Emerging Developments and Applications of Low Temperature Plasma

Aamir Shahzad
Government College University, Faisalabad, Pakistan

Maogang He
Xi'an Jiaotong University, China

A volume in the Advances in Bioinformatics and Biomedical Engineering (ABBE) Book Series

IGI Global
PUBLISHER of TIMELY KNOWLEDGE

Published in the United States of America by
IGI Global
Engineering Science Reference (an imprint of IGI Global)
701 E. Chocolate Avenue
Hershey PA, USA 17033
Tel: 717-533-8845
Fax: 717-533-8661
E-mail: cust@igi-global.com
Web site: http://www.igi-global.com

Library of Congress Cataloging-in-Publication Data

Names: Shahzad, Aamir (College teacher), editor. | He, Maogang, 1970- editor.
Title: Emerging developments and applications of low temperature plasma / Aamir Shahzad and Maogang He, editors.
Description: Hershey, PA : Engineering Science Reference, [2022] | Includes bibliographical references and index. | Summary: "This book offers the foundation for the education and research of medical physicists starting their university studies in the field of physics of nuclear medicine, low temperature plasma applications in industries and daily life"-- Provided by publisher.
Identifiers: LCCN 2021036279 (print) | LCCN 2021036280 (ebook) | ISBN 9781799883982 (hardcover) | ISBN 9781799883999 (paperback) | ISBN 9781799884002 (ebook)
Subjects: LCSH: Low temperature plasmas. | Medical physics. | Nuclear medicine.
Classification: LCC QC718.5.L6 E44 2022 (print) | LCC QC718.5.L6 (ebook) | DDC 530.4/42--dc23
LC record available at https://lccn.loc.gov/2021036279
LC ebook record available at https://lccn.loc.gov/2021036280

This book is published in the IGI Global book series Advances in Bioinformatics and Biomedical Engineering (ABBE) (ISSN: 2327-7033; eISSN: 2327-7041)

British Cataloguing in Publication Data
A Cataloguing in Publication record for this book is available from the British Library.

All work contributed to this book is new, previously-unpublished material.
The views expressed in this book are those of the authors, but not necessarily of the publisher.

For electronic access to this publication, please contact: eresources@igi-global.com.

Advances in Bioinformatics and Biomedical Engineering (ABBE) Book Series

Ahmad Taher Azar
Prince Sultan University, Riyadh, Kingdom of Saudi Arabi and Benha University, Egypt

ISSN:2327-7033
EISSN:2327-7041

MISSION

The fields of biology and medicine are constantly changing as research evolves and novel engineering applications and methods of data analysis are developed. Continued research in the areas of bioinformatics and biomedical engineering is essential to continuing to advance the available knowledge and tools available to medical and healthcare professionals.

The **Advances in Bioinformatics and Biomedical Engineering (ABBE) Book Series** publishes research on all areas of bioinformatics and bioengineering including the development and testing of new computational methods, the management and analysis of biological data, and the implementation of novel engineering applications in all areas of medicine and biology. Through showcasing the latest in bioinformatics and biomedical engineering research, ABBE aims to be an essential resource for healthcare and medical professionals.

COVERAGE

- Molecular Simulations
- Finite Elements
- Protein Engineering
- Applied Biosystems
- Bayesian Methods
- Therapeutic Engineering
- Artificial Organs
- Databases
- Drug Design
- Robotics and Medicine

IGI Global is currently accepting manuscripts for publication within this series. To submit a proposal for a volume in this series, please contact our Acquisition Editors at Acquisitions@igi-global.com or visit: http://www.igi-global.com/publish/.

The Advances in Bioinformatics and Biomedical Engineering (ABBE) Book Series (ISSN 2327-7033) is published by IGI Global, 701 E. Chocolate Avenue, Hershey, PA 17033-1240, USA, www.igi-global.com. This series is composed of titles available for purchase individually; each title is edited to be contextually exclusive from any other title within the series. For pricing and ordering information please visit http://www.igi-global.com/book-series/advances-bioinformatics-biomedical-engineering/73671. Postmaster: Send all address changes to above address.

Titles in this Series

For a list of additional titles in this series, please visit: http://www.igi-global.com/book-series/

Biomedical Computing for Breast Cancer Detection and Diagnosis
Wellington Pinheiro dos Santos (Universidade Federal de Pernambuco, Brazil) Washington Wagner Azevedo da Silva (Universidade Federal de Pernambuco, Brazil) and Maira Araujo de Santana (Universidade Federal de Pernambuco, Brazil)
Medical Information Science Reference • © 2021 • 357pp • H/C (ISBN: 9781799834564) • US $295.00

Deep Neural Networks for Multimodal Imaging and Biomedical Applications
Annamalai Suresh (Department of Computer Science and Engineering, Nehru Institute of Engineering and Technology, Coimbatore, India) R. Udendhran (Department of Computer Science and Engineering, Bharathidasan University, India) and S. Vimal (Department of Information Technology, National Engineering College (Autonomous), Kovilpatti, India)
Medical Information Science Reference • © 2020 • 294pp • H/C (ISBN: 9781799835912) • US $275.00

Biomedical and Clinical Engineering for Healthcare Advancement
N. Sriraam (Ramaiah Institute of Technology, India)
Medical Information Science Reference • © 2020 • 275pp • H/C (ISBN: 9781799803263) • US $285.00

Attractors and Higher Dimensions in Population and Molecular Biology Emerging Research and Opportunities
Gennadiy Vladimirovich Zhizhin (Russian Academy of Natural Sciences, Russia)
Engineering Science Reference • © 2019 • 232pp • H/C (ISBN: 9781522596516) • US $165.00

Computational Models for Biomedical Reasoning and Problem Solving
Chung-Hao Chen (Old Dominion University, USA) and Sen-Ching Samson Cheung (University of Kentucky, USA)
Medical Information Science Reference • © 2019 • 353pp • H/C (ISBN: 9781522574675) • US $275.00

IGI Global
PUBLISHER of TIMELY KNOWLEDGE

701 East Chocolate Avenue, Hershey, PA 17033, USA
Tel: 717-533-8845 x100 • Fax: 717-533-8661
E-Mail: cust@igi-global.com • www.igi-global.com

Table of Contents

Detailed Table of Contents

The basic properties of classical and quantum plasmas are discussed. Quantum plasmas behave differently due to high densities and low temperatures at nanometer scale in contrast to classical ones which are characterized by low densities and high temperatures. A literature survey is made to investigate the plasma phenomenon with quantum mechanical effects. Classical and quantum viewpoints are also presented to understand the free electron gas in metals. In particular, the excitation of stable plasmon wakefield is studied due to a short electron pulse propagating in axial direction of nanowire. The latter contains degenerate electrons and classical static ions. By using the Trivelpiece-Gould configuration and Fourier transform techniques, a general dispersion is obtained for the electrostatic plasmons and analyzed numerically. Nevertheless, an evolution equation for the wakefield is derived and carried out the stability analysis. In a gold nanowire, the amplitudes of wakefield become significantly modified by the variation of quantum diffraction, quantum exchange-correlations and mode quantization in the radial direction. The present findings may prove useful for investigating new radiation sources in the extreme-ultraviolet range.

molecular dynamics (NEMD) simulations. The new simulation result was obtained by employing the integral formula of velocity autocorrelation functions (VACF) using the Green-Kubo relation. The normalized self-diffusion coefficient (D*) is investigated for different combinations of plasma coupling (Γ) and Debye screening (κ) parameters. The simulation outcome shows that the decreasing position of D* shifts toward Γ and also increased with the increase of κ. The D* linearly decreased with Γ and increased when applied external E increases. It is observed that the increasing trend of D* depends on the E strength. These investigations show that the present algorithm provides precise data with fast convergence and effects of κ, Γ, E. It is shown that the current NEMD techniques with applied external E can be employed to understand the microscopic mechanism of dusty plasmas.

Adeel Aabir, University of Agriculture, Faisalabad, Pakistan
Muhammad Yasin Naz, University of Agriculture, Faisalabad, Pakistan
Shazia Shukrullah, University of Agriculture, Faisalabad, Pakistan

CNTs are the element that exists with predominant physio-chemical properties, which have been extensive researched today. These properties make carbon nanotubes (CNTs) valuable in a wide potential range of applications. The production of high-quality carbon nano-tubes (CNTs) via different precursors has been reported for many years. The arc discharge is a pristine technique to form CNTs with a high-quality yield. This technique has been elucidated for a long time, but the growth condition and mechanism of affected synthesized parameters and coorelation between synthesized parameters and nucleation of carbon have not been explored. In this chapter, the authors present the factors affecting temperature, geometry, grain size, electrodes, pressure, catalyst, arc current, power supply, and growth mechanism of CNTs. The variation in parameters has been elicited along with challenges and gaps.

Linda Agun, Universiti Teknologi Malaysia, Malaysia
Chang Shu Ting, Universiti Tunku Abdul Rahman, Malaysia
Norizah Redzuan, Universiti Teknologi Malaysia, Malaysia
Santhana Krishnan, Universiti Teknologi Malaysia, Malaysia
Siti Sarah Safaai, Universiti Teknologi Malaysia, Malaysia
Zarita Zakaria, Universiti Teknologi Malaysia, Malaysia
Muhamad Nor Firdaus Zainal, Universiti Teknologi Malaysia, Malaysia
Mohd Fadthul Ikmal Misnal, Universiti Teknologi Malaysia, Malaysia
Norhayati Ahmad, Universiti Teknologi Malaysia, Malaysia

Cold plasma is implemented in the food industry for protecting the agricultural product from foodborne pathogens. In this case, dielectric barrier discharge cold plasma pen (DBD-CP) was applied to study its efficiency in inactivation of bacterial on oyster mushroom. The surface of the fresh oyster mushroom was treated with 5 kV of AC voltage with variable of treatment times (0-4 min). Data showed sufficient energy by DBD-CPP has inactivated the existence of bacterial on the oyster mushroom surface with undetectable of bacteria colony. The reactive species generated by cold plasma undoubtedly irreversibly damage the deoxyribonucleic acid, ribonucleic acid, and enzymes of gram bacterial, which eventually causes cell death. Above all, an understanding of the microorganism cell structure, the food surface types, and roughness is an essential in manipulating cold plasma processing parameters to achieve the maximum rate of microbial inactivation.

Cold plasma has attracted lots of attention among researchers because it has a wide range of applications, such as the automotive industry, textile industry, microelectronics, packaging, biomedical technology, food preservation, and agricultural sectors. Scientists have shown a great interest in non-thermal plasma because of its advantages such as low temperature, scalable size, low operation cost, flexible operation, and high electron and reactive specie density. Also, non-thermal plasma can be operated at atmospheric pressure, which is an advantage in the agriculture industry rather than operating in a vacuum. Recently atmospheric cold plasma pressure was selected as one of the plasma technologies applied in the agricultural industry for treating the surface of the seed with environmentally friendly technology that produces no hazardous waste. DBD plasma is one of the cold plasma techniques, which can be easily triggered at atmospheric pressure and room temperature.

Dielectric barrier discharges (DBD) are the configurations for the production of electrical discharges using a dielectric medium between the metallic electrodes.

Preface

Principal intention of this book is to present a fundamental and applied research for the researchers to study emerging developments and applications of Low Temperature Plasma (EDA-LTP) researches. This also provides latest development and innovative techniques of plasma technology and its practical uses in control ways in agricultural, medical, energy production and industries. Book is designed based on novel outcomes for graduate students and advanced postgraduate researcher work those have been offered at global-wise universities, including in south Asian counties. The challenge in exploring emerging technology and indirect plasma physics is its obvious complexity. The significance of plasma phenomena found in different directions formulates it quite diverse from other branches of physics. Development of clean and environment-friendly plasma technology at large scale becomes global challenges for plasma science and engineering. To achieve plasma applications it becomes very significant, how we can employ plasma technology in agriculture, medical, alternative energy source and food processing. Low temperature plasma is a key scientific task involves how one can understand the anomalous transport processes observed in agricultural, industrial and scientific environments in a cost-effective fashion.

This book establishes up-to-date developments of the applications of low temperature plasmas in different field of interest. The EDA-LTPs of emerging fields are an outstanding question and a new frontier in applied science and technology that have applications in both basic and applied research. Recently, EDA-LTPs have been shown that modern advances in material development, experiment and theory tools have suggested that low temperature plasma may complete with conventional refrigeration and power generation technologies. The device performance is the major task for physical properties control design and how plasma parameters can control devices. Different attempts have been made to increase the performance of devices with physical applications and new improved techniques are required for application efficiency from material parameters optimization. Considerable advances in this area have been obtained over the last decades using a diverse set of theoretical, experimental, and computational techniques. This covers a wide spectrum of research areas in low temperature plasams, and is mainly based on the dusty plasmas and nano-scale of studies in global universities. The preferred research topics show a large range and well matched to syllabi that are presently being studied in different countries. The book is involved nine chapters and each chapter beings with fundamental ideas involving low temperature applications, properties at nano-level and complex plasmas, followed by a number of research issues and their complete clarifications. Developments in low temperature plasma are provoked by improvements in multidisciplinary areas of science and technology including physics, chemistry, applied biology, engineering and applied disciplines with daily used applications (for instance, plasmas oncology, semiconductor, and agricultural and food processing's, etc.).

ORGANIZATION OF THE BOOK

Chapter 1: The basic properties of classical and quantum plasmas are discussed. Plasmas at nanometer scale behave differently due to high-particle densities and low-temperatures in contrast to classical ones which are characterized by low-densities and high-temperatures. A literature survey on the plasma phenomena is made to account for quantum mechanical effects. Classical and quantum viewpoints are presented to understand a free electron gas in metals. In particular, the excitation of stable plasmon wakefield is studied due to a short electron pulse propagating in axial direction of nanowire. The latter contains degenerate electrons and classical static ions. By using the Trivelpiece-Gould configuration and Fourier transform technique, a general dispersion is obtained for electrostatic plasmons and analyzed numerically. Nevertheless, an evolution equation for the wakefield is derived with stability analysis. In a gold metallic nanowire, the amplitudes of wakefield become significantly modified by the variation of quantum diffraction, quantum exchange-correlations

Chapter 2: Equilibrium molecular dynamics (EMD) simulation has been employed to explore the dynamical structure factors (DSFs) of two dimensional (2D) dusty plasma systems for a wide domain of plasma parameters of Coulomb coupling (Γ) and Debye screening strength (κ).The influence of varying wave vectors (k) on plasma DSFs S(k, ω) have been reported with different combinations of plasma state points (Γ, κ). New simulations has been tested for the influence of different wave vectors on plasma density S(k, ω) in addition to different combinations of plasma state points. New results of plasma density S(k, ω) show that amplitude of oscillation and frequency will vary with increasing value of Coulomb coupling parameter (Γ) and Debye screening strength (κ). These simulation techniques show that transient behavior has been reported for frequency (ω) with various values of Debye screening strength (κ) and number of particles (N). Moreover, EMD simulation has been checked in order to investigate the behavior of plasma DSFs with increasing number of particles (N).

Chapter 3: Atmospheric-pressure plasmas have a wide variety of potential industrial applications. They are used in extractive metallurgy; metal recovery; novel nanomaterial synthesis; refractory and wear resistant coatings deposition; chemical synthesis; energy conversion; industrial, medical, and nuclear waste destruction; engine combustion enhancement; and exhaust gas pollutants clean up. Atmospheric plasmas are produced by applying DC or AC high voltage between two electrodes designed as cylindrical in shape for jets and planar for the dielectric barrier discharge systems. This review presents an overview of the use of atmospheric-pressure plasma dedvices and industrial processes carried out in several of these areas.

Chapter 4: The effects of external electric field (E) on the diffusion coefficient of dust particles in low-temperature dusty plasmas (LT-DPs) have been computed through Nonequilibrium molecular dynamics (NEMD) simulations. The new simulation result was obtained by employing the integral formula of velocity autocorrelation functions (VACF) using the Green-Kubo relation. The normalized self-diffusion coefficient (D*) is investigated for different combinations of plasma coupling (Γ) and Debye screening (κ) parameters. The simulation outcome shows that the decreasing position of D* shifts toward Γ and also increased with the increase of κ. The D* linearly decreased with Γ and increased when applied external E increases. It is observed that the increasing trend of D* depends on the E strength. These investigations show that the present algorithm provides precise data with fast convergence and effects of κ, Γ, E. It is shown that the current NEMD techniques with applied external E can be employed to understand the microscopic mechanism of dusty plasmas

Chapter 5: CNTs are the element that exists with predominant physio-chemical properties which has been extensive research today. These properties make carbon nanotubes (CNTs) valuable in a wide potential range of applications. The production of high-quality carbon nano-tubes (CNTs) via different precursors has been reported for many years. The arc discharge is a pristine technique to form CNTs with a high-quality yield. This technique has been elucidated for a long time, but the growth condition and mechanism of affected synthesized parameters and correlation between synthesized parameters and nucleation of carbon are not explored. In this chapter, we present the factors affecting such as temperature, geometry, grain size, electrodes, pressure, catalyst, arc current, power supply, and growth mechanism of CNTs. The variation in parameters has been elicit along with challenges and gaps.

Chapter 6: Cold plasma implement in the food industry for protecting the agricultural product from foodborne pathogen. In our case, dielectric barrier discharge cold plasma pen (DBD-CP) was applied to study its efficiency in inactivation of bacterial on oyster mushroom. The surface of the fresh oyster mushroom was treated with 5 kV of AC voltage with variable of treatment times (0-4 min). Data showed sufficient energy by DBD-CPP has inactivated the existence of bacterial on the oyster mushroom surface with undetectable of bacteria colony. The reactive species generated by cold plasma undoubtedly irreversible damage of deoxyribonucleic acid, ribonucleic acid, and enzymes of gram bacterial that eventually causes cell death. Above all, an understanding of the microorganism cell structure, the food surface types, and roughness is an essential in manipulating cold plasma processing parameters to achieve the maximum rate of microbial inactivation.

Chapter 7: Cold plasma has attracted lots of attention among researchers because it has a wide range of applications, such as the automotive industry, textile industry, microelectronics, packaging, biomedical technology, food preservation, and agricultural sectors. Scientist has shown a great interest in the non-thermal plasma because of its advantages such as low temperature, scalable size, low operation cost, flexible operation, and high electron and reactive specie density. Also, non-thermal plasma can be operated at atmospheric pressure, which is an advantage in the agriculture industry rather than operating in a vacuum Recently atmospheric cold plasma pressure was selected as one of the plasma technologies applied in the agricultural industry for treats the surface of the seed with environmentally friendly technology that produces no hazardous waste. DBD plasma is one of the cold plasma techniques, which can be easily triggered at atmospheric pressure and room temperature.

Chapter 8: Dielectric barrier discharges (DBD) are the configurations for the production of electrical discharges using a dielectric medium between the metallic electrodes. Plasma treatment produces negative radicals which increase the adhesion of fabric for nanoparticles. The plasma treatment made the fabric surface rougher because of the etching effect. UV-vis spectra of the Plasmon resonance band observed at 253-400 nm. X-ray diffraction results showed that AgNPs has a cubical structure, and the average crystalline size is 25 nm. SEM results determined that the morphology of the silver nanoparticles flowers shaped. The energy bandgap of AgNPs was observed at 2.59 eV. The silver nanoparticles found to have enhanced antimicrobial properties and showed better zone of inhibition against isolated bacteria (Escherichia coli). DBD plasma treatment changed the chemical as well as physical properties of the cotton fabric. FTIR spectrum revealed that oxygen-containing groups, such as C-O, C=O, O-C-O as well as O-C=O increased on DBD treatment of cotton samples.

Chapter 9: In the last 20 years, A rapidly developing technique over the last two years is laser induced break-down spectroscopy (LIBS) also be known as laser-induced plasma spectroscopy (LIPS), a non-destructive spectroscopy technique that is mostly be used for the analytical study of the samples in last 20 years. With this technique, multiple elemental composition of elements are analyzed simultaneously without considering the form of sample that maybe in solid, liquid or gas . Moreover, economically it is very beneficial and valuable technique because no sample preparation is required, and sample consumption is very small. This technique is powerful enough that it can bore a microscopic crater usually in the solid samples to target. This technique has great sensitivity to find out the resolution down to a single grain it has variety of applications in the field of science for past two decades such as archeology, soils, environmental protection and so on.

Aamir Shahzad
Molecular Modeling and Simulation Laboratory, Department of Physics, Government College University, Faisalabad, Pakistan

Maogang He
School of Energy and Power Engineering, Xi'an Jiaotong University, China

Introduction

Plasma is a partially or fully ionized gas containing charged and neutral particles showing collective behavior. Plasma is the study of non-linear collective interactions of electrically charged particles (electrons and ions) not only with each other but also with the neutral particles like atoms and molecules. Plasma has many significant applications in addition to material processing, agricultural, medical/ clinical and semiconductor industries. Plasmas are scientifically productive due to two important characteristics: (1) it has high energy density, and (2) it generates various kinds of active species of different energies than those produced in chemical reactors. These characteristics of plasmas make it productive for material processing. Such type of active species of different kinds and different energies and fluxes are frequently produced by Magnetron Sputtering System.

Emerging Developments and Applications of Low Temperature Plasma (EDA-LTP) provides the multidiscipline probes that fundamental thermophysical behaviors, interaction physics and reaction chemistry of partially ionized gases and their application to technologies ranging from plasma refrigeration, lighting sources, lasers and sensors to materials processing, agriculture and food processing, biotechnology, healthcare, microelectronics fabrication/ semiconductor industries and space and environmental sciences. This book will discuss the introductory and research contents on EDA-LTP trained by different world-wide universities and institutes. Nowadays, the developments in industrial fabrication technology and the desires on performance and feature in advanced manufacture processes have become in complex and incorporated innovation technology. This book gives amazingly wide and planned impact on national, health sectors and economic security, and providing societal advantage. EDA-LTP quantifies the information's how modern semiconductor devices and thin film solar cell technologies can be fabricated through plasma processes and EDA-LTP to be economically viable. EDA-LTP introduces the manufacturing of bioengineering devices that depend on plasma processes to harden artificial joints and plan biocompatible surfaces on tissue scaffolding. This will probe the fundamental mechanisms in interplanetary that are powered by plasma thrusters. In addition, this text provides the fundamentals and applications and other several new topics related to plasma based energy devices and low temperature plasma for complex systems. These contents indentify the key challenges including basic interactions of plasma with organic materials, living tissues, plasmas that contain chaotic and stochastic processes, very high density plasma and microplasmas, high temperature and pressure plasmas. Advancement in life sciences is gradually more sourced by deployment of discrete technologies and information and plasma technology becomes a key advancement in modern medicine and health care.

Low temperature plasma in medicine is rather new field that rose from the research in application of cold plasmas (CPs) in bioengineering. Plasma medicine is innovative and promising multidisciplinary novel field of research covering plasma physics, life sciences and clinical medicine to apply physical plasma for therapeutic applications. This book offers the foundation for the education and research of medical physicists starting their university studies in the field of physics of nuclear medicine, low temperature plasma applications in industries and daily life. This book gives equally beneficial for those wish to advance their knowledge of physics of plasma medicine, plasma agriculture and industrial applications. It provides, in the form of a syllabus, a comprehensive overview of the basic Emerging Developments and Applications of Low Temperature Plasma (EDA-LTP) physics knowledge required for the practice of medical physics in modern medicine. This book provides the guide of nuclear medicine that is the exercise of radionuclides in medicine for diagnosis, staging of disease, therapy and monitoring the response of a disease process. This book comprehensively covers a broad range of topic including but not limited to field of Plasma Oncology, Plasma Medicine many applications including, agriculture, plasma processing, catalysis, and aerospace engineering.

Here is a quick run through some basics, which enclosed in more detail in succeeding chapters: First Chapter provides the discussion of the most fundamental properties of classical and quantum plasmas. In quantum regimes, the length and time scales are seen quite different due to high equilibrium densities and low-temperatures in contrary to classical ones, which are characterized by low-densities and high-temperatures. The second Chapter explains the modeling of dynamical structure factors (DSFs) of two dimensional (2D) dusty plasma systems for a wide domain of plasma parameters of Coulomb coupling (Γ) and Debye screening strength (κ). This chapter also gives the fluctuation of dynamical density increases at intermediate to higher values of coupling parameter. However, it shows less fluctuation at higher values of Debye screening strength (κ). The third Chapter incorporates the atmospheric pressure plasma generation, thermal atmospheric plasma generation, arc and torches, nonthermal atmospheric plasma, dielectric barrier discharges and atmospheric pressure plasma in Medical fields. In fourth chapter, the authors elaborate the numerical analysis of diffusion coefficients in low-temperature dusty plasmas in the presence of external electric field. The fifth chapter gives the fabrication of carbon nanotubes via plasma arc discharge method and the arc discharge is a pristine technique to form CNTs with a high-quality yield. In the sixth chapter, the authors tackle the cold plasma, a non-thermal emerging technology implement in the food industry. This chapter provides the detailed information of foodborne pathogen inactivation by low temperature plasma species. In the seventh chapter, researchers discuss the dielectric barrier discharge plasma (DBD) as the application for agricultural purposes as in this study for seed treatment. Moreover, the optimization of the plasma reactor has been successfully determined using air as carrier gas with different flow rates. Low temperature plasma treatment at different times and voltage proves that the best treatment time for sample. The chapter eight provides the biosynthesis of silver nanoparticles for study of their antimicrobial effect on plasma treated textiles. The silver coating of plasma treated cotton fabric is used this chapter. The final ninth chapter presents the plasma formation and its parameters used in calibration free laser induced breakdown spectroscopy.

CONCLUSION

This book highlights the needs for alternative plasma technology in different fields that discuss theoretical, numerical and experimental techniques, and can be available for all countries. One of the most significant problems facing civilization today is how to generate alternative clean technology sustainably for health care and in sufficient quantity to supply a growing global production. Conventional fuel burning machines pollute the atmosphere resulting in global warming and cause a health risk. Plasma energy does not produce any carbon dioxide and this book will provide the number challenges that have to be solved on before it may become the alternative source of technology and health care solutions in future. This book includes the theoretical and experimental observations of plasma technology for agricultural, medical, and industrial applications.

Chapter 1
Wakefield Formation Due to a Short Electron Beam in Quantum Nanowires:
Plasma Oscillations with Quantum Effects

Shahid Ali
National Centre for Physics, Pakistan

Ioannis Kourakis
Khalifa University of Science and Technology, UAE

ABSTRACT

The basic properties of classical and quantum plasmas are discussed. Quantum plasmas behave differently due to high densities and low temperatures at nanometer scale in contrast to classical ones which are characterized by low densities and high temperatures. A literature survey is made to investigate the plasma phenomenon with quantum mechanical effects. Classical and quantum viewpoints are also presented to understand the free electron gas in metals. In particular, the excitation of stable plasmon wakefield is studied due to a short electron pulse propagating in axial direction of nanowire. The latter contains degenerate electrons and classical static ions. By using the Trivelpiece-Gould configuration and Fourier transform techniques, a general dispersion is obtained for the electrostatic plasmons and analyzed numerically. Nevertheless, an evolution equation for the wakefield is derived and carried out the stability analysis. In a gold nanowire, the amplitudes of wakefield become significantly modified by the variation of quantum diffraction, quantum exchange-correlations and mode quantization in the radial direction. The present findings may prove useful for investigating new radiation sources in the extreme-ultraviolet range.

DOI: 10.4018/978-1-7998-8398-2.ch001

FUNDAMENTALS OF PLASMAS AND NANOSTRUCTURES

Classical plasmas are low-density and high-temperature plasmas, occur in many laboratory and space environments, e.g., plasmas in glow discharges and laser laboratories, in industrial processing plants and tokamaks, in magnetosphere, in ionosphere, in solar wind, in stellar interiors, in interstellar medium, etc. An ionized gas can be treated as *classical plasma* (Chen, 1984) if it holds certain criteria that impose specific conditions. These conditions not only rely on the high-temperature and low-particle density but also on the characteristic Debye length for maintaining the quasi-neutrality and collective behavior of plasma particles. First, the plasma must be a *quasi-neutral gas* on the global scale such that the dimension of a plasma system should be much larger than the characteristic Debye length, viz., $L \gg \lambda_D$, otherwise, the shielding phenomenon microscopically becomes ineffective and there would be just an ionized gas rather than a plasma gas. An immobile positive test charge in a plasma attracts electrons and gives rise to the formation of Debye-Hückel potential, i.e., $\phi_{DH} = \frac{q_T}{R}\exp\left(-R / \lambda_D\right)$, where q_T is the charge of the test particle, R is the distance of test charge from an observation point and $\lambda_D = \left(\varepsilon_0 k_B T / N_0 e^2\right)^{1/2}$ is the Debye shielding length. It is important here to note that for a distance much smaller than the Debye length, the potential is reduced to Coulomb potential while in opposite case, the potential decreases exponentially with the distance from the test charge. Second, it is the Debye shielding length, which basically characterizes a length scale at which the thermal particle energy is balanced with the electrostatic potential energy, helping us to determine the Debye sphere containing large number of plasma particles, viz., $N_D = \left(\frac{4\pi}{3} N_0 \lambda_D^3\right) >> 1$. This further implies that a plasma is generally dominated by long-range collective interactions or forces that prevail over the short-range individual interactions (i.e., the binary interactions or collisions). If this condition violates, the plasma will simply correspond to *strongly coupled state*, which behaves differently from a conventional plasma state. Third, the collisional time between the charged particles and neutrals must be larger than the inverse of plasma oscillation frequency, viz., $\tau_{en} >> \omega_p^{-1}$.

To understand the physics of a collective phenomenon that involves a plasma oscillation frequency, we zoom out a small portion of the plasma gas, as shown in Figure 1. Observe that an electric field is established if we slightly displace the electrons from their mean positions, where ions are fixed, providing a neutralized plasma background. Let's say the displacement of electrons is along the x-axis and these electrons are literally come back towards the ions due to their restoring force and do not stop at the mean position owing to their inertia. They eventually execute a back-and-forth motion about the mean position and oscillate with a distinct frequency known as *plasma oscillation frequency* ω_p.

Figure 1. The diagram interprets specific features of a classical plasma whose constituents are the electrons, positive ions and neutrals.

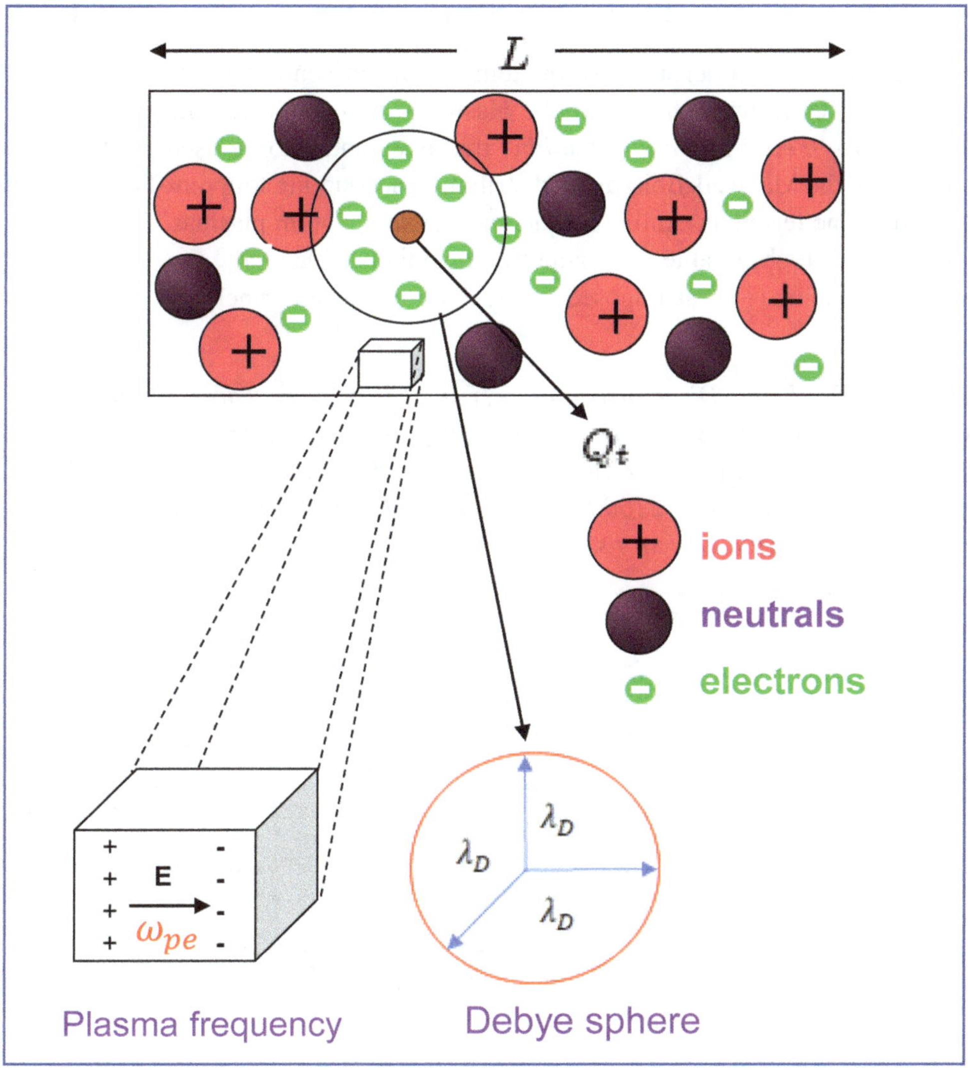

On the other hand, quantum plasmas are highly dense at very low temperatures. New quantum forces and degenerate laws (Haas et al., 2003, Manfredi, 2005) are utilized, playing a vital role in the dynamics of charge carriers and plasma modes. In quantum dense plasmas (Manfredi & Haas, 2001, Manfredi, 2005), the quantum scales are distinctly modified due to strong dependence of the Fermi speed $u_F\left[=(2\varepsilon_F/m_e)^{1/2}\right]$, the Fermi length $\lambda_F\left(=u_F/\omega_p\right)$ and plasma oscillation frequency $\omega_p\left[\equiv\left(N_0e^2/\varepsilon_0m_e\right)^{1/2}\right]$. These parameters are also, in turn affected by the high equilibrium density and low-temperature in contrast to classical plasma parameters. Quantum behavior of the plasma particles is often modelled by the well-known "*Fermi-Dirac distribution*" that is essential to be used for fermions rather than the Maxwell-Boltzmann distribution accounting for point-like classical particles. The fermions exhibit unique features in quantum plasmas, for example, they cannot be observable and cannot be distinguishable, and carry half-integral multiple of spins $(\frac{1}{2}\hbar)$ as imposed by the Pauli-exclusion principle.

Figure 2. The Fermi-Dirac distribution [$f_{FD}(\varepsilon)$] is plotted against the parameter x(=ε/kB) for different values of absolute temperature, i.e., T=0.01K (pink solid thin curve), T=300K (black solid thick curve), 7000K (blue dotted curve), 14000K (green dashed curve) and 21000K (red solid thick curve) with Fermi temperature T_F=6.3×10⁴K for gold electrons.

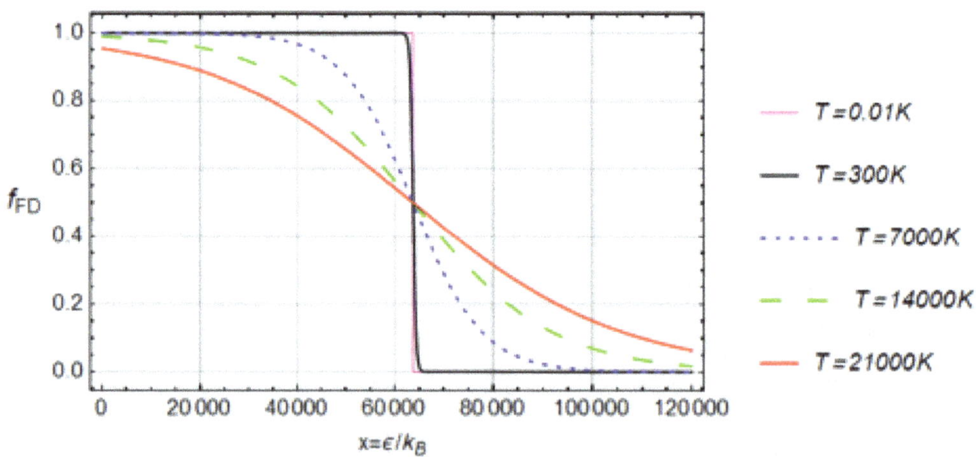

The Fermi-Dirac distribution function, essentially represents the probability for occupancy of the energy levels caused by the fermions, can be given, as

$$f_{FD}(\varepsilon) = \frac{1}{1 + \exp[(\varepsilon - \varepsilon_F) / k_B T]}. \tag{1}$$

In Figure 2, we see that how different values of the absolute temperature *(T)* affect the Fermi-Dirac distribution function. As, the temperature increases, the Fermi-Dirac distribution approaches the Maxwell-Boltzmann distribution. A partially degenerate plasma is achieved as long as the probability strictly holds a limit $T < T_F$ and in turn, the distribution takes the form of a step curve. At exactly absolute zero temperature *(T = 0),* the probability of fermions becomes unity under the regime $\varepsilon < \varepsilon_F$ and the plasma is then regarded as the fully degenerate plasma. It is clear to note that lower and upper edges of the step curve are sharper at very low temperature T = *0.01K a*s compared to the step curve at T = *300K.* Furthermore, the assumptions $\varepsilon > \varepsilon_F$, $\varepsilon = \varepsilon_F$ and $\varepsilon \gg \varepsilon_F$ in the Fermi-Dirac distribution function will lead to the probabilities in the order of magnitude of zero, half, and $\exp(-\varepsilon/k_B T)$, respectively. Observe that probability appeared in the form of exponential factor exactly confirms the Maxwell-Boltzmann distribution as a function of energy. The electron Fermi energy (ε_F) and electron Fermi temperature (T_F) strongly depend on the equilibrium density, which can be expressed, as:

$$\varepsilon_F = \frac{\hbar^2}{2m_e k_B}(3\pi^2 N_0)^{2/3} \text{ and } T_F = \frac{m_e u_F^2}{2k_B} \equiv \frac{\varepsilon_F}{k_B}, \tag{2}$$

where k_B ($\hbar$) is the Boltzmann (scaled Planck) constant and m_e the electron mass.

The electron de Broglie wavelength (λ_B) is another important parameter in quantum plasmas that roughly exhibits the spatial extension of a state function caused by the quantum uncertainty. If λ_B is so small compared to the average inter-particle spacing $d\left(= 3N_0^{-1/3} / 4\pi\right)$, then the plasma charges are assumed to be point-like and consequently we do not have any overlapping of waves and quantum interference. If the de Broglie wavelength is comparable to or larger than the inter-particle spacing, i.e., $\lambda_B \geq d$, then quantum corrections play a significant role in studying the collective behavior in dense quantum plasmas. The particles of large mass will have small quantum effects as compared to lighter species. Similarly, an interesting parameter (Γq) which indicates the ratio of the interaction energy to the average kinetic energy, is known as 'quantum *coupling parameter' can be* read, as

$$\Gamma_q \sim \left(\frac{1}{N_0 \lambda_F^3}\right)^{2/3} \sim \left(\frac{\hbar\omega_p}{k_B T_F}\right)^2. \tag{3}$$

It may be noted that coupling parameter determines the square of plasmon energy $\left(\hbar\omega_p\right)$ to Fermi energy (ε_F). For $\Gamma q < 1$, the Fermi plasma is considered to be a collision-less quantum plasma and collective forces become more effective over the collisional forces. This also suggests that all lower energy states are filled in by the fermions in accordance to the Pauli`s exclusion principle below the Fermi energy in a fully degenerate plasma. If a fermion is further added to the plasma, then it occupies a higher energy state, which will increase the plasma density and reduce the value of quantum parameter Γq, such that quantum plasma exhibits a collective behavior at large density. It is worth expressing here that if the Fermi length tends to the classical Debye length $\lambda_F \rightarrow \lambda_{D,}$ then the corresponding quantum coupling parameter will also approach to the classical counterpart, i.e., $\Gamma_q \rightarrow \Gamma_c = \left(1/N_0\lambda_D^3\right)^{2/3}$.

Numerous dispersive wave modes (Shukla et al., 2006, Ali et al. 2007) and instabilities (Ali, Nitin & Shukla 2007, Moslem et al., 2008) have been identified in quantum plasmas by taking into account the quantum corrections. These corrections have been caused by the degenerate electron pressure, quantum Bohm potential, electron exchange-correlations, electron spin-1/2 dynamics and have many implications in intense laser-solid density plasma experiments (Kremp et al. 1999, Kremp et al. 2005), in micro- and nano-electronic devices (Markowich et al. 1990) [i.e., semiconductors, quantum dots, quantum wells, quantum nanowires (Shpatakovskaya, 2006), quantum nanotubes (Wei & Wang, 2007), and quantum diodes (Ang, Kwan & Lau 2003)], in exterior of Jupiter, in superdense astrophysical systems [i.e., white dwarf stars, neutron stars, magnetars, etc.] (Jung, 2001, Chabrier et al. 2002).

Moreover, in recent years, low-dimensional systems (LDs) have attracted a considerable interest in the fields of nano-electronics and nano-technology. The LDs are basically the confined systems, which are made of different nanostructures, in which the charge carriers can move freely in one, two, and zero dimensions to form a quantum nanowire, quantum well, and quantum dot, respectively. The reduction in one or more dimensions substantially changes the physical and chemical properties of nanostructures in comparison with the bulk material. When the size of a device reduces more and more, the electrons can get more confined and consequently, the LDs can function even at a voltage less than 0.2 Volts and currents in nano or pico Amps. In such a situation, the spatial dimension of the device is comparable to the thermal de-Broglie wavelength of charge carriers, and hence the electron quantum effects and their correlations cannot be ignored. The confined devices (Duan et al. 2001, Lee et al. 2009, Johnson et al. 2002, Huang et al. 2001, Xiang et al. 2006) are effectively used in fabrication of transistors, light emitting diodes, memory devices, electronic circuits, biomolecular nano-sensors, photo-detectors, nanowire lasers, solar cells, etc.

Nanowires are often regarded as *quantum nanowires b*ecause the fermions behave quantum mechanically and need to be confined in the transverse direction while allowing them to be free along the longitudinal direction for novel peculiarities in contrary to three-dimensional bulk material. Thus, quantum nanowires essentially require a nanometer dimension, viz., 10^{-9} m *(*=1nm*)* with a diameter of the order of 10 nm or less without any constraints on longitudinal size. Nanowires can be found in the form of different metals, e.g., Ni, Pt, Au, etc., in semiconducting materials, e.g., InP, Si, GaN, etc., and in insulating materials like SiO2 and TiO2. Nanostructures with quantum settings have also opened up a new field of plasmonics, where plasmon may act as quasiparticle - the quantization of collective phenomena, profoundly affected by the high density and low temperature to account for electron exchange-correlations. In addition, quantum recoil effects via the quantum Bohm potential appear as higher order correction to the QHD model in the long wavelength limit. The neglect of quantum diffraction effects in the non-stationary Thomas-Fermi model, one can eventually arrive at the well-known QHD model, which plays a major role in the physics of superconductors (Feynman, 1972), superfluids (Loffredo & Morato 1993), metal clusters, thin films, etc.

HISTORICAL BACKGROUND

Several new analytical and computational techniques have been developed to explain particle transport, collective modes and instabilities, nonlinear structures and nonlinear wave couplings at nanometer scales. The most familiar theoretical frameworks to model the degenerate plasmas, are the Schrödinger-Poisson and Wigner-Poisson models, have often led to the simple quantum hydrodynamic (QHD) equations (Manfredi & Haas, 2001, Manfredi, 2005) in the long wavelength constraint [i.e., $\lambda_F K \ll 1$], where λ_F(K) represents the electron Fermi length (wavenumber). About two decades ago, Haas, Manfredi and Feix (2000) have reported a classic paper on multi-streaming model in an unmagnetized collision-less quantum plasma using the Schrödinger-Poisson coupled equations to examine fermionic behavior of electrons and quantum diffraction on the dispersions of two-stream instability. They obtained a new quantum branch, purely based on the quantum corrections and analyzed the growth rate instability numerically. Later, the dispersive features of quantum ion-acoustic (IA) waves were investigated in an electron-ion unmagnetized plasma by using a one-dimensional QHD equations (Haas et al. 2003). They retrieved a classical IA wave in the small quantum parameter regime and presented the propagation characteristics of IA waves in weakly and fully nonlinear regimes. Shukla &

Eliasson (2006) explained the formation of dark solitons and vortices in quantum electron plasma and Ali et al. (2007) reported the large-amplitude IA solitons in Fermi magnetoplasmas.

In the past, several nonlinear structures were studied taking into account the classical scales in multi-dimensional systems (Allen & Rowlands, 1993, Infeld, & Rowlands 2000) and obtained the Zakharov and Kunetsov (ZK) equation (Zakharov & Kunetsov, 1974) and instability (Parkes & Munro 2005) in magnetoplasmas. Moreover, Waleed et al. (2007) derived a quantum ZK equation for investigating small-amplitude IA waves with quantum settings and identified solitary, explosive, and periodic solutions by utilizing the Conte's truncation technique, as well as, a bending instability with small-k perturbation expansion method. Later on, Sabry, Moslem & Shukla (2008) employed a new direct method to solve nonlinear ZK equation in a very dense magnetoplasma by taking into account the quantum degenerated electrons and ions. However, the above investigations have ignored the quantum exchange correlations (XCs) due to electrons, which can be incorporated by relying on the density functional theory (DFT) and QHD model.

The concept of DFT was originally presented by two great scientists, Thomas (1927) and Fermi (1927), whereas Dirac (1930) and Weizsäcker (1935) subsequently interpreted the electron-exchange term in relation to energy functional and nonlocal gradient with kinetic energy. Kohn & Sham (1965) have based their model on the studies of Hohenberg & Kohn (1964) and developed approximate techniques for studying slowly varying and high-density systems to account for XCs as an unknown functional of density. Isihara (1968) have examined the dispersive properties of Langmuir waves with electron XCs in quantum plasmas. The impact of XCs is mainly caused by the electrons, generally taken to be very weak and often neglected. However, if the density (temperature) is enough high (low) in degenerate plasmas, then this effect is no longer ignorable, especially in the study of plasmons in nanostructures. Hence, the XCs potential is added up in the equation of motion for electrons, which results in the form of an additional force to modify the dielectric and dispersive properties in nanostructures. A comprehensive review containing both analytical and simulation findings, has been reported (Crouseilles, Hervieux & Manfredi 2008), studying the nonlinear ultrafast electron dynamics in thin metallic films with QHD treatment. They analyzed the electron XCs effects phenomenologically and compared their findings with Sham and Kohn's ground state calculations involving the DFT model, finding a complete analogy between the two approaches. In particular, Shukla & Akbari (2013) employed the QHD model to express the ion structure and its stopping power with XCs effects. Subsequently, Ekman & Zamanian (2015) investigated the dispersive IA and Langmuir modes with electron exchange effects only using the kinetic model.

With advent of nanotechnology and nano-electronics, many new avenues have been opened up for the research and development projects, playing a crucial role in fabricating nanostructures, in characterizing nano-materials and in developing new technologies. Long ago, Anderegg, Feuerbacher & Fitton (1971) experimentally verified the optically excited longitudinal plasmons (Jones, Kliewer & Fuchs, 1969) in alkali metallic thin potassium films and evaluated a dispersion completely in agreement with theoretical results of Boltzmann equation. Nanostructures can be composed of ten-to-hundred thousand atoms and if ultrafast femtosecond/attosecond lasers are irradiated upon them, it could lead to significant modifications in the basic properties of nanostructures. Spectroscopic observations (Willets & Van Duyne 2007) have been carried out to achieve spectral location of localized surface plasmon resonance in nanostructures involving their size 10 nm or less and established a relationship between the Raman excitation energy and plasmon resonance. Liu et al. (2007) studied the optical properties of arbitrarily shaped nanostructures by taking into account three-dimensional computational electrodynamics, and entirely ignored the dependence of wavenumber in collective processes to obtain similar results as observed in recent experiments. Later, McMahon, Gray & Schatz, (2009) have self-consistently solved the Maxwell's equations by utilizing the finite difference method to show the dielectric response of arbitrary shaped nanostructures and examined spatially nonlocal effects incorporating electron collisions and quantum corrections. They achieved a strong anomalous absorption in gold nanowires, completely overshadowing the plasmon resonance as predicted by the local electrodynamics. In recent years, Pamyatnykh, (2020) analytically confirmed the existence of quantum waves in metallic nanowires, and explored several aspects of nanowire and acoustic dispersion in the long wavelength limit.

In this Chapter, we discuss the basic properties of classical and quantum plasmas and consider different tools to describe dynamical behavior of conduction electrons in metallic nanowires. At nanometer scales, stable electrostatic plasmons can be excited to form wakefields behind the short electron pulse propagating with a constant speed axially along the direction of nanowire. It is found that dispersion and wakefield excitations are significantly influenced by the quantum parameter H, the electron XC speed, and finite-size mode quantization effects. The present findings are important to understand the electron trapping and acceleration caused by the plasmon wakefield, which may lead to the emission of new radiations in the extreme-ultraviolet range.

THE CONCEPT OF A FREE ELECTRON GAS IN METALS

Metals are the most common materials, used for different purposes depending on their specific characteristics, hardness and stability. Nowadays, they have plenty of implications, directly or indirectly affecting our lives, for example, metals are used in making surgical and electrical tools, in household utensils, in industries, and in modern structures like ceilings and flyovers, in big machines, cars, buses, trains, railway lines, warships, airplanes, power transmission lines, traffic signals, opto-electronics, nano-structures, etc. As a matter fact, we can differentiate various metals and make their adequate utility. For this, we need to understand their physical, chemical and electrical properties. By composition, metals can be composed of electrons and static ions. It is interesting to see how a free electron gas evolves in metals due to the presence of interatomic forces [see Figure 3]. So, as an example, we first look into the electronic configuration of single sodium atom (Na) [cf. Figure 3(a)], where electrons revolve around the nucleus. Note that core electrons are relatively tightly bound to the nucleus in contrast to a loosely bound single valence electron in its outermost shell. If there are so many sodium atoms that can be brought

Figure 3. Energy levels and bands in sodium metal.

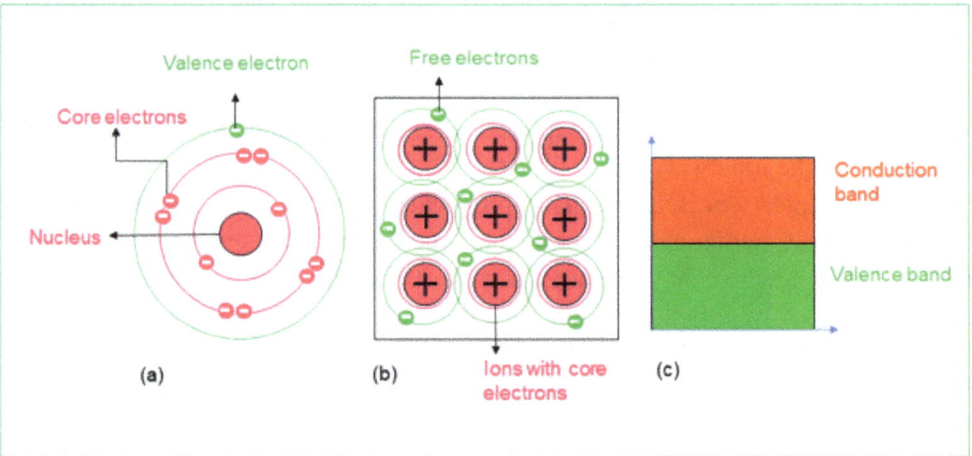

closer to each other, such that the outermost orbits of sodium atoms start to overlap in metallic solid. In this situation, the inter-atomic forces play their crucial role in making the valence electrons to be free to move from an atom to nearby atoms. The valence electrons thus can wander away from parent atoms and move freely in metallic volume, resulting in the electrical flow/conduction.

At present, there are three important theories to understand the dynamics of electrons in metals, for instance, the classical theory, quantum theory and band theory of electrons. In 1900, Drude and Lorentz introduced a model to investigate transport properties of free electrons in metals by imposing the laws of classical mechanics. This theory could not succeed in verifying the experimental results on temperature dependence of the resistivity and heat capacity. Shortcomings of the Drude's theory were later satisfactorily addressed by the Sommerfeld in 1928. He essentially made use of quantum mechanical laws to explain the behavior of quantum free electrons that can be trapped in a potential well of a metal. He correctly achieved the values of specific heat, electrical and thermal conductivities of metals, as well as, considered the Fermi electrons to be responsible for electrical conductivity and matter waves. Nevertheless, in 1928, a band theory was also stated by Bloch to describe free electron motion in a periodic field of lattice. If the electrons in atomic orbitals have the same energy, then they may be combined to form molecular orbitals and show energy to be closer to one another to form the so called "band". Specifically, in metals, the conduction and valence bands overlap on each other [as can be seen in Figure 2(c)] and by applying the electric field, the electrons can move easily from valence band to conduction band to give rise to the flow of electricity in metals. Overall, this theory has introduced several new features and interpreted the concepts of holes, band gap and effective mass of electrons in semiconductors.

QUANTUM HYDRODYNAMIC (QHD) MODEL AND EQUATIONS

The most common approaches to describe a quantum dense plasma, are the Schrödinger-Poisson and Wigner-Poisson models. These models essentially express the fluid and statistical behaviors of fermions and are reduced to classical counterparts by simply neglecting the quantum effects through $\hbar \to 0$. In this respect, Manfredi, (2005) was the first who presented a review on quantum models and plasma modes in an unmagnetized dense plasma. In this section, our main objective is to derive QHD equations and discuss salient features of QHD model.

Let us consider a one-dimensional quantum plasma system (Manfredi, 2005) that is composed of electrons and ions. The ions being the heavier species compared to lighter electrons, are often taken as static in the background of plasma. Supposing that we have N number of electrons and each electron is associated with a wave function at time T, then one can express all the electrons in terms of N-particle wave functions. It means that we need to mathematically decompose the total wave function into the product of N-one-particle functions, i.e.,

$$\psi_N(X_1, X_2, ..., X_N, T) = \psi_1(X_1, T), \psi_2(X_2, T), \psi_3(X_3, T)........\psi_N(X_N, T)$$

for the sake of identification concerning the wave characteristics. To make this model more realistic and achievable, we neglect two-body and higher order correlations using a weak form of Pauli-exclusion principle if none of the wave functions are identical. We also observe that N electrons are involved with N particle wave functions $\psi_N(X_1, X_2, ..., X_N, T)$, occupying N number of positions $X_1, X_2, X_3...X_N$ at time T. Thus, such a system contains N electrons and each electron is denoted by an index j (where $j = 1, 2, 3...N$), representing the N-independent Schrödinger equations along with the Poisson equation, respectively, as

$$i\hbar\partial_T\psi_j + \frac{\hbar^2}{2m_e}\partial_X^2\psi_j + e\phi\psi_j - \phi_{eff}\psi_j = 0, \tag{4}$$

and

$$\frac{\partial^2\phi}{\partial X^2} = 4\pi e\left(\sum_{j=1}^{N} p_j\left|\psi_j\right|^2 - N_0\right). \tag{5}$$

Note that plasma species at equilibrium holds a condition, such as $N_{e0} \approx N_{i0} = N_0$, here ϕ is the electrostatic potential, whereas $\phi_{eff}\left(= m_e u_F^2\left|\psi_j\right|^4 / 2N_0^2\right)$ stands for the effective potential. Let us consider a complex wave solution of the form $\psi_j = A_j(X, T)\exp\left(iS_j(X, T)/\hbar\right)$ in Eq. (4) and (5) with its real-amplitude $A_j(X,T)$ and real-phase $S_j(X,T)$. Hence, the average electron density $N_j(X,T)$ and average electron fluid speed $U_j(X,T)$ can be expressed in the real quantities as

$$N_j(X, T) = \left|A_j(X, T)\right|^2 = \left|\psi_j(X, T)\right|^2$$

and $m_e U_j = \partial_X S_j(X,T)$. The use of Madelung representation

$$\left[viz., \psi_j(X, T) = \sqrt{N_j(X, T)}\exp\left(iS_j(X, T)/\hbar\right)\right]$$

into Eq. (4) then yields the following form:

$$i\hbar\left(\sqrt{N_j}\frac{i}{\hbar}\frac{\partial S_j}{\partial T}+\frac{1}{2\sqrt{N_j}}\frac{\partial N_j}{\partial T}\right)\exp\left(\frac{iS_j}{\hbar}\right)=-\frac{\hbar^2}{2m_e}\left[\frac{i}{\hbar}\left\{\frac{i}{\hbar}\sqrt{N_j}\left(\frac{\partial S_j}{\partial X}\right)^2+\sqrt{N_j}\frac{\partial^2 S_j}{\partial X^2}\right.\right.$$
$$\left.+\frac{1}{2\sqrt{N_j}}\frac{\partial S_j}{\partial X}\frac{\partial N_j}{\partial X}\right\}+\frac{1}{2}\left\{\frac{i}{\hbar}\frac{1}{\sqrt{N_j}}\frac{\partial N_j}{\partial X}\frac{\partial S_j}{\partial X}\right.$$
$$\left.\left.+\frac{1}{\sqrt{N_j}}\frac{\partial^2 N_j}{\partial X^2}-\frac{1}{2}\frac{\partial N_j}{\partial X}N_j^{-3/2}\frac{\partial N_j}{\partial X}\right\}\right]\exp\left(\frac{iS_j}{\hbar}\right)$$
$$-e\phi\sqrt{N_j}\exp\left(\frac{iS_j}{\hbar}\right)+\frac{m_e u_F^2}{2N_0^2}N_j^{5/2}\exp\left(\frac{iS_j}{\hbar}\right). \tag{6}$$

Differentiating the real part of Eq. (6) with respect to X and cancelling out the factor $\exp\left(iS_j/\hbar\right)$ from both of its sides, we eventually arrive at the momentum equation for jth electrons, as

$$m_e\left(\frac{\partial U_j}{\partial T}+U_j\frac{\partial U_j}{\partial X}\right)=e\frac{\partial \phi}{\partial X}-\frac{1}{N_j}\frac{\partial p_{Fj}}{\partial X}-\frac{\partial \phi_{Bj}}{\partial X}, \tag{7}$$

where the fermionic pressure and quantum Bohm potential are, respectively, given by

$$p_{Fj}=2k_B T_F N_j,\ \text{and}\ \phi_{Bj}=-\frac{\hbar^2}{2m_e}\left(\frac{\partial_X^2\sqrt{N_j}}{\sqrt{N_j}}\right).$$

The corresponding imaginary part of Eq. (6) is then simplified to get a continuity equation for electrons, as

$$\frac{\partial N_j}{\partial T}+\frac{\partial}{\partial X}\left(N_j U_j\right)=0. \tag{8}$$

The Poisson equation finally closes the system of equations to give

$$\frac{\partial^2 \phi}{\partial X^2}=4\pi e\left(N_j-N_0\right). \tag{9}$$

In deriving (9), we have used the probability $\sum_{j=1}^{N} p_j = 1$. The coupled set of Eqs. (7)-(9) are called the quantum hydrodynamic (QHD) equations, representing the macroscopic electron fluid in terms of macroscopic quantities such as the electron number density, the electron fluid velocity and electron current density. In other words, QHD equations describe the transport of charge density and momentum taking into account the electron diffraction and statistical pressure effects through the self-consistent Hartree potential ϕ. Alternatively, in the long wavelength limit [viz., $K\lambda_F \ll 1$], Eqs. (7) and (8) can be obtained by solving the moments of Wigner equation in velocity space. The Fermi pressure (p_{Fj}) actually represents the statistical effects of fermions, plays a crucial role in LDs, e.g., in ordinary metals, in metallic clusters, and in nanoparticles. In such nano-devices, the Fermi temperature is often taken much larger than the room temperature. It is the degenerate Fermi pressure that provides a balance against the gravitational pull to prevent the collapsing of superdense white dwarf stars. Interestingly, the fermions describe a wavelike behavior associating with the wave functions in quantum mechanics, which is evident from the relation $\psi_j(x,t) = \sqrt{N_j(x,t)}$, leading to the existence of quantum Bohm potential. It introduces a new interaction length scale to give rise to the existence of nonlocal effects in many quantum plasmas and has attracted significant attention in semiconductors (Gardner 1994) for studying the resonant tunneling diode and negative differential resistance. The latter has also interpreted the time taken by the electrons in quantum wells.

PLASMON WAKEFIELDS IN METALLIC NANOWIRES

Let us consider a short electron beam propagating in a cylindrical metallic nanowire (Ali, Tercas & Mendonca 2011, Ali, Shafiq-ur-Rehman, & Ding, 2018), whose spatial dimension is much larger than its radius, i.e., $L \gg a$ and contains conduction electrons with a neutralized background of positive ions. Since the electron de Broglie wavelength is much larger compared to ion de Broglie wavelength, therefore, immobile ions are assumed as classical species and conduction electrons follow the laws of quantum mechanics.

The electrons in quantum nanowires follow a quasi-one-dimensional QHD model, which reveals a dimensional quantization due to electron motion in radial direction and allows the electron motion to be free in longitudinal direction. So, we need to express accordingly the Laplacian operator in terms of perpendicular and parallel components, as $\nabla^2 = \nabla_\perp^2 + \partial_z^2$. The total electron density in such a nanowire is then given by $N = N_0 + N_1 + N_{b1}$, with N_1 being the perturbed electron density and $N_0(=N_{e0} \approx N_{i0})$ the unperturbed density, while N_{b1} denotes the electron beam density.

Figures 4. A short electron beam enters into a cylindrical metallic nanowire to excite plasmon modes.

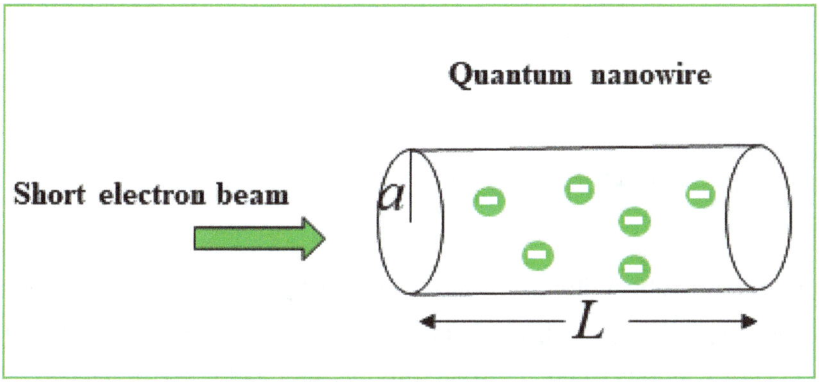

Thus, the dynamics of plasmons in a metallic nanowire can be governed by the following set of linearized QHD equations (Crouseilles 2008, Ali, Shahfiq-ur-Rehman, & Ding 2018, Cabral, Lapao & Mendonca 1993) including the continuity, momentum, and Poisson equations for electrons:

$$\frac{\partial N_1}{\partial T} + N_0 \frac{\partial U_1}{\partial Z} = 0, \tag{10}$$

$$\left(\frac{\partial}{\partial T} + \nu_c\right)U_1 - \frac{e}{m_e}\frac{\partial \phi_1}{\partial Z} - \frac{\hbar^2}{4m_e^2 N_0}\frac{\partial^3 N_1}{\partial Z^3} + \left(\frac{u_F^2 + u_{XC}^2}{N_0}\right)\frac{\partial N_1}{\partial Z} = 0, \tag{11}$$

and

$$\left(\frac{\partial^2}{\partial Z^2} + \nabla_\perp^2\right)\phi_1 - \frac{e}{\varepsilon_0}\left(N_1 - N_{b1} + N_{i0}\right) = 0, \tag{12}$$

where U_1 (ϕ_1) denotes the electron fluid velocity (self-consistent electrostatic potential), v_c stands for the electron-crystal lattice collisional frequency and ε_0 is the electric constant in SI units. In deriving Eq. (11), we have utilized the relations $\frac{N_1}{N_0 m_e}\frac{\partial \phi_{XC}}{\partial Z} = \frac{u_{XC}^2}{N_0}\frac{\partial N_1}{\partial Z}$ and $p_F = m_e u_F^2 N^3 / 3N_0^2$, where the electron Fermi speed, the electron XC speed and the electron XC potential are explicitly (Ali, Shahfiq-ur-Rehman, & Ding 2018, Khan & Hassan 2014) given by

$$u_F = \left(\frac{2k_B T_F}{m_e}\right)^{1/2}, u_{XC} = \left(\frac{0.985 e^2 N_0^{1/3}}{3m_e \varepsilon} + \frac{0.03349 e^2}{3m_e \varepsilon}\frac{18.37 N_0^{1/3}}{1+18.37 a_B^* N_0^{1/3}}\right)^{1/2},$$

and

$$\phi_{XC} = \frac{0.985 e^2}{\varepsilon}\left\{\frac{0.034}{a_B^* N^{1/3}}\ln\left(18.37 a_B^* N^{1/3}+1\right)+1\right\} N^{1/3}.$$

The effective Bohr atomic radius is denoted by $a_B^* \left(=\varepsilon\hbar^2 / m_e e^2\right)$ with $\varepsilon(= 4\pi\varepsilon_0)$ being the relative dielectric constant of metallic material. Notice that 4th term in Eq. (11) corresponds to quantum Bohm potential as higher order correction to the QHD model in the long wavelength limit, showing the quantum diffraction due to density fluctuations. The fifth and sixth terms in Eq. (11) purely determine the quantum statistical and electron exchange-correlation effects. In general, it may be noted that electron Fermi temperature in metal clusters, in ordinary metals, and in nanoparticles is strongly affected by the unperturbed electron density and is often taken much greater than the room temperature. If electron density (electron temperature) is high (low) enough then one cannot ignore the interactions between identical electrons because of overlapping of state functions. Hence, the XCs term is added in momentum equation in the adiabatic local-density approximation to account for XC potential acting as functional of electron density.

Next, we transform all the variables from Eqs. (10)-(12) in terms of Bessel function by utilizing the cylindrical coordinates (R,Θ,Z) and properly apply the Trivelpiece-Gould (TG) configuration (Cabral, Lapao & Mendonca 1993, Ali, Shahfiq-ur-Rehman, & Ding 2018) by following

$$\begin{aligned}
N_{b1}(R,\Theta,Z,T) &= \sum_{l,m} N_{bl,m}(Z,T) J_m(K_{\perp l,m}R)\exp(im\Theta),\\
N_1(R,\Theta,Z,T) &= \sum_{l,m} N_{l,m}(Z,T) J_m(K_{\perp l,m}R)\exp(im\Theta),\\
U_1(R,\Theta,Z,T) &= \sum_{l,m} U_{l,m}(Z,T) J_m(K_{\perp l,m}R)\exp(im\Theta),\\
\phi_1(R,\Theta,Z,T) &= \sum_{l,m} \phi_{l,m}(Z,T) J_m(K_{\perp l,m}R)\exp(im\Theta),
\end{aligned} \tag{13}$$

obtaining

$$\frac{\partial N_{l,m}(Z,T)}{\partial T} + N_0 \frac{\partial U_{l,m}(Z,T)}{\partial Z} = 0, \tag{14}$$

$$\left(\frac{\partial}{\partial T} + \nu_c\right) U_{l,m}(Z,T) - \frac{e}{m_e}\frac{\partial \phi_{l,m}(Z,T)}{\partial Z} - \frac{\hbar^2}{4m_e^2 N_0}\frac{\partial^3 N_{l,m}(Z,T)}{\partial Z^3} + \frac{u_F^2 + u_{XC}^2}{N_0}\frac{\partial N_{l,m}(Z,T)}{\partial Z} = 0, \tag{15}$$

and

$$\left(\frac{\partial^2}{\partial Z^2} - K_{\perp l,m}^2\right)\phi_{l,m}(Z,T) - \frac{e}{\varepsilon_0}\left\{N_{l,m}(Z,T) + N_{bl,m}(Z,T)\right\} = 0. \tag{16}$$

The physical quantities $U_{l.m}(Z,T)$, $\phi_{l,m}(Z,T)$, and $N_{l,m}(Z,T)$ represent the arbitrary amplitudes of the perturbations. The electron motion is confined in the radial direction denoted by the transverse wavenumber, $K_{\perp l,m}\left(=\frac{\alpha_{l,m}}{a}\right)$, specifying the finite-size effects of the system. Here $\alpha_{l,m}$ is the lth zero of the Bessel function of order m with l (m) being the radial (azimuthal) index, satisfying the condition $l \geq m$ [i.e., l is greater than or equal to m]. One can easily derive the basic low-lying modes propagating along the longitudinal direction in a nanowire, just by assuming the values $m = 0$ and $l = 1$, which may ultimately result in the form of quantized wavenumber, $K_{\perp 1,0}$.

Combining Eqs. (14), (15) and (16) together, we eventually obtain an evolution equation as

$$\left\{\frac{\partial^2}{\partial T^2} + \nu_c \frac{\partial}{\partial T} + \frac{\hbar^2}{4m_e^2}\frac{\partial^4}{\partial Z^4} - \left(u_F^2 + u_{XC}^2\right)\frac{\partial^2}{\partial Z^2}\right\}\left(\frac{\partial^2}{\partial Z^2} - K_{\perp l,m}^2\right) N_{l,m}(Z,T) + \omega_p^2 \frac{\partial^2}{\partial Z^2}\left\{N_{l,m}(Z,T) + N_{bl,m}(Z,T)\right\} = 0 \tag{17}$$

This the required equation to describe plasmon excitations caused by the propagation of a short electron pulse in a quantum nanowire. In order to look for the linear plasmon modes, we neglect the pulse density, i.e., $N_{bl,m}(Z,T) = 0$ in (17) and seek a plane wave solution to the perturbed quantities by substituting $\partial_T = -i\omega_{l,m}$ and $\partial_Z = iK$. Hence a more simplified form of the TG plasmons can be given as

$$\left\{(\omega_{l,m} + i\nu_c)\omega_{l,m} - \frac{\hbar^2 K^4}{4m_e^2} - \frac{\omega_p^2}{K^2 + K_{\perp l,m}^2} K^2 - \left(u_F^2 + u_{XC}^2\right) K^2\right\} N_{l,m} = 0. \tag{18}$$

This yields a distinct solution, provided the electron number density $N_{l,m} \neq 0$, as

$$\omega_{l,m} = -\frac{i\nu_c}{2} \pm \left(-\frac{\nu_c}{4} + \frac{\hbar^2 K^4}{4m_e^2} + \left(u_F^2 + u_{XC}^2\right) K^2 + \frac{\omega_p^2 K^2}{K^2 + K_{\perp l,m}^2}\right)^{1/2} = 0. \tag{19}$$

See that plasmon modes propagate along the axial direction with a wavenumber (K) and wave frequency ($\omega_{l,m}$), and are strongly modified by the dimensional quantization of electron motion in radial direction and quantum mechanical effects. For neglecting the XCs speed [viz., $u_{XC} = 0$] only, the TG plasmon modes with distinct finite-size effects can be reproduced (Ali, Tercas & Mendonca 2011). For neglecting the collisions and finiteness of the system, i.e., $\nu_c = K_{\perp l,m} = 0$, a well-known linear dispersion is retrieved for quantum Langmuir waves (Ren, Wu & Chu 2007) without taking the effects of Fermi pressure and XCs. However, the dispersion relation (19) in classical limits exactly leads to the earlier findings of TG oscillations in the plasma column.

To proceed further, let us assume that electron beam moves with a uniform constant velocity u_0 in a quantum nanowire along the longitudinal direction. So, we introduce a new moving frame with its spatial and temporal coordinates (Ali, Shahfiq-ur-Rehman, & Ding 2018) $\xi = Z - u_0 T$ and $\tau = T$, respectively, and use the scaling of different parameters, as

$$\tilde{\xi} = \xi / \lambda_F, \tilde{\tau} = \omega_p \tau, \tilde{N}_{bl,m} = N_{bl,m} / N_0, \tilde{K}_{\perp l,m} = K_{\perp l,m} \lambda_F, \tilde{u}_0 = u_0 / u_F, \tilde{u}_{XC} = u_{XC} / u_F,$$

and $\tilde{u}_F = 1$ in (17) to eventually derive the following result:

$$\begin{aligned} &\left\{\frac{\partial^2}{\partial \tilde{\tau}^2} + (\tilde{u}_0^2 - \tilde{u}_F^2 - u_{XC}^2)\frac{\partial^2}{\partial \tilde{\xi}^2} + \frac{H^2}{4}\frac{\partial^4}{\partial \tilde{\xi}^4} - 2\tilde{u}_0^2 \frac{\partial^2}{\partial \tilde{\tau} \partial \tilde{\xi}}\right\} \\ &\times \left(\frac{\partial^2}{\partial \tilde{\xi}^2} - \tilde{K}_{\perp l,m}^2\right) \tilde{N}_{l,m}(\tilde{\xi}, \tilde{\tau}) + \frac{\partial^2}{\partial \tilde{\xi}^2}\left(\tilde{N}_{l,m} + \tilde{N}_{bl,m}\right) = 0. \end{aligned} \tag{20}$$

Here the new quantum parameter $H[=\hbar\omega_p/(2k_BT_F)]$ is introduced, showing the ratio of plasmon energy-to-Fermi energy. We shall be omitting Tilda notations (~) from the normalized quantities in the upcoming equations for simplicity. It is clear that a short electron beam looks immobile in the moving frame and leaves somehow a static wake-field behind it. This helps us to assume a stationary solution [i.e., $\partial_\tau \approx 0$], which in other words gives rise to the utilization of quasi-static approximation in Eq. (20) to obtain a more simplified form

$$N_{bl,m}(\xi) + \left(\frac{H^2}{4}\frac{\partial^4}{\partial\xi^4} + A^2\frac{\partial^2}{\partial\xi^2} + B^4\right)N_{l,m}(\xi) = 0, \tag{21}$$

with

$$A = \left(u_0^2 - \frac{H^2}{4}K_{\perp l,m}^2 - 1 - u_{XC}^2\right)^{1/2} \text{ and } B = \left(1 - (u_0^2 - 1 - u_{XC}^2)K_{\perp l,m}^2\right)^{1/4}.$$

It is pertinent to express that Eq. (21) is the fourth order inhomogeneous partial differential equation due to the presence of a finite source and can be solved by the Green function methodology.

Applying the Fourier transform to Eq. (21), we get the Fourier transformed electron density, as

$$\hat{N}_{l,m}(K) = -\hat{N}_{bl,m}(K)\hat{G}(K), \tag{22}$$

where the Fourier transformed Green function is **defined** by

$$\hat{G}(K) = \left(\frac{H^2K^4}{4} - A^2K^2 + B^4\right)^{-1},$$

or

$$\hat{G}(K) \equiv \frac{1}{K_+^2 - K_-^2}\left(\frac{1}{K^2 - K_+^2} - \frac{1}{K^2 - K_-^2}\right), \tag{23}$$

with wavenumbers

$$K_{\pm}^{2} = \frac{2}{H^{2}}\left\{A^{2} \pm \left(A^{4} - B^{4}H^{2}\right)^{1/2}\right\}.$$

In deriving Eq. (23), we have made the use of partial fraction technique. The inverse Fourier transformed Green function can be expressed, as

$$G(\xi - \xi_{0}) = \tfrac{1}{K_{+}^{2} - K_{-}^{2}} \int_{-\infty}^{\infty} \frac{dK}{2\pi}\left(\tfrac{1}{K^{2} - K_{+}^{2}} - \tfrac{1}{K^{2} - K_{-}^{2}}\right)\exp[-iK(\xi - \xi_{0})] \tag{24}$$

Equation (24) is solved by using the residue theorem, which eventually leads to this form

$$G(\xi - \xi_{0}) = \tfrac{1}{K_{+}^{2} - K_{-}^{2}}\left(\frac{\sin K_{+}(\xi - \xi_{0})}{K_{+}} - \frac{\sin K_{-}(\xi - \xi_{0})}{K_{-}}\right) \tag{25}$$

However, applying the convolution theorem, Eq. (22) immediately becomes as

$$N_{l,m}\left(\xi\right) = \int_{-\infty}^{\infty} d\xi_{0} N_{bl,m}(\xi_{0}) G(\xi - \xi_{0}). \tag{26}$$

Now combining Eqs. (25) and (26) together and introducing a unit step function θ, we can finally come up with the following expression (Ali, Shahfiq-ur-Rehman, & Ding 2018):

$$N_{l,m}(\xi) = \frac{-1}{K_{+}^{2} - K_{-}^{2}} \int_{-\infty}^{\infty} N_{bl,m}(\xi_{0})\left(\frac{\sin K_{+}(\xi - \xi_{0})}{K_{+}} - \frac{\sin K_{-}(\xi - \xi_{0})}{K_{-}}\right)\theta(\xi_{0} - \xi)d\xi_{0}. \tag{27}$$

This is the plasmon wake-field equation indicating the electron density excitations caused by a short electron pulse moving in a quantum metallic nanowire along the axial direction. Let us assume that this electron beam follows the simplest normalized Gaussian pulse of the form $N_{bl,m}(\xi_{0}) = \exp\left(-\xi_{0}^{2} / \sigma^{2}\right)$ with a pulse width σ and dummy variable $\xi 0$. To find out whether the plasmon wake-field is stable or unstable, we need to solve inequalities like $A^2 > 0$, $B^4 > 0$, and $A^4 > H^2B^4$ simultaneously, which in turn impose certain constraints on the lower and upper bounds of the pulse speed as

$$u_{0\min} \le u_0 \le u_{0\max} \tag{28}$$

or

$$\left(1 + u_{XC}^2 + H - H^2 K_{\perp l,m}^2 / 4\right)^{1/2} \le u_0 \le \left(\left\{K_{\perp l,m}^2 + u_{XC}^2 K_{\perp l,m}^2 + 1\right\} / K_{\perp l,m}^2\right)^{1/2}. \tag{29}$$

Note that lower and upper limits of the pulse speed are significantly altered by the impact of quantum corrections including the electron XCs speed, the electron Fermi speed, the quantum dimensionless parameter, and finiteness of the system. Nevertheless, the condition $u_{0\min}=u_{0\max}$ can be used to determine a critical value of the transverse wavenumber [i.e., $K_{\perp l,m}^{cr} = \left(2 / H\right)^{1/2}$] in terms of quantum parameter H. Thus, the transverse wavenumber originally indicates the finite-size effect of the system along the radial direction and plays a key role in stabilizing the electrostatic wake-field excitations in quantum metallic nanowires.

Solution of the Problem and Implications

In this section, we solve numerically the solutions to our main findings in this Chapter, as given by Eqs. (19), (27) and (28) by using different metallic based quantum nanowires. For this purpose, we first take into account typical electron density $N_0 = 5.85 \times 10^{28}\text{m}^{-3}$ and electron Fermi temperature $T_F = 6.4 \times 10^4$K. These values are consistent to gold nanowires (Manfredi, 2005) at room temperature $T =$ 300K and help us in finding other relevant plasma parameters, e.g., the plasma oscillation frequency $\omega_p = 1.36 \times 10^{16}$/s, the plasmon energy $\hbar\omega_p \sim 8.97$ eV, the electron Fermi energy $\varepsilon_F = 5.49$ eV the electron Fermi length $\lambda_F = 0.102$nm, the electron Fermi speed $u_F = 1.39 \times 10^6$ m/s and the electron XCs speed $u_{XC} = 6 \times 10^5$ m/s. The dimensionless quantum parameter, which is the plasmon energy-to-Fermi energy ratio, yields a numerical value H = 0.817 for quantum electrons in gold nanowire.

A parametric variation of quantum parameters is shown in Table 1 for different metals at room temperature that may be helpful in understanding the plasmon's wakefield excitations and generating new radiation sources in nanowires of different metals.

The present model basically demonstrates stable wake-field excitations of the plasmon modes caused by a short electron pulse at quantum scales. The wave frequency of plasmon modes is denoted by $\omega_{l,m}$ with corresponding wavelength as

Table 1. Typical quantum parameters of different metals at room temperature

Metals	Ne0 (10^{28} m^{-3})	ω_p ($10^{16} s^{-1}$)	T_F (10^4K)	λ_F (nm)	u_F (10^6 m/s)	u_{XC} (10^5 m/s)	H (Scaled)	$K^c_{\perp 0} = \sqrt{2/H}$
Au	5.85	1.36	6.37	0.102	1.39	6.0	0.817	1.56
Li	4.7	1.223	5.511	0.105	1.29	5.845	0.847	1.53
Na	2.65	0.902	3.675	0.116	1.06	5.335	0.937	1.46
K	1.4	0.667	2.458	0.123	0.863	4.877	1.037	1.38
Rb	1.15	0.604	2.155	0.133	0.808	4.737	1.071	1.36
Cs	0.91	0.538	1.844	0.138	0.747	4.576	1.114	1.33
Cu	8.45	1.639	8.148	0.095	1.57	6.388	0.768	1.61
Ag	5.85	1.365	6.384	0.101	1.391	6.043	0.816	1.56
Be	24.2	2.775	16.432	0.080	2.231	7.510	0.644	1.76
Mg	8.6	1.654	8.244	0.095	1.580	6.406	0.766	1.61
Ca	4.6	1.209	5.432	0.106	1.283	5.826	0.850	1.53
Sr	3.56	1.064	4.579	0.117	1.178	5.60	0.887	1.50
Ba	3.2	1.009	4.265	0.126	1.137	5.516	0.903	1.48
Zn	13.1	2.041	10.914	0.089	1.818	6.832	0.714	1.67
Cd	9.28	1.718	8.673	0.094	1.621	6.480	0.756	1.62
Al	18.06	2.391	13.520	0.084	2.024	7.178	0.677	1.71
Ga	15.3	2.206	12.105	0.086	1.915	6.996	0.696	1.69
In	11.49	1.912	10.001	0.091	1.741	6.696	0.730	1.65
Pb	13.2	2.049	10.970	0.089	1.823	6.839	0.713	1.67
Sn	14.48	2.146	11.467	0.087	1.880	6.937	0.702	1.68

$\lambda_{\omega_{l,m}} \left(= 2\pi / K\right)$. Notice that this wavelength must be much longer compared to the electron Fermi length (λ_F) for the validation of QHD equations. For gold electrons, the Fermi length is found to be in the order of magnitude of an Ångstrom with a time-period of one-femtosecond, which is, of course, much shorter than the collisional timescale owing to electron-lattice collisions. Consequently, collisional effects (Manfredi, 2005) are neglected in the specific region of the Fermi distribution,

Figure 5. Normalized TG plasmon modes ($\tilde{\omega}_{l,m} = \omega_{l,m} / \omega_p$) as function of normalized wave number ($\tilde{K} = K\lambda_F$), where the radial and azimuthal indices are taken as l (=1,3) and m=0 with nanowire's radius a=4λF

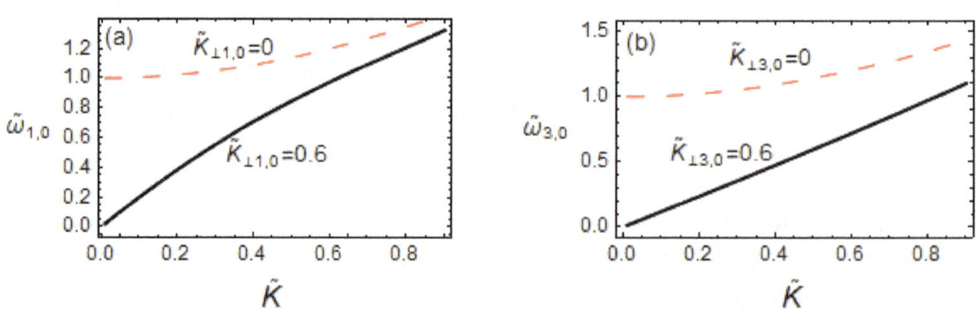

where quantum condition $T < T_F$ holds, even with large quantum coupling parameter $\Gamma_q > 1$.

The two normalized TG plasmon modes with their corresponding wave frequencies $\tilde{\omega}_{1,0}$ and $\tilde{\omega}_{3,0}$ are plotted against the normalized wave number ($\tilde{K}$) in a quantum nanowire [see Figure 5] with other fixed parameters $H = 0.81$, $\tilde{u}_F = 1$, and $\tilde{u}_{XC} = 0.42$. These modes are assumed to be quantized through the transverse wavenumber $\tilde{K}_{\perp l,m}$ which indicates the finiteness of the system in radial direction. The presence of mode quantization in the TG plasmon modes [as given by Eq. (19)]

Figure 6. Contours of the normalized TG plasmon modes ($\tilde{\omega}_{l,m} = \omega_{l,m} / \omega_p$) are displayed against the normalized longitudinal wavenumber ($\tilde{K} = K\lambda_F$) for varying the electron XCs speed $0u_F \leq u_{XC} \leq 1u_F$ and quantum parameter $0.01 \leq H \leq 0.91$. The radial and azimuthal indices l (=1,4) and m=0 are taken into account with a nanowire radius $a=4\lambda_F$ and electron Fermi speed $\tilde{u}_F = 1$.

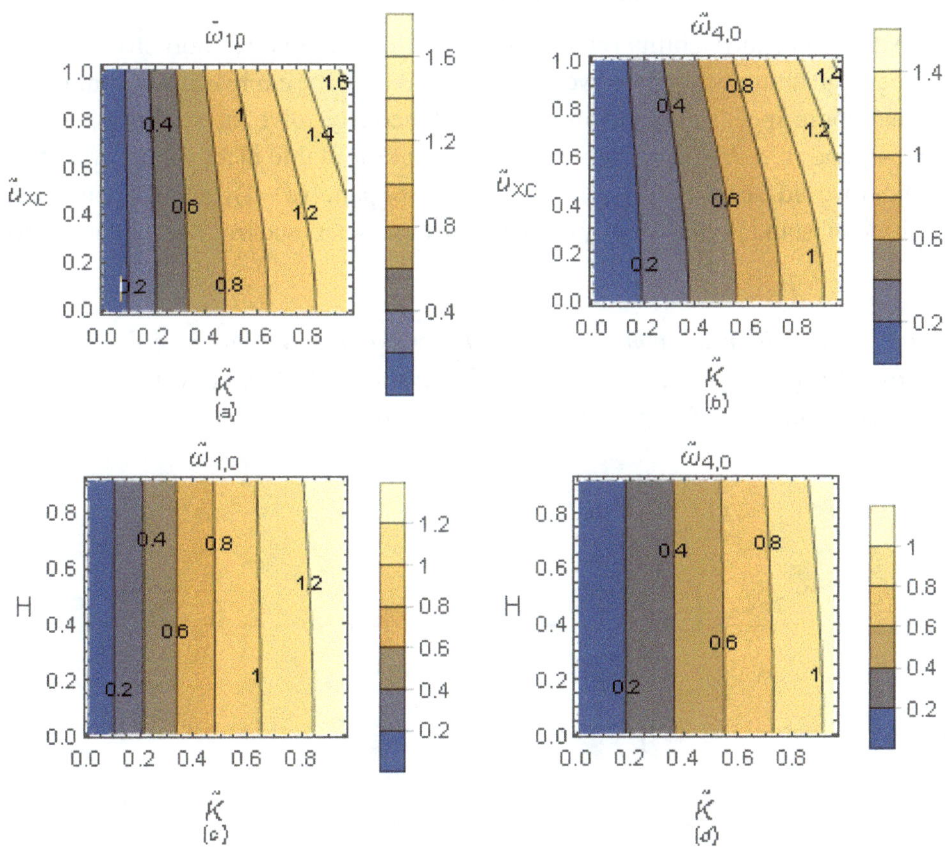

leads to an acoustic curve (Moradi, 2015) in similar way as that of shielding term in the linear dispersion relation of the usual ion-acoustic waves. For null modes quantization, i.e., $\tilde{K}_{\perp 1,0} = 0$ and $\tilde{K}_{\perp 3,0} = 0$, the frequencies of TG plasmon modes start at corresponding asymptotic values ($\tilde{\omega}_{1,0} = \tilde{\omega}_{3,0} = 1$) and tend to increase with varying the longitudinal wavenumber in the axial direction of nanowire.

Figure 6 displays how the mode quantization and quantum mechanical effects modify the contours of the linear TG plasmon modes ($\tilde{\omega}_{1,0}, \tilde{\omega}_{4,0}$) as a function of longitudinal wavenumber. Observe that electron XCs and quantum parameters affect the magnitudes of these frequencies at fixed mode quantization $\tilde{K}_{\perp 1,0} = 0.6$ [Figure 6 (a) and (c)] and $\tilde{K}_{\perp 1,0} = 2.94$ [Figure 6 (b) and (d)]. One can clearly see that lighter regions in the graphs determine higher magnitudes of frequencies, whereas darker regions identify the lower magnitudes of wave frequencies. Quantum corrections start playing a significant role in modifying the contours for larger values of the wavenumbers under an approximation $\tilde{K} << 1$. The magnitude of the low-lying TG plasmon mode at $\tilde{K}_{\perp 1,0} = 0.6$ is larger as compared to TG plasmon mode evaluated at $\tilde{K}_{\perp 1,0} = 2.94$ with fixed $\tilde{u}_F = 1$ and $\tilde{a} = 4$.

To determine the stability region for wakefields caused by a short electron pulse moving in quantum nanowires, we first need to derive the expressions for minimum and maximum speeds and for this, we simultaneously solve inequalities, i.e., $A^2 > 0$, $B^4 > 0$ and $A^4 > H^2B^4$. In Figure 7, we present the stable electrostatic wakefield regimes for gold [see panel (a)-(c)] and sodium [see panel (d)-(f)] metallic nanowires, which are substantially affected by the quantum parameters and finite-size quantization

Figure 7. The diagrams compare the stable wakefield regions of gold nanowire [Figure 7 (a)-(c)] with sodium nanowire [Figure 7 (d)-(f)] for varying the electron XCs effects with H=0.817 and H=0.932, respectively.

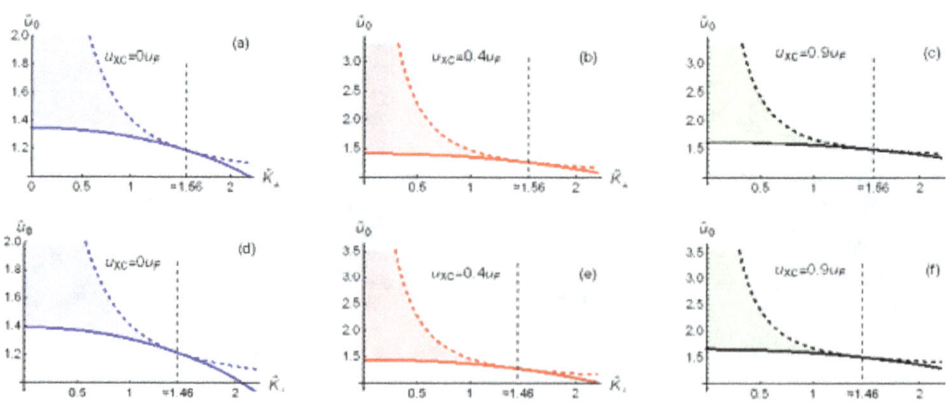

effects in the transverse direction. The shadowed areas correspond to stable regions for wakefield excitations and un-shadowed areas represent dynamical instability regions, which are separated by a vertical dashed line at $\tilde{K}_{\perp}^{cr} = \left(2 / H\right)^{1/2} \sim 1.56$ for gold metal and $\tilde{K}_{\perp}^{cr} = \left(2 / H\right)^{1/2} \sim 1.46$ for sodium metal.

The stability diagram shown in Figure 7(a) exactly matches the earlier investigation (Ali, Tercas & Mendonca 2011) to account for null electron XC effects. However, as the electron XCs effects increase, the wakefield regimes shift to the higher values of the pulse speed. The stability regimes and critical wavenumbers may vary quantitatively from metal to metal, depending on the electron equilibrium

Figure 8. Normalized density perturbations $\tilde{n}_{1,0} \left(= n_{1,0} / N_0\right)$ of the plasmon mode are displayed against the normalized axial distance $\tilde{\xi} \left(= \xi / \lambda_F\right)$ for changing (a) the XCs speed $u_{XC}=0u_F$ (blue solid curve), $u_{XC}=0.4u_F$ (red solid curve), and $u_{XC}=0.7u_F$ (green solid curve) with H=0.817, (b) the quantum parameter H(=0.417, 0.617, 0.917) with $u_{XC}=0.5u_F$, (c) the pulse speed u_0(-2.15, 2.19, 2.23)u_F with H=0.817, $u_{XC}=0.5u_F$, and (d) the mode quantization effect $K_{\perp}(= 0.267, 0.400, 0.480) / \lambda_F$ with H=0.817, and $u_{XC}=0.5u_F$. The electron pulse width and pulse speed are, respectively, taken as $\sigma=0.1\lambda_F$ and $u_0=2.15u_F$.

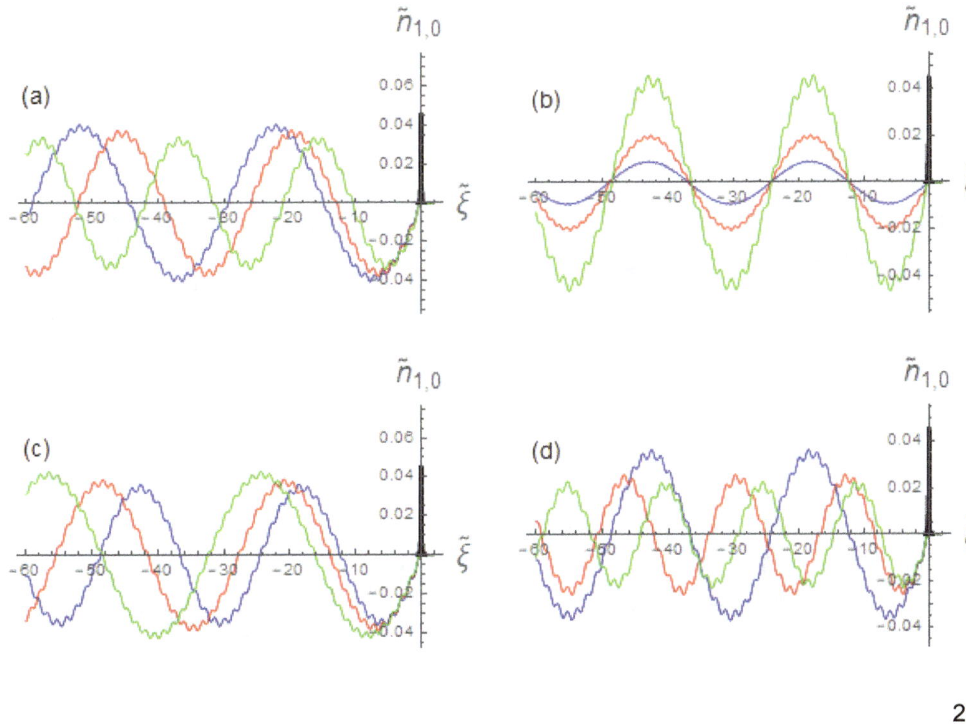

densities and Fermi energies. Hence, the transverse wavenumber plays a major role in stabilizing the wakefield in quantum nanowires.

It is now observed that electron density excitations appear in the form stable oscillatory wakefields along the longitudinal direction behind the short electron pulse in a quantum gold nanowire. The effect of quantum XCs reduces the magnitude of the wakefield and quantum diffraction leads to enhancement of amplitudes of plasmon excitations as displayed in Figures 8(a) and (b), respectively. In addition, the pulse speed also matters a lot to affect the plasmon's wakefield as well as the finiteness of the system significantly modifying the existence of wakefield at nanometer scales [see Figures 8(c) and (d), respectively].

The wakefield formation essentially provides a possibility for generating new radiation sources in quantum metallic nanowires, where conduction electrons can be trapped and accelerated by plasmon excitations to emit electromagnetic radiations, i.e., $\omega_{\pm} = cK_{\pm}$, with c the normalized speed of light in vacuum. In Figure 9, the

Figure 9. The variation of new wave numbers ($\tilde{K}_{\pm} = K_{\pm}\lambda_F$) is displayed as function of pulse speed ($\tilde{u}_0 = u_0 / u_F$) for varying the electron XC effects u_{XC}=0u_F (black solid - dashed curves), u_{XC}=0.4u_F (red solid - dashed curves), and u_{XC}=0.8u_F (blue solid - dashed curves) with mode finiteness effects (a) $K_{\perp} = 0.3 / \lambda_F$ and (b) $K_{\perp} = 0.8 / \lambda_F$. Similarly, the quantum parameter also varies through H=0.417, 0.617, 0.917 with finiteness effects (c) $K_{\perp} = 0.3 / \lambda_F$ and (d) $K_{\perp} = 0.8 / \lambda_F$.

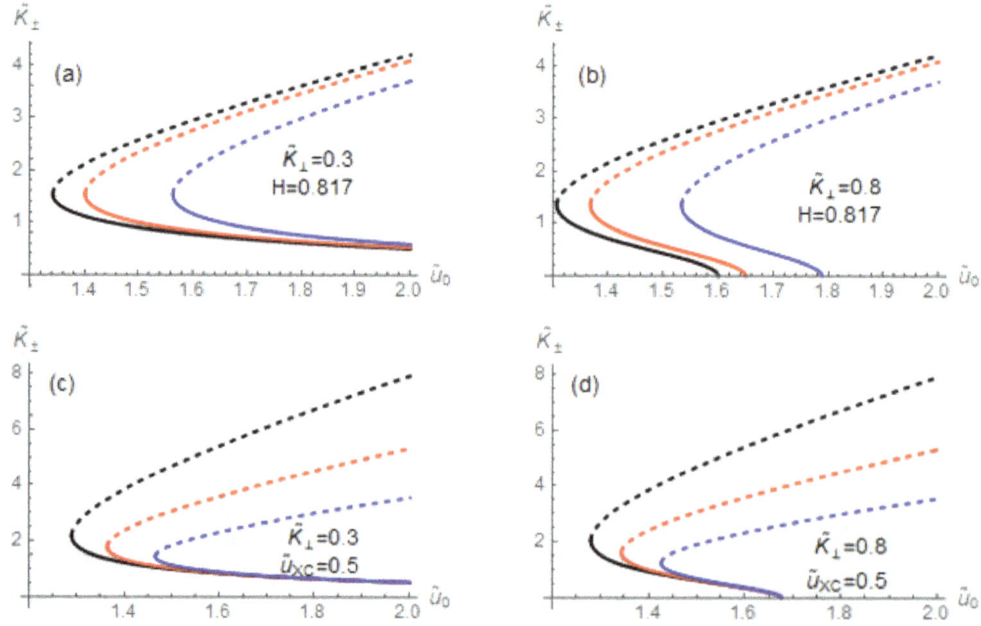

Figure 10. The normalized wavelengths ($\tilde{\lambda}_{\pm} = \lambda_{\pm} / \lambda_F$) are plotted against the quantum parameter H for varying the finite-size effects (a) $K_{\perp} = 0.267 / \lambda_F$ and (b) $K_{\perp} = 0.480 / \lambda_F$.

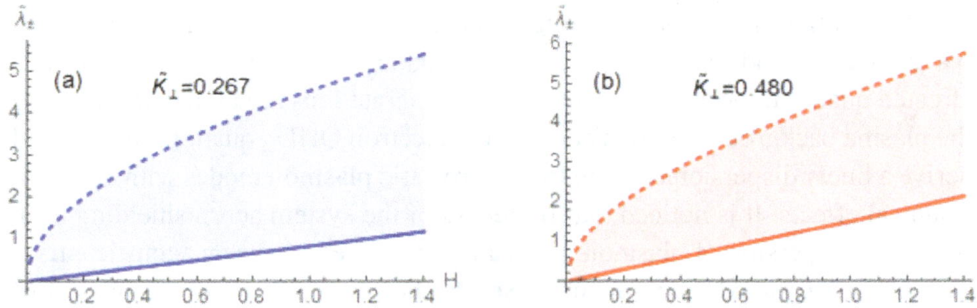

solid curves correspond to the wavenumber K_- and dashed curves are related to the wavenumber K_+. See that both wavenumbers $K_{\pm}$ have strong dependence on quantum corrections including the electron XCs and quantum diffraction, as well as, mode quantization effects.

It is now well-known (Ali, Shahfiq-ur-Rehman, & Ding 2018) that finiteness of the system squeezes the accessible wavelength range [see Figure 10] but these wavelength regimes are independent of the electron XCs effects in view of stability condition (29), where the XCs speeds are cancelled out in the definition of $K_{\pm}$. Thus, a short electron beam can excite the plasmon modes with a plasmon energy ~8.97eV in gold quantum nanowire and generate signals involving new radiations in the range [viz., $\lambda_{\pm}$~1-200 nm]. It is observed that these wavelengths are strongly depended on quantum H-parameter, the lower and upper beam speeds, as well as the finite-size effects of the system in the radial direction.

CONCLUSION

To conclude, we have studied the most fundamental properties of the classical and quantum plasmas. In quantum regimes, the length and time scales are seen quite different due to high equilibrium densities and low-temperatures in contrary to classical ones, which are characterized by low-densities and high-temperatures. We have made a literature survey to present the current status of the field on plasma collective modes and instabilities with quantum degenerate laws and forces, and explained novel features of quantum nanowires. In addition, the concept of free electron gas in metals is briefly discussed, relying mainly on the viewpoints of Drude, Sommerfeld and Bloch. We have also derived QHD equations for electrons by utilizing the Madelung representation in the Schrödinger-Poisson model and simplified the real and imaginary parts in terms of electron continuity and electron momentum equations.

Specifically, we have aimed at providing a theoretical model, which shows stable plasmon excitations in the form wakefields due to a short electron pulse propagating into a quantum cylindrical nanowire. In such a nanowire, the conduction electrons are treated quantum mechanically and non-degenerate ions are assumed immobile in the plasma background. In this context, the electron QHD equations are utilized to derive a linear dispersion relation for electrostatic plasmon modes with quantum mechanical effects. It is noticed that finiteness of the system acts a shielding term in the linear dispersion of plasmon mode and gives rise to have an acoustic curve/ behavior, in contrast to Langmuir curve. So, the dielectric and dispersive properties significantly depend on the mode quantization effects through the variation of radial and azimuthal indices.

The application of Fourier transform technique further leads to the derivation of a wake-field equation. The latter is analyzed to find out the stability conditions and solved it numerically at nanometer scales by choosing typical quantum parameters of metals. It is found that wakefield oscillates behind a short electron pulse moving in the axial direction of a quantum nanowire. Since these oscillations appear in terms of plasmon's wakefield excitations and therefore significantly modified by the variation of quantum statistics, quantum diffraction, and quantum exchange-correlations, as well as, the finite-size effects associated with transverse wavenumber. The present findings have relevance to metallic nanowires, where plasmon's wakefield is excited to generate new sources of radiation in the extreme-ultraviolet range.

ACKNOWLEDGMENT

The authors (S. Ali and I. Kourakis) gratefully acknowledge financial support from the project FSU-2021-012/8474000352 entitled "Modeling of Nonlinear Waves and Shocks in Space and Laboratory Plasmas" internally funded by Khalifa University of Science and Technology, Abu Dhabi, UAE. Funding is gratefully acknowledged from Abu Dhabi Department of Education and Knowledge (ADEK), currently ASPIRE UAE, through AARE-2018 (Abu Dhabi Award for Research Excellence) research grant ADEK/HE/157/18.

REFERENCES

Ali, S., Moslem, W., Shukla, P., & Kourakis, I. (2007). Fully nonlinear ion-sound waves in a dense Fermi magnetoplasma. *Physics Letters. [Part A], 366*(6), 606–610. doi:10.1016/j.physleta.2007.05.073

Ali, S., Shukla, N., Shukla, P. K., & Kourakis, I. (2007). Instability of drift-like waves and cross-field charged particle transport in a nonuniform collisional quantum magnetoplasma. *Europhysics Letters, 78*(4), 45001–p4. doi:10.1209/0295-5075/78/45001

Ali, S., Terças, H., & Mendonça, J. T. (2011). Nonlocal plasmon excitation in metallic nanostructures. *Physical Review. B, 83*(15), 153401–153404. doi:10.1103/PhysRevB.83.153401

Ali, S., Ur Rehman, S., & Ding, Z. J. (2018). Stable plasmon excitations in quantum nanowires. *Physics of Plasmas, 25*(8), 082115–082117. doi:10.1063/1.5041297

Allen, M. A., & Rowlands, G. (1993). Determination of the growth rate for the linearized Zakharov-Kuznetsov equation. *Journal of Plasma Physics, 50*(3), 413–424. doi:10.1017/S0022377800017244

Anderegg, M., Feuerbacher, B., & Fitton, B. (1971). Optically excited longitudinal plasmons in potassium. *Physical Review Letters, 27*(23), 1565–1568. doi:10.1103/PhysRevLett.27.1565

Ang, L. K., Kwan, T. J. T., & Lau, Y. Y. (2003). New scaling of child-Langmuir law in the quantum regime. *Physical Review Letters, 91*(20), 208303. Advance online publication. doi:10.1103/PhysRevLett.91.208303 PMID:14683407

Cabral, J. A. C., Lapão, L. M., & Mendonça, J. T. (1993). Nonlinear Trivelpiece–Gould waves: Recurrence, harmonic cascade, and sidebands. *Physics of Fluids. B, Plasma Physics, 5*(3), 787–795. doi:10.1063/1.860975

Chabrier, G., Douchin, F., & Potekhin, A. Y. (2002). Dense astrophysical plasmas. *Journal of Physics Condensed Matter, 14*(40), 9133–9139. doi:10.1088/0953-8984/14/40/307

Chen, F. F. (1984). *Introduction to Plasma Physics and Controlled Fusion*. Springer. doi:10.1007/978-1-4757-5595-4

Crouseilles, N., Hervieux, P. A., & Manfredi, G. (2008). Quantum hydrodynamic model for the nonlinear electron dynamics in thin metal films. *Physical Review. B, 78*(15), 155412. Advance online publication. doi:10.1103/PhysRevB.78.155412

Dirac, P. A. M. (1930). Note on exchange phenomena in the Thomas atom. *Mathematical Proceedings of the Cambridge Philosophical Society*, *26*(3), 376–385. doi:10.1017/S0305004100016108

Duan, X., Huang, Y., Cui, Y., Wang, J., & Lieber, C. M. (2001). Indium phosphide nanowires as building blocks for nanoscale electronic and optoelectronic devices. *Nature*, *409*(6816), 66–69. doi:10.1038/35051047 PMID:11343112

Ekman, R., Zamanian, J., & Brodin, G. (2015). Exchange corrections in a low-temperature plasma. *Physical Review. E*, *92*(1), 013104. Advance online publication. doi:10.1103/PhysRevE.92.013104 PMID:26274291

Fermi, E. (1927). Un metodo statistico per la determinazione di alcune priorietà dell'atome. *Rend. Accad. Naz. Lincei*, *6*, 602–607.

Feynman, R. P. (1972). *Statistical Mechanics, A Set of Lectures*. Benjamin.

Gardner, C. L. (1994). The quantum hydrodynamic model for semiconductor devices. *SIAM Journal on Applied Mathematics*, *54*(2), 409–427. doi:10.1137/S0036139992240425

Haas, F., Garcia, L. G., Goedert, J., & Manfredi, G. (2003). Quantum ion-acoustic waves. *Physics of Plasmas*, *10*(10), 3858–3866. doi:10.1063/1.1609446

Haas, F., Manfredi, G., & Feix, M. (2000). Multistream model for quantum plasmas. *Physical Review. E*, *62*(2), 2763–2772. doi:10.1103/PhysRevE.62.2763 PMID:11088757

Hohenberg, P., & Kohn, W. (1964). Inhomogeneous electron gas. *Physical Review*, *136*(3B), B864–B871. doi:10.1103/PhysRev.136.B864

Huang, Y., Duan, X., Cui, Y., Lauhon, L. J., Kim, K.-H., & Lieber, C. M. (2001). Logic gates and computation from assembled nanowire building blocks. *Science*, *294*(5545), 1313–1317. doi:10.1126cience.1066192 PMID:11701922

Infeld, E., & Rowlands, G. (2000). *Nonlinear Waves, Solitons and Chaos*. Cambridge University Press. doi:10.1017/CBO9781139171281

Isihara, A. (1968). *Electron Liquids*. Springer Verlag.

Johnson, J. C., Choi, H. J., Knutsen, K. P., Schaller, R. D., Yang, P., & Saykally, R. J. (2002). Single gallium nitride nanowire lasers. *Nature Materials*, *1*(2), 106–110. doi:10.1038/nmat728 PMID:12618824

Jones, W. E., Kliewer, K. L., & Fuchs, R. (1969). Nonlocal theory of the optical properties of thin metallic films. *Physical Review*, *178*(3), 1201–1203. doi:10.1103/PhysRev.178.1201

Jung, Y. D. (2001). Quantum-mechanical effects on electron–electron scattering in dense high-temperature plasmas. *Physics of Plasmas*, *8*(8), 3842–3844. doi:10.1063/1.1386430

Khan, S. A., & Hassan, S. (2014). Effects of electron exchange-correlation potential on electrostatic oscillations in single-walled carbon nanotubes. *Journal of Applied Physics*, *115*(20), 204304. doi:10.1063/1.4878936

Kohn, W., & Sham, L. J. (1965). Self-consistent equations including exchange and correlation effects. *Physical Review*, *140*(4A), A1133–A1138. doi:10.1103/PhysRev.140.A1133

Kremp, D., Bornath, T., Bonitz, M., & Schlanges, M. (1999). Quantum kinetic theory of plasmas in strong laser fields. *Physical Review. E*, *60*(4), 4725–4732. doi:10.1103/PhysRevE.60.4725 PMID:11970338

Kremp, D., Schlanges, M., & Kraft, W. D. (2005). *Quantum Statistics of Nonideal Plasmas*. Springer.

Lee, M. R., Eckert, R. D., Forberich, K., Dennler, G., Brabec, C. J., & Gaudiana, R. A. (2009). Solar power wires based on organic photovoltaic materials. *Science*, *324*(5924), 232–235. doi:10.1126cience.1168539 PMID:19286521

Leemans, W. P., Nagler, B., Gonsalves, A. J., Tóth, C., Nakamura, K., Geddes, C. G. R., Esarey, E., Schroeder, C. B., & Hooker, S. M. (2006). GeV electron beams from a centimeter-scale accelerator. *Nature Physics*, *2*(10), 696–699. doi:10.1038/nphys418

Lindau, I., & Nilsson, P. (1970). Experimental evidence for excitation of longitudinal plasmons by photons. *Physics Letters. [Part A]*, *31*(7), 352–353. doi:10.1016/0375-9601(70)90978-3

Liu, M., Guyot-Sionnest, P., Lee, T. W., & Gray, S. K. (2007). Optical properties of rodlike and bipyramidal gold nanoparticles from three-dimensional computations. *Physical Review. B*, *76*(23), 235428. Advance online publication. doi:10.1103/PhysRevB.76.235428

Loffredo, M. I., & Morato, L. M. (1993). On the creation of quantized vortex lines in rotating He II. *Il Nuovo Cimento B*, *108*(2), 205–215. doi:10.1007/BF02874411

Manfredi, G. (2005). How to model quantum plasmas. *Fields Inst. Commun.*, *46*, 263–287.

Manfredi, G., & Haas, F. (2001). Self-consistent fluid model for a quantum electron gas. *Physical Review. B*, *64*(7), 075316. Advance online publication. doi:10.1103/PhysRevB.64.075316

Markowich, P. A., Ringhofer, C. A., & Schmeiser, C. (1990). *Semiconductor Equations*. Springer-Verlag. doi:10.1007/978-3-7091-6961-2

McMahon, J. M., Gray, S. K., & Schatz, G. C. (2009). Nonlocal optical response of metal nanostructures with arbitrary shape. *Physical Review Letters*, *103*(9), 097403. Advance online publication. doi:10.1103/PhysRevLett.103.097403 PMID:19792829

Moradi, A. (2015). Plasmon modes of metallic nanowires including quantum nonlocal effects. *Physics of Plasmas*, *22*(3), 032112–032115. doi:10.1063/1.4916059

Moslem, W. M., Ali, S., Shukla, P. K., & Eliasson, B. (2008). Three-dimensional electrostatic waves in a nonuniform quantum electron-positron magnetoplasma. *Physics Letters. [Part A]*, *372*(19), 3471–3475. doi:10.1016/j.physleta.2007.09.080

Moslem, W. M., Ali, S., Shukla, P. K., Tang, X. Y., & Rowlands, G. (2007). Solitary, explosive, and periodic solutions of the quantum Zakharov-Kuznetsov equation and its transverse instability. *Physics of Plasmas*, *14*(8), 082308. doi:10.1063/1.2757612

Ourabah, K., & Tribeche, M. (2013). Quantum ion-acoustic solitary waves: The effect of exchange correlation. *Physical Review. E*, *88*(4), 045101. doi:10.1103/PhysRevE.88.045101 PMID:24229311

Pamyatnykh, E. A. (2020). Electron quantum waves in metal nanowires. *The Physics of Metals and Metallography*, *121*(5), 405–407. doi:10.1134/S0031918X20050099

Parkes, E. J., & Munro, S. (2005). The stability of obliquely-propagating solitary-wave solutions to Zakharov–Kuznetsov-type equations. *Journal of Plasma Physics*, *71*(05), 695. doi:10.1017/S0022377805003727

Petersilka, M., Gossmann, U. J., & Gross, E. K. U. (1996). Excitation Energies from Time-Dependent Density-Functional Theory. *Physical Review Letters*, *76*(8), 1212–1215. doi:10.1103/PhysRevLett.76.1212 PMID:10061664

Sabry, R., Moslem, W., & Shukla, P. (2008). Explosive and solitary excitations in a very dense magnetoplasma. *Physics Letters. [Part A]*, *372*(35), 5691–5694. doi:10.1016/j.physleta.2008.06.090

Shpatakovskaya, G. V. (2006). Semiclassical model of a one-dimensional quantum dot. *Journal of Experimental and Theoretical Physics*, *102*(3), 466–474. doi:10.1134/S1063776106030095

Shukla, P., Brodin, G., Marklund, M., & Stenflo, L. (2009). Excitation of multiple wakefields by short laser pulses in quantum plasmas. *Physics Letters. [Part A]*, *373*(35), 3165–3168. doi:10.1016/j.physleta.2009.06.053

Shukla, P. K., & Akbari-Moghanjoughi, M. (2013). Hydrodynamic theory for ion structure and stopping power in quantum plasmas. *Physical Review. E*, *87*(4), 043106. Advance online publication. doi:10.1103/PhysRevE.87.043106 PMID:23679529

Shukla, P. K., Ali, S., Stenflo, L., & Marklund, M. (2006). Nonlinear wave interactions in quantum magnetoplasmas. *Physics of Plasmas*, *13*(11), 112111–112116. doi:10.1063/1.2390688

Shukla, P. K., & Eliasson, B. (2006). Formation and Dynamics of Dark Solitons and Vortices in Quantum Electron Plasmas. *Physical Review Letters*, *96*(24), 245001. Advance online publication. doi:10.1103/PhysRevLett.96.245001 PMID:16907249

Terças, H., Mendonça, J. T., & Shukla, P. K. (2008). Quantum Trivelpiece–Gould waves in a magnetized dense plasma. *Physics of Plasmas*, *15*(7), 072109. doi:10.1063/1.2947235

Thomas, L. H. (1927). The calculation of atomic fields. *Mathematical Proceedings of the Cambridge Philosophical Society*, *23*(5), 542–548. doi:10.1017/S0305004100011683

von Roos, O., & Zmuidzinas, J. S. (1961). Effect of electron exchange on the dispersion relation of plasmons. *Physical Review*, *121*(4), 941–942. doi:10.1103/PhysRev.121.941

Wei, L., & Wang, Y.-N. (2007). Quantum Ion-Acoustic Waves in Single-Walled Carbon Nanotubes Studied with a Quantum Hydrodynamic Model. *Physical Review. B*, *75*(19), 193407. doi:10.1103/PhysRevB.75.193407

Weizsäcker, C. F. V. (1935). Zur Theorie der Kernmassen. *Zeitschrift für Physik*, *96*(7–8), 431–458. doi:10.1007/BF01337700

Willets, K. A., & Van Duyne, R. P. (2007). Localized surface plasmon resonance spectroscopy and sensing. *Annual Review of Physical Chemistry*, *58*(1), 267–297. doi:10.1146/annurev.physchem.58.032806.104607 PMID:17067281

Xiang, J., Lu, W., Hu, Y., Wu, Y., Yan, H., & Lieber, C. M. (2006). Ge/Si nanowire heterostructures as high-performance field-effect transistors. *Nature*, *441*(7092), 489–493. doi:10.1038/nature04796 PMID:16724062

Zakharov, V. E., & Kuznetsov, E. A. (1974). Three-dimensional solitons. *Soviet Physics, JETP*, *39*, 285–286.

Chapter 2
Dynamical Spectra in Two-Dimensional Dusty Plasmas

Aamir Shahzad
Government College University, Faisalabad, Pakistan

Zakia Rafiq
Government College University, Faisalabad, Pakistan

Alina Manzoor
Government College University, Faisalabad, Pakistan

Muhammad Kashif
Government College University, Faisalabad, Pakistan

ABSTRACT

Equilibrium molecular dynamics (EMD) simulation has been employed to explore the dynamical structure factors (DSFs) of two dimensional (2D) dusty plasma systems for a wide domain of plasma parameters of Coulomb coupling (Γ) and Debye screening strength (κ). The influence of varying wave vectors (k) on plasma DSFs $S(\mathbf{k}, \omega)$ have been reported with different combinations of plasma state points (Γ, κ). New simulations has been tested for the influence of different wave vectors on plasma density $S(\mathbf{k}, \omega)$ in addition to different combinations of plasma state points. New results of plasma density $S(\mathbf{k}, \omega)$ show that amplitude of oscillation and frequency will vary with increasing value of Coulomb coupling parameter (Γ) and Debye screening strength (κ). These simulation techniques show that transient behavior has been reported for frequency (ω) with various values of Debye screening strength (κ). Moreover, EMD simulation has been checked in order to investigate the behavior of dusty plasma to evaluate DSFs. The outcomes of EMD simulations are matched to earlier known numerical and experimental data. It has been shown that fluctuation of dynamical density increases at intermediate to higher values of coupling parameter. However, it shows less fluctuation at higher values of Debye screening strength (κ).

DOI: 10.4018/978-1-7998-8398-2.ch002

1. INTRODUCTION

The dynamic structure factors (DSFs) is mainly generalization of structure factor which tell us about correlation in time and space. A mathematical function called DSFs gives information about correlation of inter-particles and time evolution. In simple and complex systems, the information about dynamic and static properties is taken through DSFs. The measurable quantities such as ratio of specific heats, adiabatic sound velocity and thermal diffusivity are given by DSFs in hydrodynamic condition. The various techniques such as inelastic neutron scattering, x-rays and light scattering are mostly used for the evaluation of dynamic structure factor of glasses, liquids and dense plasmas. The evolution of structural behavior of different materials in material science has become tremendous task and also shows different applications in complex (dusty) plasmas. Various approaches like theoretical, experimental and computational have been used to attain various developments in science and technology. Complex plasma usually consists of dust particles of various size (nano to millimeter) and shapes usually circular/ spherical. Particles (dust) in plasma establish their pattern in given system. The interaction of plasma and dust particles has helped effectively to investigate new region of research and developments. In condensed matter physics, the crystalline structure of the complex plasma has proved to be an authentic tool to determine physical process. Various experimental and theoretical (and or simulation) techniques are efficiently used to examine the structural behavior of different materials (including liquid materials). The numerically investigation of structural behavior of the various complex systems has become a significant task in material science. In thermophysics and allied sciences, the exploration of a new system such as non- ideal (complex) plasma usually referred as dusty plasmas helped to discover another research gate. The dust particles in plasma show interaction with the plasma particles (electrons and positive ions).

Dusty plasma usually in extreme limits considered as a many body system for both strong and weak interactions. One is observed gaseous behavior for weak interactions. Reliable and authentic theoretical approaches are basically used based on LRT (linear response theory) to determine dynamical structural properties of the different materials in the absence of effects of the correlation. If interactions are strong then particles get localization and systems turned into crystallize form (Shahzad & He, 2012c). The liquid and solid conditions of the complex plasma (dusty) usually strongly coupled complex (dusty) plasmas have great importance in this regard. The coupling parameter of dusty plasma usually related with temperature (1/T) and describes states in this plasma region. Therefore, the values of coupling parameter help to illustrate the plasma behavior from non-ideal nature to crystal-structure states and ordered state also called crystal state of plasma.

1.1 Plasma

Plasma is basically the 4th state of matter and it was firstly introduced by Sir William Crooks. It was introduced in 1879. Irving Langmuir was the first scientist who recommended the name of "Plasma" for this state of matter. For this achievement "Irving Langmuir" was also awarded Nobel Prize. About 99% matter in our universe is in the plasma phase whereas remaining 1% is either gas or liquid or solid. It is an ionized gas composed of electrons, ions and neutral particles. In a well-organized way, "Plasma is an ionized gas that must have two characteristics, collective behavior and quasi-neutrality. The word quasi neutral shows that plasma possesses equal density of positive charge particles and negative electrons such as ($ni \approx ne$). Low temperature plasma also exists in flames. We can define plasma as "plasma is quasi-neutral gas of charged and neutral particles which exhibits collective behavior". It is such an electrically conductive state of matter that responds collectively to electromagnetic forces. Plasma is obtained at a very temperature nearly a few tens of thousands of degrees which is the lowest temperature at which plasma is usually observed. Plasma is an electrically conducting fluid. Plasma is described as a gas having high temperature and it consists of charged species which show classical or quantum mechanical behavior. Moreover, the word plasma is combination of Greek letters τλᾰμα meaning something which is fabricated, molded or composed (Chen, 2006).

1.2 Dusty Plasma

Dusty plasma was introduced by Mendis in 1997. For the study of dusty plasma, the existence of bright comet provided an efficient cosmic laboratory to study plasma particles interactions as well as their dynamical and physical consequences. Other manifestations provided by this plasma type by ancient were the existence of zodiacal light, Nebula, Notilucent clouds, planetary rings etc. (Shahzad, Shakoori, Mao-Gang et al, 2019). Hannes Alfven along with Spitzer investigated that dust was not the cause of hindrance in universe, but it is the essential part of our universe. It is also observed from last survey that dust particles in shape of spokes are rotating around Saturn Rings. These described observations was improved and made by Cassini space-craft in 2005. A European spacecraft Ulysses in 1992 also detected these particles and measured their speed and masses efficiently. Meanwhile, NASA space-craft Galileo in (1995) give the verification of the existence of the dust particles around Jupiter. An enormous interest has been started from the mid of 1980 to study this plasma type along with the laboratory investigations of DSFs (dynamical structure factor). Now this plasma (dusty) is very significant in medical science, industries as well as in energy sectors etc.

1.3 Plasma Classifications

The plasma may be categories based on electrons or Ions temperature, degree of ionizations or density.

1.3.1 Ideal Plasma

The Coulomb coupling parameter basically differentiate these two types of plasma (ideal and non-ideal). The coupling parameter can be denoted such as Γ=P.E/K.E, it is ratio of potential to kinetic energy of dusty particles. The value of this parameter is less than '1' due to maximum value of kinetic energy of dust particles in case of ideal plasma. The charge dust particles have greater value of K.E. as contrast to P.E. This plasma parameter Γ has negligible or very small value in this type of plasmas. This plasma system does not have proper structure and interaction of particles is also less due to low density (Boulos et al., 2013).

1.3.2 Non-Ideal Plasma

In case of -deal plasmas, the particles possess greater value of potential energy as compare to kinetic energy here, the value of (Γ) is greater from one or may approaches to '1'. The presence of dust particles is responsible to make this plasma, dusty plasma.

1.4 Dusty (Complex) Plasma

A complex or dusty plasma is a type of plasma having electrons, neutral and positive particles as well as some other particles also exist in this plasma type and particles size vary from Nano to Millimeter range. These foreign particles are also known as dust particles. Sometimes aggregations occur between these particles and size become large. Mixture of gases also used such as argon, oxygen and silane etc. for the formation and growth of these particles. The sputtering of atoms and molecules from the walls and by electrodes of the plasma system causes the formation of the dust particles. The development of these particles in the system (plasma) is nucleation, surface growth and coagulation etc. There is vital role of Coulomb interaction and the thermal fluctuation in the description of the dusty plasmas system (strongly coupled). These dust particles are generally suspended with few electrons in gaseous system (plasma). At room temperature, the dusty clouds are also found in plasma system (dusty). The dust particles possess large mass comparatively to ions and electrons resulting in the slowness of down time scale, which is why the dynamical behavior could be studied properly in real time and space system (Shahzad, Shakoori, He et al, 2019; Shahzad, Shakoori, Mao-Gang et al, 2019). Complex plasmas have very low value of temperature and its components are neutral atoms, electrons, ions and dust particles etc. In different fields such as (material science, plasma chemistry, plasma physics, space & astrophysics, etc.), the properties of dusty grains are studied. The examples of dusty plasmas are planetary rings, comets, circumstellar and interstellar clouds, interplanetary space etc. The presence of these dusty particles alters the equality of plasma system leading to possess the wonderful and interesting phenomenon. This plasma system has applications in different fields such as fluid dynamics, acoustics, plasma physics, material science, optics, Nano physics and Nanotechnology etc. Some properties of dusty particles are given below:

a) Charge on Dust Particle

The presence of dust particles in plasma as well as charge on them has introduced an interesting field in science and technology. The value of charge differs from (zero to thousands of electronic charge) totally depending on the shape and size of the dusty particles. The dust particles get charged on immersing in plasma system due to surface contact with plasma. The determination of charge is usually done by balancing positive and negative charges in plasma. Both currents produced by electrons and positive particles in plasma system must have equal values in stable state. As electrons possess more mobility than positive ions, the surfaces get negative charge and strongly attracts positive ions. The repulsion of electrons by negatively charged dust particles produces a motionless state. So, by adjusting the fluxes between positive ions and electrons the charge on dust particles is obtained. Due to associations among these particles, the phenomenon of waves, instabilities and collective behavior are observed.

b) Size of Dust Particle

Dust particle possess large size than positive ions and electrons and size of these particles vary from millimeter to nanometer range. The diameter of these particles usually varies from 3.50 to 0.05μm. The dust particles have any shape mostly spherical, so it is easily observed in laboratory with the help of any microscope. It is made through dielectric material or conducting material (Shukla & Mamun, 2015).

c) Nature of Dust Particle

Dust particles show interaction with other particles (electrons and positive ions) present in plasma system and gets charge. As we know that size of dust particle is much larger than electrons and positive ions inside plasma so, dust particle can acquire charge 10^4e approximately. In this type of plasma, the momentum, mass and energy etc. are shifted through these dusty particles. Dusty particle may exist in gaseous, liquid and solid form.

1.5 Applications of Plasma

There are number of applications of plasmas both in industries and technologies. Plasma is available in naturally and manmade plasma applications. Here, there are application of plasmas in different ways.

1.5.1 Plasma in Industry

The surfaces of different contact lenses, optics are clearly cleaned by using different plasma using tools. The organic-contaminated surfaces are properly cleaned through these tools in different ways. The adhesion ability of the automobile tools in industry can be increased externally and internally using authentic plasma treatments without heating effect. The work of Henniker is very famous in automobile industry due to controlling of plasma systems in authentic and good way (Shahzad & He, 2012a). His mission was improvement of adhesion property of different automobiles. These plasma treatments greatly enhanced the adhesion property and cleaning of great materials. The dust particles also used in the developing of semiconductor chips causing efficient flow of current (Shahzad & He, 2012b; Shahzad & He, 2014). This Plasma property also play significant role in aerospace industry to manufacture polymers and composite materials. This treatment of plasma is very efficient in order to increasing adhesion and cleaning of different materials.

1.5.2 Plasma in Daily Life

As plasma is a forth state of matter, this state also greatly influenced human lives additionally. It has a significant rule in energy productions, intense beam of laser, in development of fusion reactions, high power radiation source, lightening, sterilizations of various surgical and medical tools, purification of water plants, as well as for controlling radiations etc. X-rays and different types of radiations which are emitted through plasma sources possess different applications in different fields of life.

1.5.3 Plasma in Textile

The fourth state of matter (plasma) also plays a vital role in development of textile industry greatly and in an excellent way. It is greatly used in surface treatments and a lot of benefits have been found by using plasma in this industry. By using various techniques of plasma, like techniques of cold plasma widely used to vary the surface of fabrics properly. Different techniques exhibit different and excellent positive effects in dying, printing and shrinking of wool properly. For proper repelling of dirty things on fabrics hydrophilic treatments are properly utilized (Sparavigna,

2008). It presence is excellent for environment due to adaptation of the upper layer surface without changing bulk properties. It has great importance to enhance various properties like; in repelling of liquid, in excellent surface making and for greatly increasing wet ability properly.

1.6 Dynamical Structure Factors (DSFs)

The dynamic structure factors (DSFs) is mainly generalization of structure factor which tells us about correlation in time and space. A mathematical function called DSFs gives information about correlation of inter-particles and time evolution. In simple and complex systems, the information about dynamic and static properties is taken through DSFs. The measurable quantities such as ratio of specific heats, adiabatic sound velocity and thermal diffusivity are given by DSFs in hydrodynamic condition. The various techniques such as inelastic neutron scattering, x-rays and light scattering are mostly used for the evaluation of dynamic structure factor of glasses, liquids and dense plasmas. The many body systems such as strongly coupled complex dusty plasmas (SCCDPs) which show various collective excitations, the various properties of such system are investigated through theoretical approaches and mathematical simulations. The pair correlation function could not be evaluated directly in condensed matter physics, however DSFs is determined directly through light scattering or x-rays. The DSFs is basically Fourier Transformation of pair correlation function. The extreme states of warm dense matter are diagnosed through DSFs (Shahzad, Shakoori, He et al, 2019). The fluctuation in dynamical density $S(\boldsymbol{k}, \omega)$ leads to spectra of current correlation, like longitudinal and transverse currents. DSFs show association with the density fluctuation thus giving determination of transport properties, stopping power and electron-ion temperature properly. Dust particles are active species (chemically) and they also possess either positive or negative charge depending upon situation.

2. MODEL AND NUMERICAL TECHNIQUE FOR DSFS

In this section, we described about mathematical model and simulation tool to investigate about dynamic structure factors (DSFs). In equilibrium molecular dynamics (EMD) technique, Newton's second law of motion is used to calculate the interaction force and $n = N/V$ gives the number density of dust particles in given plasma system. The EMD simulation is employed in computational box of volume V. The simulations are performed for Yukawa system having number of particles i.e. (i=1, 2, 3…N) interacting dust grains. The summation of forces is calculated due to dust grains as $Fi = \sum_j F_{ij}$ here, F_i is derivative form of Yukawa potential, given in equation (1). First potential energy function (Yukawa potential) that explains interacting force among dust particles within system is chosen. For strongly coupled Yukawa liquids (SCYL), the interaction potential between dust particles is given;

$$\phi(|r|) = \frac{Q^2}{4\pi\varepsilon_0} \frac{e^{-|r|/\lambda_D}}{|r|} \tag{1}$$

In above equation (1), Q is charge on particle, **r** is interparticle distance of dust particle, $\in_0$ permittivity for free space and λD is Debye screening length. An atomistic depiction of system, a function such as number density at a point r and time t is given as:

$$\int \rho(r,t)dr = N \tag{2}$$

The equation (2) explains the condition for conservation of matter and then given as:

$$\rho(k,t) = \int \rho(r,t)e^{-ik.r}.dr \tag{3}$$

In this case, k is along z-direction. The above data gives information for a range of wave numbers which are multiple of k. Here $k=2\pi/L$, where L is length of simulation box. The Fourier transform for van Hove function

$$F(k,t) = \int G(r,t)e^{ik.r}dr \tag{4}$$

Here $F(\boldsymbol{k}, t)$ is the intermediate function (scattering), and can be evaluated through neutron spin-spectroscopy. Similarly, $G(\boldsymbol{r},t)$ is Van Hove (scattering) function.

$$G(r,t) = 1/_N \left\langle \int \rho(r' + r,t)\rho(r',0)dr' \right\rangle \tag{5}$$

The DSFs is temporal and spatial Fourier transform of van- Hove time dependent pair correlation function. Therefore, F(r, t) is Fourier components of density such as;

$$F(k,t) = \frac{1}{N}\langle \rho(k,t)\rho(-k,t)\rangle \tag{6}$$

So, the DSF is given as;

$$S(k,\omega) = \frac{1}{2\pi}\int F(k,t)e^{i\omega t}dt \tag{7}$$

$$S(k,t)dt = S(k) \tag{8}$$

The dynamic structure factor is linked to longitudinal current correlation

$$c_l\left(k,t\right) = \frac{k^2}{N_m}\left\langle \pi_z\left(k,t\right)\pi_z\left(-k,0\right)\right\rangle \tag{9}$$

Here k is wave number.

3. PARAMETERS AND SIMULATION TECHNIQUES

In this part, we choose a system having number of particles N=400 with varying values of plasma parameters (Γ, κ), wave number k and particles of this system pair-wise interact with each other through a potential called Yukawa potential as given in equation (1). EMD simulation has been used to evaluate dynamic structure factor of strongly coupled dusty plasmas which is given in equations (7) and (8). The dimensions of the simulation box are represented as $L_{x,}$ L_y and L_z. Periodic boundary condition are adopted to reduce surface size effect and applied for the simulation box including minimum image convection. In this case, main calculations are performed for N=400 particles at κ =1.4, 2.5 and 3.5 with various values of Coulomb coupling parameter (1, 5, 10, 20, 50,100) at waves numbers k (≡ 0, 1, 2, and 3). EMD technique is reported to evaluate DSFs for SCCDPs over suitable range of plasma parameters. The time step for this simulation technique is given as dt = $0.005/\omega_p$ that help us to evaluate sufficient data for long simulation run. The reported simulations are performed between $5.0\times10^5/\omega_p$ and $3.0\times10^5/\omega_p$ time units (Shahzad & He, 2013; Shukla & Mamun, 2015).

4. COMPUTATIONAL RESULTS

Figure 1 illustrates the numerical results obtained using EMD simulations for strongly coupled plasma system to evaluate dynamical structure factor, DSFs $S(k, \omega)$ as a

Figure 1. Variation of dynamical structure factor S(k, ω) as a function of frequency for strongly coupled complex plasma at system size N = 400, wave numbers k= 0, 1, 2 and 3 for six values of Coulomb coupling parameter (a) Γ =1, (b) Γ =5, (c) Γ =10, (d) Γ =20, (e) Γ =50 and (f) Γ =100

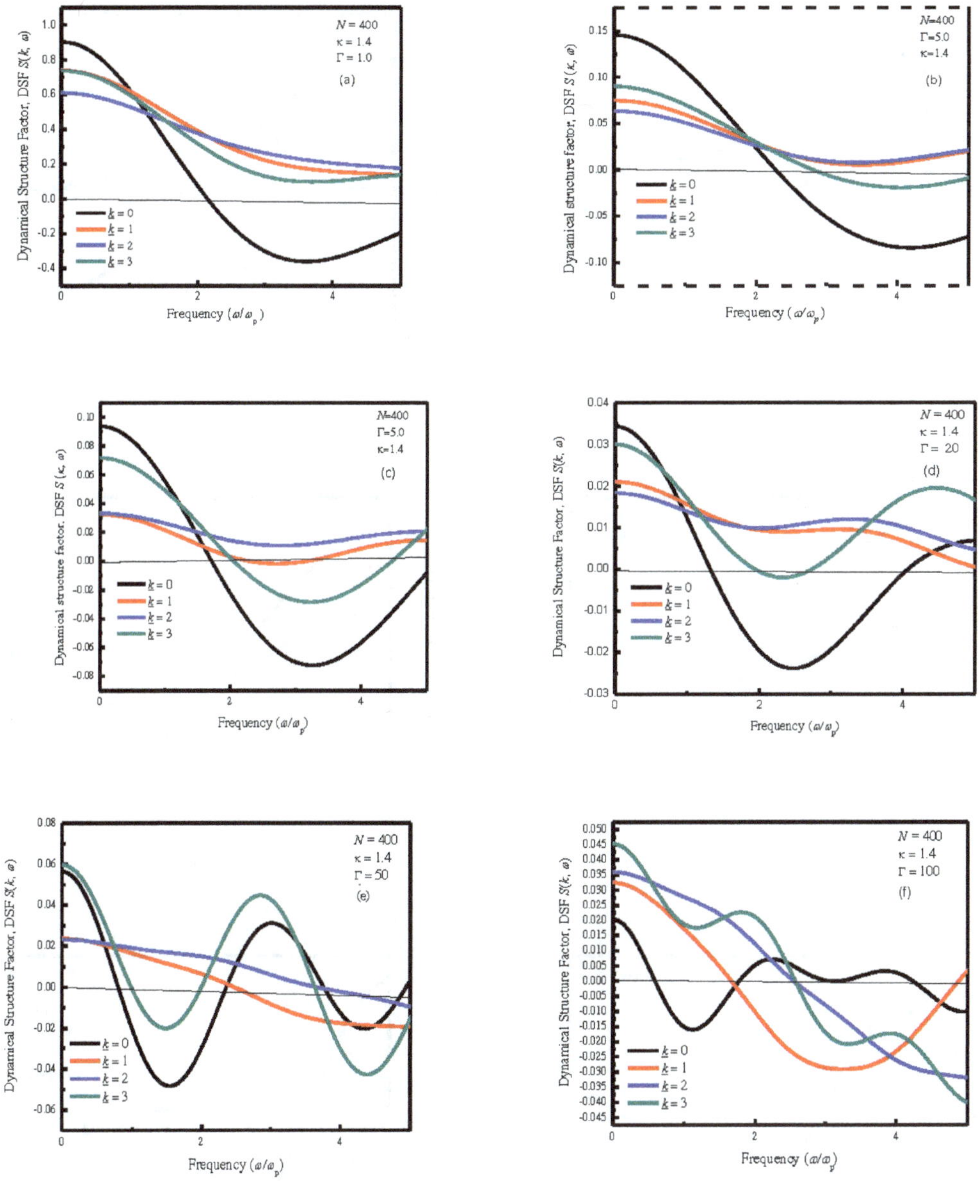

Figure 2. Variation of dynamical structure factor S(k, ω) as a function of frequency for strongly coupled complex plasma at κ=2.5, system size N = 400, waves numbers k= 0, 1, 2 and 3 for six values of coulomb coupling parameter (a) Γ =1, (b) Γ =5, (c) Γ =10, (d) Γ =20, (e) Γ =50 and (f) Γ =100.

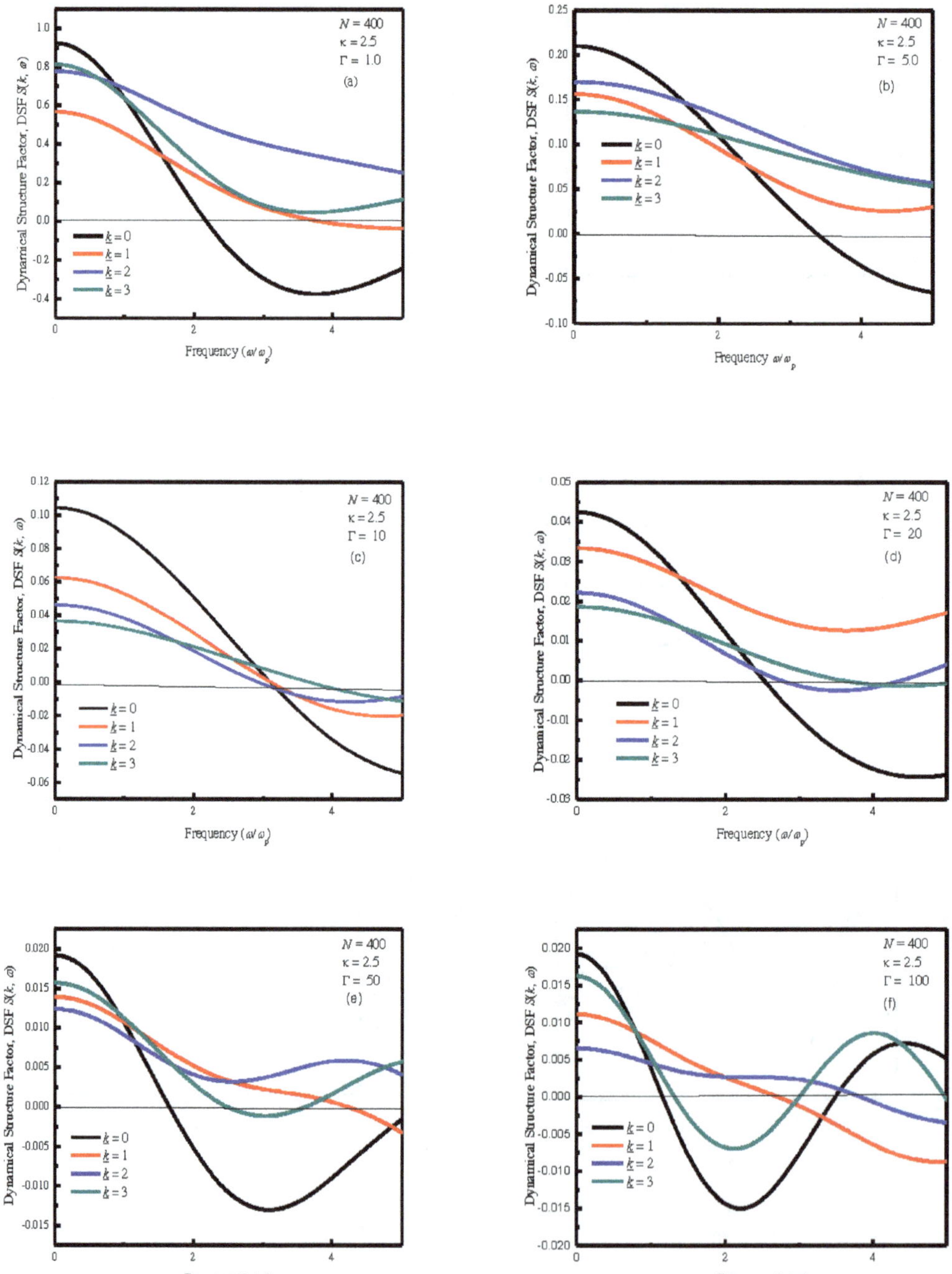

Figure 3. Variation of dynamical structure factor S(k, ω) as a function frequency of strongly coupled complex plasma at κ=3.5, system size N = 400, waves numbers k= 0, 1, 2 and 3 for five plasma states (a) Γ =1, (b) Γ =5, (c) Γ =10, (d) Γ =20, (e) Γ =50 and (f) Γ =100.

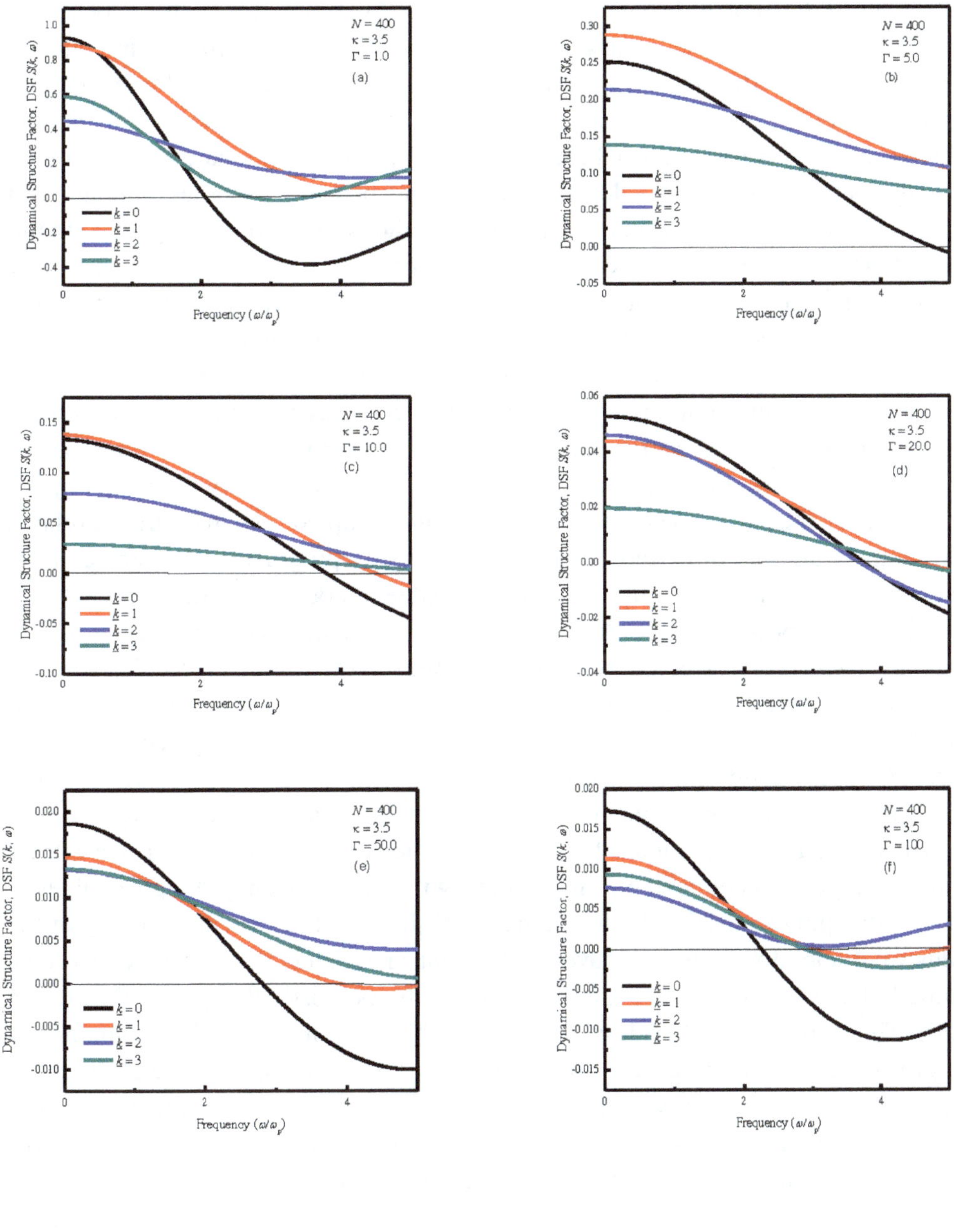

function of frequency. In present EMD simulation method, there are various limitations which may affect the outcomes effectively. These limitations are screening strength (κ), system temperature ($1/\Gamma$), system frequency (ω), wavenumber (k), system size (N), time (t), and total run time. These limitations vary to determine how these limitations can control dynamical density, DSFs $S(k, \omega)$ of Yukawa systems. It was observed from the above six panels that the amplitude of $S(\boldsymbol{k},\omega)$ shows maximum value at wave number $k(\equiv 0, 3)$ as compare to $k(\equiv 1, 2)$. At low value of Coulomb coupling parameters Γ(°1, 20) the damping of wave spectrum is maximum showing that the kinetic energy (average thermal energy) of dusty particles dominated over potential energy, but with increasing value of Γ, the damping behavior becomes minimum and damping mode shifts into sinusoidal mode. In other words, we can say a maximum oscillation in wave spectra is examined with increasing value of plasma parameter Γ. It is further examined that peaks of wave spectrum progressively becomes large and damping becomes lower with increasing value of frequency (ω/ω_p).

It has been shown that frequency of wave spectra increases by increasing the value of Coulomb coupling parameter Γ and it decreases by increasing the value of screening strength κ. The $S(\boldsymbol{k},\omega)$ amplitude decreases as the value of wave number (k) increases from zero to three values. It is summarized from above two panels (Figures 1 and 2) that the plasma density (DSFs) in strongly coupled system is directly related to temperature and amplitude of DSFs is inversely proportional to temperature, showing that damping is very low at higher value of coupling parameter Γ.

The wave spectrum of the plasma density $S(\boldsymbol{k}, \omega)$ oscillates with higher frequency at higher values of Γ. But this behavior is changed as the value of screening strength is increased. In this panel an abrupt behavior of wave spectrum is observed due to higher value of screening strength. The state of the given plasma system is changed from (liquid to solid) as the value of this parameter κ is increased. The frequency of the spectra also decreases with increasing value of this parameter (screening strength). As mentioned earlier in Figure 1 and Figure 2 at lower values of Γ peaks of spectrum show damped behavior and plasma oscillate with high frequency at higher values of Γ and damping also shifts higher values of Γ showing that potential energy of the dusty particles dominated over average thermal energy. It is also observed that wave length decreases with increasing Γ. Wave spectra is mostly damped at Γ ($\equiv$1, and 20). This damping mode is less prominent at higher values of Γ. It is summarized from above simulation results, Figure 1-Figure 3, that wavelength and DSFs of the plasma system are directly proportional to temperature (T). This is the main reason that the damping is less at the higher value of Coulomb coupling parameter (Γ) because coupling parameter inversely related to temperature.

5. SUMMARY

Frequency dependent dynamical structure factors DSFs $S(\boldsymbol{k}, \omega)$ has been evaluated for strongly coupled Yukawa system (SCYs) over a range of plasma parameters like (Γ, κ and k) etc. through EMD simulation. To understand the behavior of wave spectrum in dusty plasma four hundred number of particles are used i.e. (N = 400). It has been shown that previous outcomes and new outcomes are strongly comparable. New obtained results give reliable and wide data for DSFs $S(\boldsymbol{k}, \omega)$. It has been observed that the damping in maximum at lower value of Γ and becomes minimum as value of this parameter Γ is increases and frequency of wave spectra increases by increasing the Γ and decreases by increasing κ. The $S(\boldsymbol{k}, \omega)$ amplitude decreases as the wave number k(° 0,1,2,3) increased. We observed that the increasing and decreasing behaviors of frequency and amplitude of $S(\boldsymbol{k}, \omega)$ in given plasma system is related to this various plasma parameter (Γ, κ and k). The EMD technique used here can easily be extended to the Yukawa system, increasing the screening strength and varying wave numbers. This model can also be used to 3-D Yukawa system with simple modifications.

ABBREVIATIONS

NICDPs: Non-ideal complex (dusty) plasmas
EMD: Equilibrium molecular dynamics
Γ: Coulomb coupling parameter
SCP: Strongly coupled plasma
DSFs: Dynamical structure factors
κ: Debye screening strength
SCYs: Strongly coupled Yukawa system
k: Wave vector

ACKNOWLEDGMENT

We are grateful to the National High Performance Computing Centre of Xian Jiaotong University and National Advanced Computing Centre of National Centre of Physics (NCP), Pakistan, for allocating computer time to test and run our MD code.

REFERENCES

Bellan, P. M. (2008). *Fundamentals of plasma physics*. Cambridge University Press.

Boulos, M. I., Fauchais, P., & Pfender, E. (2013). *Thermal plasmas: fundamentals and applications*. Springer Science & Business Media. doi:10.1007/978-1-1337-1

Chen, F. F. (2006). *Introduction to plasma physics and controlled fusion* (2nd ed.). New York Springer Verlag.

Sera, B., Stranák, V., Serý, M., Tichý, M., & Spatenka, P. (2008). Germination of Chenopodium album in response to microwave plasma treatment. *Plasma Science & Technology*, *10*(4), 506.

Shahzad, A., & He, M.-G. (2012a). Thermal conductivity calculation of complex (dusty) plasmas. *Physics of Plasmas*, *19*(8). doi:10.1063/1.4748526

Shahzad, A., & He, M.-G. (2012b). Thermodynamic Characteristics of Dusty Plasma Studied by Using Molecular Dynamics Simulation. *Plasma Science & Technology*, *14*(9), 771–777. doi:10.1088/1009-0630/14/9/01

Shahzad, A., & He, M.-G. (2012c). Thermal Conductivity of Three-Dimensional Yukawa Liquids (Dusty Plasmas). *Contributions to Plasma Physics*, *52*(8), 667. doi:10.1002/ctpp.201200002

Shahzad, A., & He, M.-G. (2013). Interaction contributions in thermal conductivity of three dimensional complex liquids. AIP Conf. Proc, 173-180. DOI: doi:10.1063/1.4816866

Shahzad, A., & He, M.-G. (2014). *Computer Simulation of Complex Plasmas: Molecular Modeling and Elementary Processes in Complex Plasmas* (1st ed.). Scholar's Press.

Shahzad, A., Shakoori, M. A., He, M. G., & Yang, F. (2019). Dynamical structure factor of complex plasmas for varying wave vectors. *Physics of Plasmas*, *26*(2), 023704.

Shahzad, A., Shakoori, M. A., Mao-Gang, H. E., & Feng, Y. (2019). *Numerical Approach to Dynamical Structure Factor of Dusty Plasmas. In Plasma Science and Technology-Basic Fundamentals and Modern Applications*. Intech Open.

Shukla, P. K., & Mamun, A. A. (2015). *Introduction to dusty plasma physics*. CRC Press.

Sparavigna, A. (2008). *Plasma treatment advantages for textiles*. arXiv preprint arXiv:0801.3727.

Chapter 3
Atmospheric Pressure Plasma Systems and Applications

Necdet Aslan
Yeditepe University, Turkey

ABSTRACT

Atmospheric-pressure plasmas have a wide variety of potential industrial applications. They are used in extractive metallurgy; metal recovery; novel nanomaterial synthesis; refractory and wear-resistant coatings deposition; chemical synthesis; energy conversion; industrial, medical, and nuclear waste destruction; engine combustion enhancement; and exhaust gas pollutants clean up. Atmospheric plasmas are produced by applying DC or AC high voltage between two electrodes designed as cylindrical in shape for jets and planar for the dielectric barrier discharge systems. This review presents an overview of the use of atmospheric-pressure plasma devices and industrial processes carried out in several of these areas.

INTRODUCTION

The term plasmas are used in physics and chemistry to designate an ionized gas state and are considered the fourth state of matter. The term of plasma was proposed first time by Irving Langmuir in 1926. The plasma is a chemically active gas whose atoms or molecules are excited or ionized through thermal, reactive, electrical, and magnetic energy. The collective behavior of the plasma allows an equal number of free negative and positive charges leading to local quasi-neutrality. Plasmas consist of 99% of the physical matter in the universe (ionosphere, sun, auroras, solar winds, and other stars). Plasma is also artificially produced; several of them are part of our daily lives, such as monitors, LCD, television screens, as indicators in equipment,

DOI: 10.4018/978-1-7998-8398-2.ch003

neon lamps or fluorescent tubes in lighting, etc. plasmas are also used as industrial proposals.

Plasmas are classified into two groups based on temperature: the thermal plasma (hot plasmas) and non-thermal plasma (cold) plasmas. Atmospheric pressure plasmas are classified based on local thermodynamics equilibrium (LTE). Generally, plasmas consist of electrons, ions, neutral atoms, and molecules occupying a collision-dominated region where the mean free path between the plasma species is short or large. This gives an increase to intensive chemical reactions by inelastic collisions and heavy particles heating by elastic collisions. In LTE between electrons and heavy particles to comparable temperatures, resulting in transitions and chemical reactions mainly governed by collisions. The non-LTE plasmas can be described by the difference in electron and heavy particle temperatures. This difference increases as the departure from LTE property is elevated; thus, the Boltzmann distribution for density for LTE plasma becomes non-valid (de Groot et al., 2018; Lotfy, 2017; Matra, 2016).

THE LOW-PRESSURE PLASMAS

This type of plasma is produced in a vacuum, where the mean-free paths between the plasma species are large. Consequently, the collisions between particles are not frequent. The electron temperature is higher compare to the heavy particles ($T_e >> T_h$) (Figure 2). In this type of plasmas, the density is low, a higher ionization rate, and local thermal equilibrium cannot achieve which causes cold heavy particles.

ATMOSPHERIC PRESSURE

This type of plasma produces nearly the same pressure as the surrounding pressure (normal pressure). The mean-free path among the plasma species is very short in this type of plasmas with collisions dominated regions. The LTE may be satisfied, including kinetic equilibrium between the species. This kind of plasma is also considered in chemical equilibrium such that species concentrations in LTE plasmas are only a function of temperature. The atmospheric pressure plasmas have two different types: such as thermal and non-thermal atmospheric pressure plasma. Glow discharge, high frequency with low-intensity discharge, and corona discharge are typical examples of cold plasmas. This type of plasma exists in various research fields, including physics, engineering, medicine, and biology (Shahzad, 2019; Shahzad & Bashir, 2019a; Shahzad & Bashir, 2019b; Shahzad et al., 2019).

Thermal Atmospheric Plasmas: Arcs and Torches

Thermal plasmas (hot plasmas) are generated at higher temperatures reaching up to several thousand Celsius degrees where high temperatures are required. To create this type of plasma arcs or torches are used depending on different purposes. The commercial torches use a DC power supply between electrodes as an energy source and air as a plasma medium at atmospheric pressure. If low or no oxidation is required, some other gases other than oxygen are used. In this case, only argon or nitrogen gas, or nitrogen-argon mixtures are used to activate plasma. Some arc devices also apply electrode-less RF power to initiate and maintain arc plasmas. Compared to AC discharges, DC torches are the most commonly used devices because they perform a more stable operation with lower power consumption, produce less flicker and noise, and are easy to control. Arc plasma torches utilizing DC power supplies are examples of discharge devices consisting of an electron-emitting cathode, a gas injection system, and a nozzle at the tip of the tube used as an anode.

The arc torches either have positive anodes carrying high currents or nozzles at floating plasma potential with the treated surface taken as an anode, and the arc between cathode and anode provides intensive electric fields to ionize gases introduced the system. The plasmas in these torches are very dense, highly ionized, and have a hot surface at 6-8000k and core at 15000K. This is mainly due to high collision frequencies among molecules, atoms, ions, and electrons with short mean-free paths. In such plasmas, local thermodynamics equilibrium (LTE) exists because the gas is significantly ionized. The electron density in arcs is between 10^{22}-10^{25} $m^{-3,}$ and the energy density is 10^7–10^9 Jm^{-3}. The electric fields are relatively low (500-5000V/m), but the current densities are high (10^7–$10^9 Am^{-2}$)[1a, 1b]. Another example of thermal plasma utilization is the pyrolysis of waste tires to produce carbon and other gaseous products

The arc cutting devices, which utilize a high-velocity jet of ionized plasma particles, were established around the 1950s. The oxygen arc plasmas can cut 15 mm thick carbon steels, while nitrogen or argon/hydrogen plasmas can cut 50-80 mm thick steel.

Arc Welding Technology was developed during World War II to join metals in the aerospace industry. These devices use a centered tungsten tip as an electrode near the nozzle and prevent oxidation of molten material gases such as argon, helium, and their mixture. Typical arc devices use low DC voltages, but the currents in these devices range from 10-200A, Figure 2(a) When the gas is introduced to the arc tube (region 1 in Figure 2(b)), the plasma region around the electrode tip is produced (region 2), the protection gas encapsulates plasma (region 3) to eliminate oxidation from air. The hot plasma opens a hole on the surface (region 4) where welding material is melted and deposited. With these devices, welding 10mm thick on stainless steel materials is now possible, and welding for gold, platinum, titanium, niobium, and many other non-alloyed materials can also be performed. Cutting material with thermal plasma using are is shown in Figure 3.

Figure 1. Schematic picture of thermal plasmas of DC power supply torch during operation

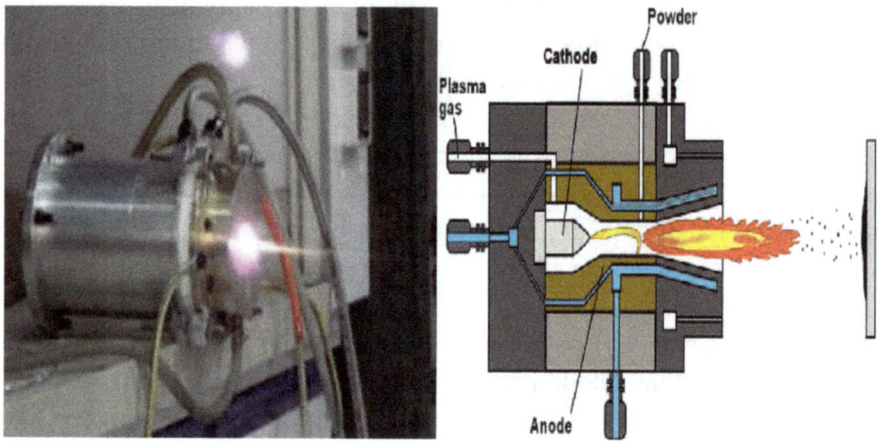

Figure 2. (a) Plasma welding device produced by Saf-Fro, a European Co., (b) the regions of plasma

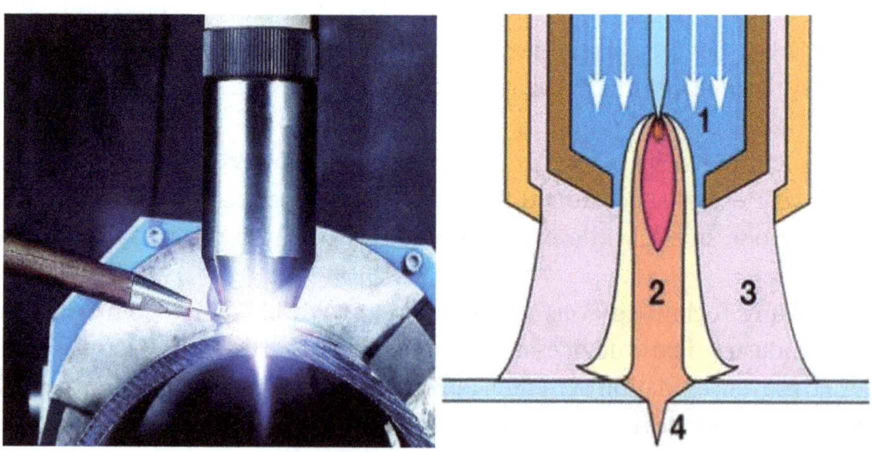

Figure 3. Picture of arc cutting device

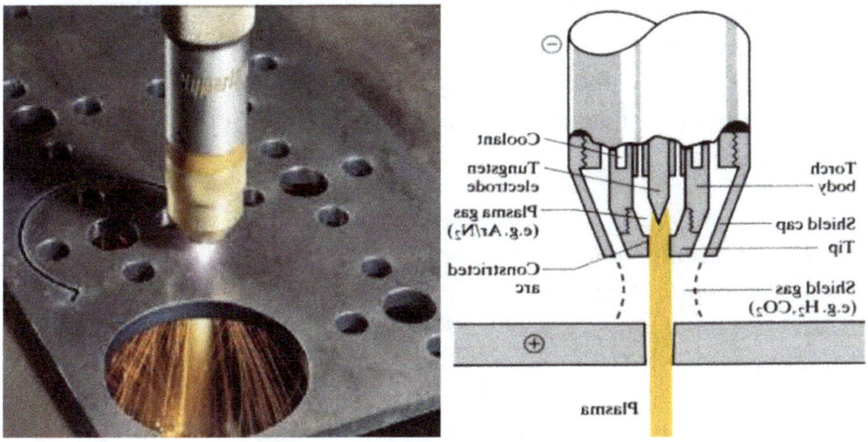

The conventional plasma arc cutting apparatus, such as plasma torch assembly, utilize a single plasma gas (nitrogen, oxygen, or a hydrogen mixture) serving both cutting and ignition gas. To reduce extra heat, air or water can be used as a purpose coolant. Plasma arc cutting employs a secondary medium, an additional gas siphoned into the plasma region to create particular characteristics for specific cutting applications. More than 500000 commercial and industrial suppliers can be found on (de Groot et al., 2018; Lotfy, 2017; Matra, 2016). The secondary gas in the plasma cutting process can increase cutting speed, power density, and cut quality. Recently attention on evolving thermal plasmas for the annihilation of hazardous waste and stabilizing nuclear waste is increasing. Nowadays many plasma waste inhalation systems with power more extensive than 0.5MW had commercialized worldwide see Figure 4

NON-LTE, NON-THERMAL ATMOSPHERIC PLASMAS

The non-thermal atmospheric plasmas (NTAPs) or cold plasmas are created with temperatures near ambient temperature. NTAPs are generated by magnetic and electric field applications to a gas. They have very high electron temperatures than other particles (atoms, molecules, ions). Atmospheric non-thermal plasmas have a low degree of ionization and the density of charged species. The atmospheric pressure discharges at room temperatures are non-LTE plasmas in which the interaction of electrons with gas atoms produces low levels of dissociation, excitation, and ionization. Since neutrals and ions are relatively cold, they do not induce thermal damages to heat-sensitive surfaces such as biological tissues and polymers. This allows the treatment of heat-sensitive materials, including polymer materials and biological tissues. Atmospheric NTAPs are very reactive because they have a high density of activated species, i.e., reactive free radicals and excited-state atoms. Typical examples of such plasmas are corona, dielectric-barrier, and glow discharges. Many plasma applications do not require high-temperature characteristics, and sometimes the heat is prohibitive.

Figure 4. Plasma torch configuration for hazardous waste removal 1: electrodes, 2:cooling water manifold, 3: magnetic field, 5: entering processing gas, 5: Plasma column, and 6: heated process gas [3bb]

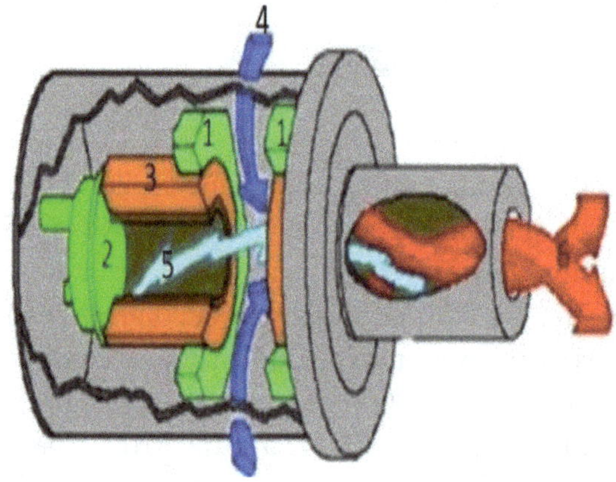

The most critical parameter of NTAPs is E / N, where E is electric field strength, and N is the gas density. These electron velocities and energy distribution ratios govern gas heating, vibration, excitation, dissociation, and ionization in NTAPs. Since electron temperature is much greater than translation and vibration temperatures, the gas displays the non-equilibrium property. The energetic electrons generate bio-chemical active species at high levels. They allow non-equilibrium plasma chemical reactions under cold temperatures. The radicals and meta stables like O, OH, N_2*, CH, He*, etc., are present in NTAPs.

ATMOSPHERIC PRESSURE PLASMA GENERATION

The gas at ambient temperature is a good insulator; the charge carrier (electrons, ions) makes a gas electrically conducting. Passing the electric current through gas leads to gas discharge. Such gaseous discharges are common to generate the plasma. The thermal plasmas are generated by passing a gas through a high-power electric arc discharge that causes heating the gas within milliseconds. Supplying a large amount of energy to the surface leads to thermionic emissions of the electrons. If the surface temperature of the emitter is not high enough for the thermionic emission of electrons, then a strong field is used. This phenomenon is known as is thermionic field emission. Plasma generators are classified as alternating current (AC), direct current (DC), microwave (MW), or radiofrequency (RF) plasma generators. The plasma produced through AC and DC generators is called electrode discharge plasmas, although RF or MW generated plasmas are inductively coupled and require no electrodes.

CORONA DISCHARGES

High pulsed DC power supply, currents in milliampere (mA), and small pulsed duration are applied to create the corona discharge. In the corona discharge process, a glowing crown appears of small diameter for this purpose; wires, sharp edges, or needles are used as electrodes [8a]. The typical configuration such as wire-to-plane, pin-to-lane, or coaxial wire in a cylindrical tube is used. A negative corona discharge is modeled when cathode wire is used near the anode surface. The cathode wire attracts positive charges (ions), and their influence creates the emission of secondary electrons having 1eV energy. The density of free electrons in corona discharges is almost 10^8electrons/cm^{3,} and the density of the gas particles at atmospheric pressure is about 10^{19} particles/cm^3. The plasmas created in corona discharges are inhomogeneous, which need a narrow space between the electrodes.

Corona discharges have industrial applications such as printers, electrostatic precipitators, electrophotography, textile processing, and powder coating. Due to inherent non-uniformity has limited industrial applications for the homogenous treatment of the large surface, coating, and deposition.

DIELECTRIC BARRIER DISCHARGES

The DBD device consists of two parallel metal plates separated by a dielectric material, or one of the plates covered up a thin layer of dielectric or high resistive material. The dielectric material plays a critical role in suppressing the current. The cathode/ anode layers charged through charges (electrons/ ions) reduce the electric field and prevent the charge transport towards the electrode. DBDs promote diffusion losses and keep the low temperature of gas because they have large surface-to-volume ratio. The DBD discharge with brush-style cathode is suitable for non-conducting sample treatment, whereas the covered cathode is applicable for organic film deposition. The first DBDs were used about 100 years ago to produce the ozone for the treatment of drinking water. After that, various industrial applications were invented for such kinds of discharges. DBDs are widely used for atmospheric pressure plasma treatment of surface modification of textile fabrics without changing bulk properties. This is also applied to flow control, conversion of CO to CO_2, modification of surface properties of polypropylene film, atmospheric pressure plasma deposition of the thin film, blood coagulation, living tissue, and bacteria sterilization, pollution control and polymer surface treatment to permute printability, wettability, and adhesion. Figure 5 shows a schematic diagram of (a) atmospheric pressure plasma torch and (b) dielectric barrier discharge.

Figure 5. Picture of (a) atmospheric pressure plasmas, (b) dielectric barrier plasmas discharge

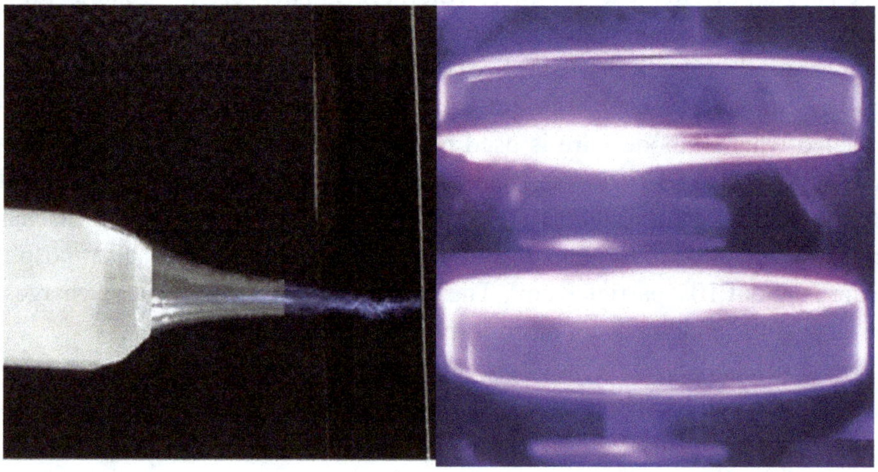

ATMOSPHERIC PRESSURE PLASMA JET (APPJ)

Atmospheric-pressure plasma jets created by DC or low frequency (kHz) high voltages consist of two concentric electrode structures through which a mixture of helium, oxygen, and other gases flow. The DC or AC voltages applied between electrodes lead to the ignition of discharge. The ionized gas from the plasma jet creates through a nozzle and is directed onto the surfaces. Under typical operating conditions, the gas velocity keeps between 2-20 m/s, with temperature changing from room temperature up to150-200 °C. Koinuma and Coworkers used a metal needle of 1 mm diameter inserted into a grounded metal cylinder which included a quartz tube between electrodes. For NTP, the air is the most frequently utilized gas for obtaining plasma jets for surface activation and modification. Mixtures of rare gases, He, Ne and Ar, or N2 and O2, or their mixtures are fed between the metal needle and quartz tube at flow velocities of about 5m/s at 200–400 °C. With this system, photoresist as hinges also achieved.

Non-thermal capacitively coupled plasmas sources usually operate in the radio frequency range (13.56-27.12MHz). This consists of two electrodes in various arrangements and setups; it can also be established with single electrodes. The distance between the electrodes is in the range of *mm*, although exposure distance must be in the range of *cm*. the distance between the electrodes depends on the electrode gap and processes gas. The ignition voltages (100V-kV) are suitable that does consume the 500W power. Generally, the process gases are noble gas such as argon or helium with few *slm* gas flow rates. The process temperature depends on power consumption, gas flow rate, and input signal. The atmospheric pressure plasma jet can operate in remote mode or direct, though most applications work in the direct mode.

Plasma jets or pencils are examples of low-temperature atmospheric plasma systems and small currents that generate low heat, allowing surface treatment of polymers and other degradable materials. When DC and low-frequency plasma jets are produced by power supplies including series-connected resistor (~100*M*) and parallel-connected small stray capacitance (~2*pF*), the discharge exiting nozzle can be touched by bare hands, without a feeling of heat and electrical shock, see Figure5. Along with producing ions and reactive species, these kinds of jets also emit electromagnetic radiation, from UV to infrared. The most important species are atomic oxygen, ozone, hydrogen peroxide, nitrite, nitrate, nitric oxide, hydroxyl radicals, superoxide anion radicals, and singlet oxygen, depending upon the gas that is introduced to the system

The main benefit of APPJ is the low plasmas dimension in grouping with the capability of penetrating into narrow gaps with high features ratio. That's why APPJ can be used at complex geometries with microstructures vessels or cavities. The APPJ is used for the applications of biomedical purposes. For metal oxide and metals, it is also used for inactivation of microbial and polymer nearly 18μm/min and 2μm/min, respectively. Plasma jet surface treatment is widely used before applying adhesives, coating, or printing to create a chemically rich environment (see Figure.6(a)). The surface activation increases the surface tension of non-polar materials, and the surface hydrophobicity increases. These treatments modify the surface properties such as conductivity, diffusion barrier against gases, low dust and carbon adherence etc.

Figure 6. Pictures of (a) the plasma jet with a spacer on skin application [89], (b) schematics of two different plasma jet systems produced by H. Koinuma [989], ELES Co., Istanbul, Turkey [6a].

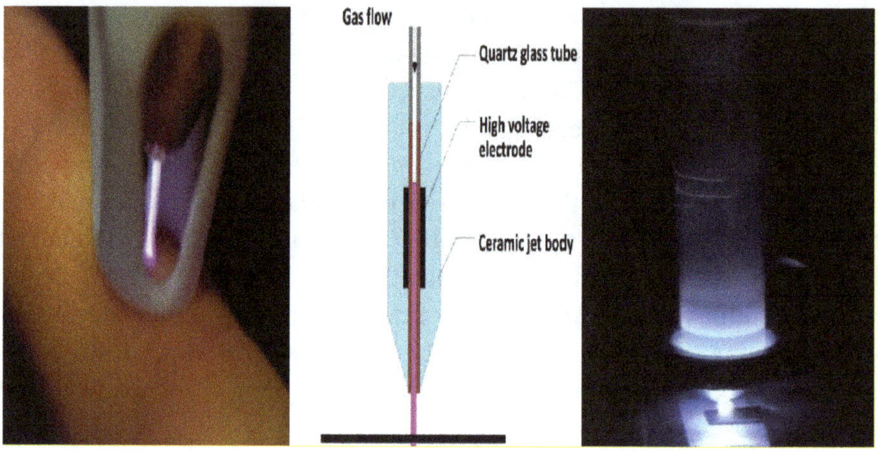

Solar cell production technology requires pre-treatment of the substrate before the thin film deposition to improve adhesion. He/O2 plasma jets were recently used to treat such substrates, resulting in increased surface energy from contact angle measurements. The reduction of surface contamination is also realized analyzed and observed by X-ray photoelectron spectroscopy measurements. Another 1-3cm range of length plasma jet creates through a stainless nozzle serving as the grounded electrode by applying 1-20 kHz high voltages (1-5 kV) to the inner tungsten electrode.

APPLICATIONS OF ATMOSPHERIC PRESSURE PLASMAS

Plasma science and technology fall in a very developing research area that has been investigated and applied on biomedical materials and devices, surface modification of textile [6zz], exclusion the chemical from the surface of heat-sensitive materials, water treatment and recently healing and food decontaminations, etc. Atmospheric pressure plasmas have many prospective applications in metallurgy, novel nanomaterial synthesis, metal recovery, and chemical synthesis, refractory and wear resistance coating deposition, energy conversion, nuclear, industrial, and industrial, medical waste destruction, exhaust gas pollutants clean up and engine combustion enhancement.

Atmospheric Plasmas in Medicine

Applications of cold plasma (NTPs) in the field of medicine are an evolving field that increased drastically last few years. The first experiment in this area was the plasma jet, successfully used to inactivating bacteria. Bacteria inactivation is one of the major challenging issues for storage, packing, contaminations, and sterilizations. Different revealed incidents demonstrated that the food safety compares food poison microorganisms (EHEC, *Listeria*, *Salmonella*) could not provide efficiently through conventional method. Low-temperature plasmas treatments are considered novel technology that can solve particular challenges.

Sterilizations and Decontamination

Mechanisms through atmospheric pressure plasma (NTAPs) are used to cell death technology. The deactivation of bacterials cell; caused several reactions, such as lipid peroxidation of polyunsaturated fatty acids and DNA oxidations of amino acids. UV radiations are produced through NTAPs that disturb microorganism's genetic material. The NTAPs can inactivate fungi, vegetative cells, yeasts, spores [8zz], and biofilm formers. The inactivation of Streptococcus mutants and Escherichia coli with the help of mycoplasmas confirms the potential of these kinds of plasma in the areas of density and microsurgery. The inactivation of Listeria monocytogenes from processed meat surfaces and on cheese through a novel atmospheric plasma jet has been added to using NTPs in food processing. Previous studies have revealed that UV produced by NTPs does not mainly affect the inactivation of the microorganism but can affect the reactive species. The reactive species in NTPs based on oxygen and nitrogen, such as atomic oxygen, hydroxyl, and nitrogen oxide. These active species play a vital role in the microbial inactivation method; they directly interact with bacterial membrane. These reactive species have short-lived in the gaseous state, disappearing within milliseconds after being produced and dissolved in the liquids. The chemicals that help to produce the reactive species in atmospheric pressure plasma are given in **Table 1**. These reactive species have advanced chemical and sterilizing potential than the conventional agents.

The plasmas discharge with higher oxygen concentrations increases microbial inhibition due to oxygen-based active species (oxygen, ozone). The active species formed during the plasma discharges such as singlet oxygen, hydroxyl, ozone, and radical can initiate lipid peroxidation, producing shorter chain fatly acyl compounds. The short chains of the charged fatty acids have a minor ability to rotate within the membrane and increase the membrane's fluidity. Gram-negative bacteria have more susceptible than gram-positive bacteria. Despite the specious perspective of atmospheric NTAPs technology, the inactivation mechanisms of different plasma systems are not yet fully understood. The difference in plasma source, process parameters, and gases results in differences in produced reactive species, making inactivation mechanisms challenging to compare. The NTAPs (cold plasma) are also used for decontamination from a microorganism located externally of the different products. Plasmas can flow around the stuff to ensure all parts of the contaminated products are treated. The NTAPs also disinfect the products before packing and can be included in the packing process. Plasma generated through electric discharge and fluorescent lighting tubes has a higher efficiency near 80% conversion rate from electricity to plasma.

Killing bacteria through NTAPs plasma was observed first time by Laroussi in 1996. In the current age, NTAPs have widely been used related contaminations of the microorganism of medical, food, oral, and agriculture products. Plasma-based processing techniques are growing over time.

Surface Modification

The surface modification of antimicrobial properties as reactive materials, possible to modify the surface of the specific substrate; this technique is identified as plasma modification or plasma activation. This technique involves depositing chemical materials such as plasma grafting and plasma polymerization to instruct required properties or remove unwanted substances (plasma etching or plasma cleaning) deposited on the substrate. Plasma surface modification technologies are used to create antimicrobial textile and heat-sensitive surfaces, such as medical equipment. This method is also used to modify the surface of several kinds of fabrics, containing nylon, woven, wool, cotton, and non-woven fabrics. Textiles are produced to have antimicrobial properties against numerous microorganisms such as E. coli and Staphylococcus aurous. The previous investigation confirmed that the copper and silver nanoparticles have antimicrobial properties when deposited (coating) on textile through different methods such as sol-gel coating, soaking, chemical washing, vacuum, and atmospheric coating. Besides, titanium oxides (TiO2) also have antimicrobial properties.

Figure 7represents electron micrographs (SEM) of cotton textile fibers coated with copper (Cu), silver (Ag), and titanium (Ti), respectively.

Sterilization of Water

The NTAPs sterilized the water by removing several kinds of bacteria's, yeasts, and fungi. The plasmas corona discharges method is performed efficiently to sterilize water from microorganisms. The pulsed plasma gas discharge and pulsed ultraviolet technology are developed to destroy unsafe chlorine-resistant cryptosporidium oocytes, affecting waterborne illness universally. One of the other water treatment methods with low energy (75 J/L) inputs is essential to obtain a one log reduction in E. coli concentration and found water treated with plasma had prolonged resistance to E. coli up to hours after treatment. The atmospheric pressure plasma with low temperature generated through a corona discharge (bot AC and DC power supply) has a potential sterilizing effect on water inoculated with different algae and microorganisms.

Figure 7. Cotton fibers examined through SEM coated by plasma coating technology (a) Cu (b) Ag and (c) Ti

Food Processing

Plasmas can inactivate and injure the pathogenic bacteria on the surface and inert food (liquid or solid foods), including decontamination of vegetables, fruit, and milk. Various researchers have focused on the cold plasma concerning contaminated pericarps of melons, mangoes, almonds, bell peppers, egg, and milk, meat, and cheese products. While these studies are preliminary, the perspective of this technology in the food industry is familiar. To date, plasma-based technology in the food industry is not working commercially due to a lack of research. The significant aspects of this technology are still underdeveloped, mainly concerning its use with food. The applications of NTPs in the food industry on commercial need a depth study and supply the basis of the technology.

ACKNOWLEDGMENT

The author thanks Aamir Shahzad (Associate Professor of Physics, Government College University Faisalabad, Pakistan) for providing his suggestions and useful discussions.

REFERENCES

Attri, P., Arora, B., & Choi, E. H. (2013). Utility of Plasma: A New Road from Physics to Chemistry. *Sensors*, *6*(207890), 7.

de Groot, G. J. J. B., Hundt, A., Murphy, A. B., Bange, M. P., & Mai-Prochnow, A. (2018). Cold plasma treatment for cotton seed germination improvement. *Scientific Reports*, *8*(1), 1–10. doi:10.103841598-018-32692-9 PMID:30258075

Lotfy, K. (2017). *Effects of Cold Atmospheric Plasma Jet Treatment on the Seed Germination and Enhancement Growth of Watermelon*. Academic Press.

Lu, P., Cullen, P. J., & Ostrikov, K. (2016). Atmospheric Pressure Nonthermal Plasma Sources. In Cold Plasma in Food and Agriculture: Fundamentals and Applications. doi:10.1016/B978-0-12-801365-6.00004-4

Matra, K. (2016). Non-thermal Plasma for Germination Enhancement of Radish Seeds. *Procedia Computer Science*, *86*, 132–135. doi:10.1016/j.procs.2016.05.033

Misra, N.N., Schlüter, O., & Cullen, P.J. (2016). *Cold plasma in food and agriculture: Fundamentals and applications*. Academic Press.

Shahzad, A. (2019). *Introductory Chapter: Role of Nuclear Medicine in Medical Science*. IntechOpen.

Shahzad, A., & Bashir, S. (2019a). *Nuclear Medicine Physics*. IntechOpen. doi:10.5772/intechopen.75255

Shahzad, A., & Bashir, S. (2019b). *Applications of Diagnostic Reference Levels of Standard Doses in Nuclear Medicine. In Nuclear medicine physics*. IntechOpen.

Shahzad, A., Bashir, S., & Anwar, A. (2019). Establishment of age-speciðc reference levels and achievable doses for children and adults undergoing nuclear medicine exams. *Radioprotection*, *54*(3), 187–194. doi:10.1051/radiopro/2019019

Chapter 4
Molecular Dynamics Study of Diffusion Coefficient for Low-Temperature Dusty Plasmas in the Presence of External Electric Fields

Muhammad Asif Shakoori
Xi'an Jiaotong University, China

Maogang He
Xi'an Jiaotong University, China

Aamir Shahzad
Government College University, Faisalabad, Pakistan

Misbah Khan
Xi'an Jiaotong University, China

Ying Zhang
Xi'an Jiaotong University, China

ABSTRACT

The effects of external electric field (E) on the diffusion coefficient of dust particles in low-temperature dusty plasmas (LT-DPs) have been computed through nonequilibrium molecular dynamics (NEMD) simulations. The new simulation result was obtained by employing the integral formula of velocity autocorrelation functions (VACF) using the Green-Kubo relation. The normalized self-diffusion coefficient (D^*) is investigated for different combinations of plasma coupling (Γ) and Debye screening

DOI: 10.4018/978-1-7998-8398-2.ch004

(κ) parameters. The simulation outcome shows that the decreasing position of D^* shifts toward Γ and also increased with the increase of κ. The D^* linearly decreased with Γ and increased when applied external E increases. It is observed that the increasing trend of D^* depends on the E strength. These investigations show that the present algorithm provides precise data with fast convergence and effects of κ, Γ, E. It is shown that the current NEMD techniques with applied external E can be employed to understand the microscopic mechanism of dusty plasmas.

INTRODUCTION

During the last few decades, the science of low-temperature plasma physics becomes a prominent research field. Plasma science and technology are serving every field of life, ranging from agriculture to nuclear fusion energy. The Irving Langmuir is the first scientist who discovered the fourth state of matter called "Plasma." From that date, plasma rapidly grows and becomes the core of numerous industries and technologies. Plasma is the fourth state of matter that contained 99% of the physical matter in the universe; and is defined as "Plasma is fully or partially ionized gas that contained neutral atoms, electrons, and ions." These particles show a collective behavior in the plasmas. Plasmas are classified based on many characteristics such as temperature, density, degree of ionization, etc. (Shahzad et al., 2018); here, we only discuss few types related to our topic.

Types of Plasma

Hot Plasma

In this type of plasma, the gas is fully ionized at high-pressure, high-temperature and acquires a local thermal equilibrium state. The temperature of the electron is equal to the ion and must satisfy the conditions ($T_e \equiv T_i$); here, T_e and T_i denote the temperature of electron and ion, respectively. The collisions between the particles are frequent. It is thermal, and equilibrium plasma associated with astrophysics, spark, flame, atmospheric, solar wind, sun, and nuclear fusion plasma cores are examples of hot plasma. It has a high density of nearly one trillion electrons per cubic centimeter and a degree of ionization near equal to 1 (100% ionized gas).

Cold Plasma

This type of plasma is composed of the same number of electrons and ions in the low-pressure ionize gas. The collisions between electrons and molecules of a gas are not frequent due to temperature differences. The electron has a higher temperature than the gas molecules. The degree of gas ionization is less than 10^{-4}—this type of plasma is known as non-thermal, non-equilibrium, and cold plasma. The condition of cold plasma based on temperature is $T_e \gg T_i \gg T_g$, here T_e, T_i and T_g are the temperatures of the electron, ion, and gas molecules, respectively, in the discharge gas. Florence and neon tubes are examples of Cold plasma.

Ultra-Cold Plasma

This type of plasma is formed at a low temperature of less than 1 K and created by photoionization laser-cooled atom. It is produced in rubidium, calcium, xenon, strontium, and cesium. In this type of plasma, the temperature of an electron in the range of *mK* ($T_e \cong$ 100mK), ion temperature in μK ($T_i \cong$ 10μK), density ($n \cong$ $2\text{x}10^9\text{cm}^{-3}$), and Debye length ($\lambda_D \cong 500nm$). It is such that cold plasma does not require a high magnetic field to confine the plasma species.

Low-Temperature Plasma (LTP)

The Ionized gas or liquid composed of electrons, ions, neutral atoms and molecules, radicals, and excited states. LTP is classified into different research areas such as electrodynamics, heat and mass transfer, fluid dynamics, statistical physics, molecular and atomic physics, thermodynamics, materials and surface science, chemistry, chemical and electrical engineering, and recently biological science(Shahzad et al., 2018, 2020a).

Dusty (Complex) Plasmas (DPs)

Dust is present everywhere in space and earth. It is a composition of different elements or single elements with different shapes and sizes. When dust particles were immersed in the plasma, they attain a charge; in most cases, they are negatively charged; sometimes, a positive charge (space plasma) also depends on the phenomena of charging. The behavior of dust particles is investigated in different science disciplines such as chemistry, physics, medicine, astrophysics, environmental science, nuclear energy, etc. it is allowing study at the microscopic and atomistic level. The existence of dust particles in the plasma modified the physical properties of plasma, such as thermophysical properties, temperatures, and potential energies. In nature, DPs are ubiquitous in astrophysical situations like comets, interplanetary, interstellar clouds, or the rings of the Giant planets, liquids in the white dwarf, interior of supernova cores, neutrons stars, etc (Shahzad et al., 2017, 2019, 2020a, 2020b). The first-time coulomb crystal of dust particles was observed in1994(Chu & Lin, 1994). After that, research on this direction increased drastically. The dust particles can be grown in a laboratory through a chemical reaction and sputtering. Dust particles are used to enhance the stability and efficiency of a solar cell, LED, thin-film deposition, and semiconductor processing techniques(Boufendi & Bouchoule, 2002).

Low-Temperature Dusty (Complex) Plasmas (LT-DPs)

The DPs are classified on the base of Plasma Coulomb coupling (Γ = P.E / K.E) parameter. Γ is the ratio of potential energy to the kinetic energy of interacting charged dust particles. If $\Gamma < 1$, than DPs are considered weakly coupled DPs (hot DPs) and have high temperatures with low density. For second conditions $\Gamma > 1$, DPs are considered strongly coupled DPs (Low-temperature DPs) and have a low temperature with higher density. The strongly coupled DPs exist in different states such as the nonideal form of gas, liquefy, liquids, solid/crystals, etc. depend on plasma Coulomb coupling values and Debye screening strengths. In recent years this type of plasma becomes a more exciting research area with different prospective. Dust particles are a highly negative charge of about 10^3 to 10^4 *e*. due to the low mass to charge ratio; these particles are coupled and arranged as liquids and solid-like structures. In this chapter, we have investigated the self-diffusion coefficients of dust particles at low temperatures (Shahzad, 2018, 2020).

Applications

LTPs in Medicine and Healthcare

Nowadays, plasma medicine has become an emerging field, which combines plasma physics, life science, and clinical medicine. The cold plasma is created easily in the atmosphere below the 40°C with reactive species (radicals, non-radicals) such as oxygen, nitrogen, photon, charged particles, and electric field. The LTPs are directly used in biological medicine, such as sterilizing surgical instruments and implants, wound healing, cancer therapy, break DNA damage for human prostate cancer. The low-temperature plasma significantly affects the biological cells on a microscopic and macroscopic scale. The LTPs were found to be helpful in killing cancer cells without affecting the healthy tissues. It is also used to inactive the bacterial in the field of health care. The LTPs is valuable as a diagnostic technique for a precise calculation of plasma parameters such as nonlinear laser spectroscopy, spectrally resolved nanosecond imaging, phase-resolved optical emission spectroscopy, and laser-induced fluorescence VUV, UV, VIS and IR, etc. (Laroussi, 2020)

LT-DPs in Manufacturing Processing Industries

LT-DPs are directly applicable in processing and manufacturing industries on a different scale with different purposes. In the semiconductor processing technique, the plasma is used to process and deposition integrated circuits, etching, masking, and stripping. The several capacitors and transistors put on the silicon wafer chip (Pentium chip) with the help of plasma without contaminations. The DPs have numerous applications in science and engineering such as thin-film deposition, ceramic processes, insulation systems, filtration processes, oil recovery, biomechanics, packed bed reactors, petroleum industry, paper industry, medicine, and biotechnology, etc. The DPs also have nuclear fusion devices or ITER (international thermonuclear experimental reactor) like Tokamak. LT-DPs science and technology significantly enhance the efficiency and stability of solar cells, LED, improving lighting source, display, and laser technology. Currently, the scientific communities are looking at plasma science's role in developing nuclear fusion reactors such as Tokamak, ITER, Z-Pinch, etc. it is hoped that in nearly future the nuclear fusion energy becomes the highest energy source in the whole of the world. Plasma is used in different diagnostic techniques in materials science such as laser-induced fluorescence, electron spectroscopy for chemical analysis, micro Raman scattering, mass spectroscopy, etc. (Adamovich et al., 2017; Boufendi & Bouchoule, 2002).

Role of Electric Fields in LT-DPs Physics

Dusty plasmas (Yukawa liquids) also have electrorheological behavior under the influence of an external electric field. This behavior of dusty (complex) plasmas was discovered by Ivlev *et al.* in 2008 through the Plasma Kristall-4 (PK-4) experiment under microgravity conditions (Ivlev et al., 2008). The role of the external electric field (EEF) in plasma science and technology is significant. In ultra-cold plasma study, the strong coupling effect sufficiently needs low temperature. DC electric field significantly affects as heating of electrons components. Due to applied EEF, the electrons accelerate and kinetic energy increased. When the external and internal electric fields of species become equal, then the electron stopped to oscillates. The oscillation of the charge component in plasma depends on the strength of the external applied electric field. The potential energy also changes with the variations of electric field strength due to the plasma component's displacement. Under the influence of DC electric field mode, attractive attractions between charged dust particles are introduced by ions streaming in the direction of the applied electric. In this way, the wake potential behind the dust grains is produced in the DC discharge plasma sheath, where charged dust particles levitated against the electric field's gravity force. In an external DC electric field, the system is known as non-Hamiltonian, and wake potential is asymmetric. For the AC electric field, the external applied AC electric field's frequency is much smaller than the ion frequency in the plasma but larger than the dust plasma frequency. In this way, the wake potential is symmetric, and a system is known as Hamiltonian. Due to the electric field, dipole-dipole interaction increased in dusty plasmas, particles are arranged in sheet, string, or crystalline structures like electrorheological fluids. It is observed during the experiments that dusty plasma changes phase as the frequency of AC was increased. The phase transition process in complex plasmas was studied by (Ivlev et al., 2008, 2011; Schwabe et al., 2020).

Thermophysical Properties of LT-DPs

Transport properties such as thermal conductivity(Shahzad et al., 2017), share viscosity(Shahzad et al., 2014), and diffusion motion(Shahzad et al., 2013), thermodynamics characteristic's (Shahzad & He, 2012a) play a significant role in science and technology for designing and modeling new systems. During this process, the chemical composition remained unaffected. The physical properties fluctuate with temperature, pressure, and composition of materials (liquids, solids, and gases). Theoretical prediction, computer simulations, and experimental work investigated these properties in various gases, liquids, and crystals. Transport properties of LT-DPs play a crucial role with the various prospectus in engineering and industries(Shahzad, 2012b, 2013, 2015a). This chapter will investigate diffusion motion behavior in LT-DPs under the influence of an external applied electric field.

Diffusion Motion in LT-DPs

Diffusion of mass and momentum is a fundamental transport property of solids, liquids, and gas. In this process, the migration of charged particles is investigated with the variations of temperature, pressure, external field (electric and magnetic), etc. Different types of diffusion motion exist in dusty plasma regimes depends on plasma parameters and physical systems. The diffusion motion of dust particles in LTPs continues as one of the active research topics, an essential consequence of understanding the transport properties. It is one of the primary sources to lose energy (stopping power). Diffusion also plays an essential role in exploring many biological, physical, and chemical system's dynamical properties (structure, waves, and instabilities). We can easily understand the microscopic phenomena of particles in the different states of complex plasmas (Nunomura et al., 2006; Ohta & Hamaguchi, 2000; Shahzad et al., 2013). An extensive amount of previous studies have been made to understand the diffusion motion in Yukawa systems. Nowadays, the effect of an electric or magnetic field is of great interest in the current research direction. The diffusion coefficient for external electric fields lacks knowledge in theories, computational work, and experiments for Yukawa liquids. The externally applied electric field has widely been investigating in many physical systems such as ionic liquids, water, biological tissues, etc.

Historical View of Diffusion Process in LT-DPs

The Self-diffusion in dusty plasmas first time was theoretically predicted by Rosenfeld in 1977(Rosenfeld, 1977). Later different computational methods such as Molecular dynamics (MD) simulations, Monte Carlo etc. algorithms, and models were developed to precisely calculate diffusion motion in material science and thermophysics. MD simulations were performed and investigated the self-diffusion coefficient in three-dimensional (3D) DPs by Ohta and Hamaguchi (Ohta & Hamaguchi, 2000). Begum *et al.* also used this technique and calculated the self-diffusion coefficient over a wide range of plasma parameters(Begum & Das, 2016). Daliguelt employed the MD simulation and computed the self-diffusion coefficient in one component plasma(Daligault, 2012b) and ionic mixtures(Daligault, 2012a). Moreover, Shahzad and He investigated diffusion motion for strongly coupled dusty plasma (Shahzad & He, 2012b) and weakly coupled dusty plasma (Shahzad et al., 2013) through molecular dynamics simulations and point out different diffusion motions in dusty plasma regimes. Langevin dynamic simulation is another computational tool. This scheme investigated diffusion in 3D dusty Plasma by .(Dzhumagulova et al., 2012). Experimental studies have been done and calculated diffusion in 2D dusty plasma systems by Vaulina *et al.* (Vaulina et al., 2001). The above literature study shows that self-diffusion is extensively investigated in 2D and 3D dusty plasma systems over wide ranges of plasma parameters through different methodologies. According to a literature survey, there is no evidence of diffusion motion in LT-DPs for the external applied electric field has been found.

The Objective of the Book Chapter

This book chapter's primary objective is to investigate the behavior of self-diffusion of charged dust particles in three-dimensional LT-DPs under the external applied electric field's influence. A well-known computational technique (molecular dynamics simulation) is used and calculated self-diffusion for different combinations of dusty plasmas parameters. The effect of DPs parameters on the diffusion motion and electric field was analyzed and discussed. This work gives more reliable data of diffusion coefficient in LT-DPs, with computational time cost-effective.

MOLECULAR DYNAMICS SIMULATION ALGORITHMS

In thermal fluids science and engineering, the experimental setups are very complex and available at a high cost, and some experiments required more time for calculations for specific investigations. In the modern age, cost-effective and low time consumption, and computational power are most prominent. Nowadays, it is a trend that before starting the experimentations, first test with computational tools than verified with experiments. There are various computer algorithms and techniques are used for the calculation of various properties in different materials. The Computer simulation provides a linkage between theoretical and experimental research work. We perform computer simulations to test a theoretical model, verify experimental data, and compare different data (Hollingsworth & Dror, 2018). Through computer simulation, the researchers investigate the structure and microscopic interactions between the particles. Various computational methods can be designed, which are very difficult or impossible (such as at high temperature and pressure) in an experiment or maybe very expensive. It also acts as a bridge between microscopic length, time scale, and macroscopic worlds. There are various standard simulation methods such as Molecular dynamics (MD) and the Monte Carlo (MC) simulation mainly used to investigate different phenomena of dusty plasmas. Additional simulation software is also available to compute multiple physical properties and build a new model following one of the simulation schemes.

The art of molecular dynamics (MD) simulations became a standard computational tool for investigating and developing new models in the field of plasma physics, polymer, materials, biological and chemical sciences, and other physical systems. This scheme became a standard computation tool to investigate various physical system's phase transition, structure, and dynamical properties. In this scheme, newton's equation of motion is used to calculate the trajectory of each particle. This method has widely been used to investigate various properties such as thermal conductivity(Shahzad & He, 2012, 2015c), share viscosity(Shahzad et al., 2014), and diffusion process (Shahzad et al., 2013), non-Newtonian behavior (Shahzad, 2017), dynamical structure factor(Shahzad et al., 2019), and waves (Shahzad et al., 2018, 2020). Also, use for phase transition of DPs in weakly and strongly coupled regimes with accuracy compared with experimental results. The calculations of interaction and forces between the charged dust particles calculate through Yukawa (screened Coulomb) potential for isotropic interactions. Yukawa is the most used potential for different physical systems such as chemical, biological, and physical, especially for the most commonly used dusty plasma potential. In this work, we also used Yukawa's potential for calculation interaction and forces of dust particles written as (Shahzad & He, 2015b).

$$\phi_Y = \frac{Q_d^2}{4\pi\varepsilon_o r_{ij}} \exp^{\frac{-r}{\lambda_D}} \tag{1}$$

In Eq. (1), the Q_d represents the amount of charge on dust particles, ϵ_o is the permittivity of free space, r_{ij} is the distance between two interaction dust particles i and j ($r_{ij}=r_j-r_i$), and λ_D is debye length. There are two fundamental dimensionless dusty plasma parameters used to characterize the dusty plasma regimes. One parameter is known as Coulomb coupling parameters same as Coulomb law can be written as

$$\Gamma = \frac{Q_d^2}{4\pi\varepsilon_o} \frac{1}{a_{ws} k_B T} \tag{2}$$

In the above equation, a_{ws} is known as interspacing between the particles or Wigner Seitz radius, can be written as $a_{ws}=(3/4n\pi)^{-1/3}$ Where n is the equilibrium dust particle density, the kB is Boltzmann constant, and T is the absolute system temperature. The plasma coupling parameters (Γ) is the ratio of neighboring dust particle's potential and kinetic energies. If the value of $\Gamma > 1$, then dusty plasma is known as strongly coupled dusty plasma and weakly coupled dusty plasma if the values of $\Gamma < 1$. The second dimensionless parameters are the Debye screening parameter

$$\kappa = \frac{a_{ws}}{\lambda_D} \tag{3}$$

The screening parameter is the ratio of the antiparticle distance to the Debye screening length. The Yukawa force interacted on dust particles can be calculated as $F_Y = -\frac{\partial \hat{O}_Y}{\partial r_{ij}}$ (Shahzad & He, 2012). In this chapter, we apply an external static electric field on the z-axis of the simulation box. the self-diffusion coefficient computed under the influence of an external electric field.

$$E = \frac{QE_z}{2\kappa T} \tag{4}$$

In Eq (4), the external applied dimensionless electric field directed along the z-axis as other interactions and forces with Yukawa interaction potential. The total electric potential accounts for the sum of Yukawa potential and potential produced due to external applied electric field. The electric force on each particle can be calculated as $\mathbf{F}_E = QE$ here; Q is the charge on the dust particles, and E is the strength of the external electric field directed to the z-axis. The external applied electric field (E) (Sukhinin et al., 2012, 2017) on each particle i and κ is Debye screening parameters ($\kappa = 1/\lambda_D$). The 3D nominal plasma frequency is $\omega_p = \left(Q_d^2 / 3\pi\epsilon_0 ma^3\right)^2$ where m is the mass of dust particles, the plasma frequency only bears in the limit of unscreened Coulomb ($\kappa \equiv 0$), and in the limit of finite screening ($\kappa > 0$), the plasma frequency (ω_p) do not bear any specific physical significance. The Einstein frequency is written as

$$\omega_E^2 = \frac{k_D^2}{3m} \sum_{i \neq j} \Phi_Y \left(r_i - r_j\right) \tag{5}$$

Where m is the mass of dust particles, Φ_Y is the Yukawa potential given in Eq (1). The Einstein frequency's numerical value depends on crystal structure types such as face-centered cubic (FCC) and body-centered cubic (BCC); for convenience, here, we only use the FCC lattice. For details values of $\omega_{E,}$ please see reference (Ohta & Hamaguchi, 2000). We considered N-identical particles confined to an FCC simulation box of edge length L in the simulation method. The velocity Varlet algorithm is used to calculate the forces on each (Spreiter & Walter, 1999). The Ewald summation method is used to calculate the long-range interaction between the particles.

Diffusion Motion

The diffusion process has characterized the motion of particles in the gas and liquid regimes. In this process, the molecules or particles migrate from a high concentration region toward a low concentration region in liquids and gas above the absolute temperature. The rate of the diffusion process depends on the mass, temperature, and viscosity of the particles. There are different methods to calculate the diffusion motion of particles of the systems. One is the Einstein relation can be written as

$$MSD(t) = \left\langle \left| r_i(t) - r_i(0) \right|^2 \right\rangle \tag{6}$$

In this equation, $r_i(0)$ is the final position of particles $i^{th,}$ and $r_i(0)$ is the initial position of ith particles over the time segment t. Using Eq. (6), we can analyze the motion of particles is diffusive, no diffusive, and sub-diffusion. The second method is the velocity autocorrelation function (VACF) of the Well-known Green-Kubo integral formula written as

$$D = \frac{1}{3N}\int_{0}^{\infty}\langle V_j(t).V_j(0)\rangle dt \tag{7}$$

In the above equation, N is the number of particles in the simulation cubic cell, and $V_j(t).V_j(0)$ is known as VACF. The V_j(t) is the velocity of jth particles at the final time (t) and initial time (0). The brackets $\langle\ldots\rangle$ represent the canonical ensemble average all-over particles. The VACF is independent of the specific particle in thermal equilibrium, so we take all particle averages to minimize the statistical errors (Ohta & Hamaguchi, 2000; Shahzad et al., 2013).

The diffusion coefficient can be normalized with plasma frequency as $D_0 = D/\omega_p a^2_{ws}$ and Einstein frequency $D^* = D/\omega_E a^2_{ws}$. The normalization diffusion coefficient, $D^* = D/\omega_E a^2_{ws}$ with D=limit$_{t\rightarrow inft}D(t)$. in this chapter present and (Ohta & Hamaguchi) results are normalized with Einstein frequency (D^*) and (Daligault, 2012a, 2012b) are normalized with the plasma frequency (D_0). For the validation of molecular simulation data, the fitting curve was calculated by using an appropriate equation given below

$$D^* = \alpha\,(T^* - 1)^{\beta} + \gamma \tag{8}$$

The α, β, and γ are fitting parameters, T^* is the ratio of system temperature (T) to the fluid-solid critical temperature (T_m), also called melting temperature, defined as $T^*=T/T_m=\Gamma_m/\Gamma$ and $\Gamma_m = Q_d^2 / 4\pi\epsilon_o a T_m$ If $T^* > 1$considered as a fluid phase for each κ also $\Gamma < \Gamma_m$ considered in the liquid phase, for $\Gamma > \Gamma_m$ is taken in solid phase for the small values of κ. It is exercised that the fitting equation is an excellent formula for the molecular dynamics simulation results of the diffusion coefficient. With the variation of κ all the fitting parameters vary in order of magnitude. Relations were used for the computation self-diffusion coefficient. We also verified our simulation data for the values of fitting parameters using Eq. (8).

RESULTS AND DISCUSSION

The NEMD simulations have been used to investigate the self-diffusion coefficient (D^*) for a wide range of plasma coupling parameter ($\Gamma \equiv 2, 500$), screening strength ($\kappa \equiv 1.4, 3$), constant number of particles ($N \equiv 500$), and for near-equilibrium strength of external applied electric field ($E = 0.10$) The authors have calculated the D^* by integrating the VACF using the Green-Kubo relation is given in Eq. (7). The simulation time step was chosen small enough ($\Delta t = 0.001\omega_p^{-1}$). To maintain the system temperature applies the Gaussian thermostat in the employed NEMD simulation method. To check the accuracy runs the MD code run-up to $2.5x10^5\omega_p^{-1}$. NEMD simulation executed and prepared the results, $N = 500$ numbers of particles. The forces (Yukawa and electric), acceleration, momentum, and positions of particles are updated at every time step. After reaching the system in thermal equilibrium, we record the data of D^*. The NEMD results of D^* for 3D SC-CDPs are optimized by varying the electric field strength and simulation time step (Δt). It is confirmed the accuracy and consistency of the used NEMD simulation method by conservations of energies (potential and kinetic. It is confirmed that this scheme can be used for the computation of D^* for E. The results are optimized at equilibrium strength of electric field ($E = 0.10$) and $\Delta t = 0.001\omega_{pd}^{-1,}$ which makes a benchmark against the previous work of (Ohta & Hamaguchi, 2000), (Daligault, 2012a, 2012b). Every data set have separate runs for each system temperature ($\equiv 1/\Gamma$) corresponding to κ at constant strength of dimensionless external electric field ($E = 0.10$).

*Figure 1. Self-diffusion coefficient (D^*) as a function of plasma Coulomb coupling parameter (Γ) for 3D SC-CDPs at screening parameter (κ = 1.4), N = 500 and for Near-equilibrium strength of external applied electric field (**E=0.10**) along the z-axis. Simulations outcomes of D^* are compared with MD simulation results in 3D DPs (Ohta & Hamaguchi, 2000), MD results in Yukawa one-component plasma obtained at κ =1.5 (Daligault, 2012b) and κ =1.0 (Daligault, 2012a).*

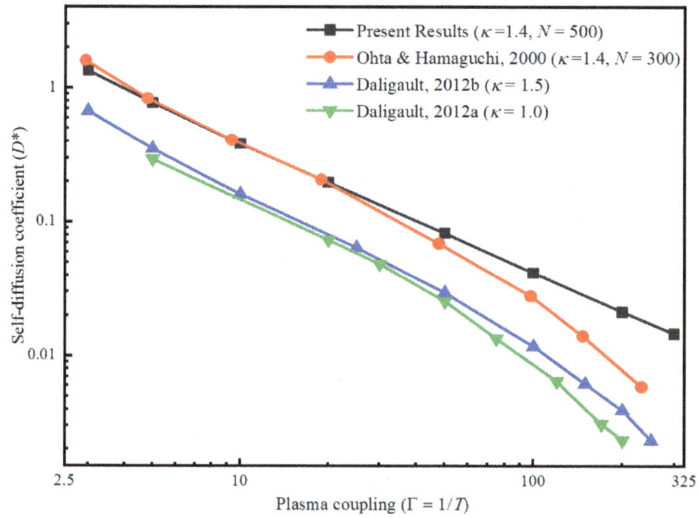

NEMD simulation outcomes of D^* in the strongly coupled regimes of dusty plasma under the constant strength of applied $E = 0.10$, for $\kappa = 1.4$, $\Gamma = 3\text{-}300$, and $N = 500$ are shown in Figure 1. We take the log scale along the x-axis and y-axis for a straightforward comparison. The filled black square (-■-) demonstrated the presented NEMD results of D^* for 3D SC-CDPs. Results are compared with previous MD simulation data of D^* for DPs and OCPs. The filled red line circle shows the EMD results for 3D DPs calculated by (Ohta & Hamaguchi, 2000), nearly the same plasma parameters for the absence of electric field. The blue and green line-triangle presented the MD results of D^* in Yukawa OCP calculated at $\kappa = 1.5$ (Daligault, 2012b) and $\kappa = 1.0$ (Daligault, 2012a), respectively. The simulation outcomes of D^* in SC-CDPs also agree well with results obtained through computer simulation in 3D Yukawa liquids, in strongly correlated liquids type structure, and 3D Dusty Plasmas over Γ and κ. It is observed that the D^* are monotonically reduced with increasing Γ which are similar to the DPs and YOCPs under the influence of near-equilibrium strength of $E = 0.10$. The presented NEMD results show a good agreement with previous results. It is observed from Fig. 1 that D^* linearly decreased with increasing Plasma coupling parameters under the constant strength of E. It is observed that the plasma D^* fast decays at higher plasma temperature (low Γ) and gradually decays at low plasma temperature (higher Γ). The system has a high temperature for low plasma coupling values; the charged dust particles also have higher kinetic energy due to high temperature. The charged dust particles have high collision rate at lower strength of Γ so that the plasma D^* decay fast. In the higher value of Γ (low temperature), collisions between the particles decreased, and D^* gradually decay. In the solid-like phase, the D^* increased compared with (Ohta & Hamaguchi, 2000) due to E. However, with increasing Γ, increased collision in dusty plasma regimes and diffusion coefficient attain a sizeable fraction of its field-free value. When the value of Γ is decreased, the friction of dust particles decreased. It was observed that $\Gamma = 50$ was nearly the same as (Ohta & Hamaguchi, 2000). For the region $50 < \Gamma \leq 300$, the D^* was slightly decayed, and the values of D^* definitely higher than the one-component plasma. The results demonstrated linear responses under the action of constant.

Figure 2 demonstrated the obtained NEMD simulation results of D^* in 3D SC-CDPs as a function of plasma Coulomb coupling ($\Gamma \equiv 5, 400$) for $\kappa = 2.0$ for the number of particles ($N = 500$) and low strength of external $E = 0.10$ comparison purposes. The simulation outcomes of D^* for different values of Γ are compared with previous data of MD in 3D DPs (Ohta & Hamaguchi, 2000) and OCP (Daligault, 2012b). It was observed that the MD simulation outcomes of SC-CDPs for low external electric filed have good agreement with prior MD data of D^* for 3D DPs and OCPs. The obtained results are more closed with 3D DPs (Ohta & Hamaguchi, 2000) and slightly higher than the OCPs (Daligault, 2012b). In Fig. 3, the results of

D^* at $\kappa = 3.0$ for $\Gamma = 5\text{-}500$ under the influence of $E = 0.10$ for $N = 500$ are display in Fig. 3. The numerical outcomes compared with (Daligault, 2012a, 2012b; Ohta & Hamaguchi, 2000). From the results, one can observe that presented D^* data in-between the 3D DPs and OCPs. The possible reason is the difference in numerical schemes, input parameters such as the number of simulated particles, total simulation run, time step, etc. Overall for the given plasma parameters, the presented algorithm shows good accuracy. The obtained simulation results of plasma D^* as a function of plasma coupling (temperature) for constant electric field shows that plasma have linear behaviors instead of a curve. The effect of constant E was observed as an increasing trend for $\kappa = 1.4$, $\Gamma > 50$, $\kappa = 2.0$, $\Gamma > 70$ and $\kappa = 3.0$, $\Gamma > 100$ shows in Figs. 1-3, respectively. So what the external applied electric field are mainly affects liquids and solid-like regimes of SC-CDPs.

Figure 2. Self-diffusion coefficient (D^*) as a function of plasma Coulomb coupling parameter (Γ) for 3D SC-CDPs at screening parameter (κ = 2), N = 500 and for near-equilibrium strength of external applied electric field (E = 0.10) along the z-axis. Simulations outcomes of D^* are compared with MD simulation results in 3D DPs (Ohta & Hamaguchi, 2000) and Yukawa one-component plasma obtained (Daligault, 2012b).

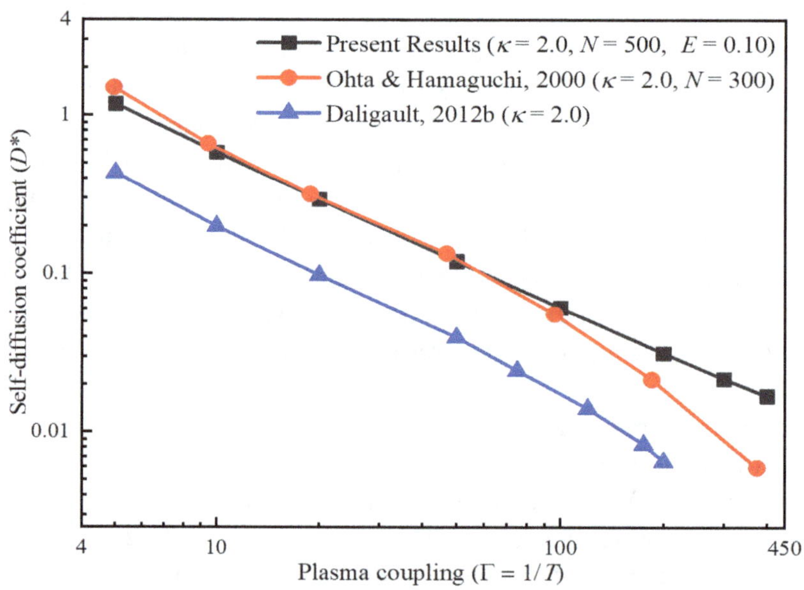

Figure 3. Self-diffusion coefficient (D^*) as a function of plasma Coulomb coupling parameter (Γ) for 3D SC-CDPs at screening parameter (κ = 3), N = 500 and for near-equilibrium strength of external applied electric field (E = 0.10) along the z-axis. Simulations outcomes of D^* are compared with MD simulation results in 3D DPs (Ohta & Hamaguchi, 2000) and Yukawa one-component plasma obtained (Daligault, 2012b).

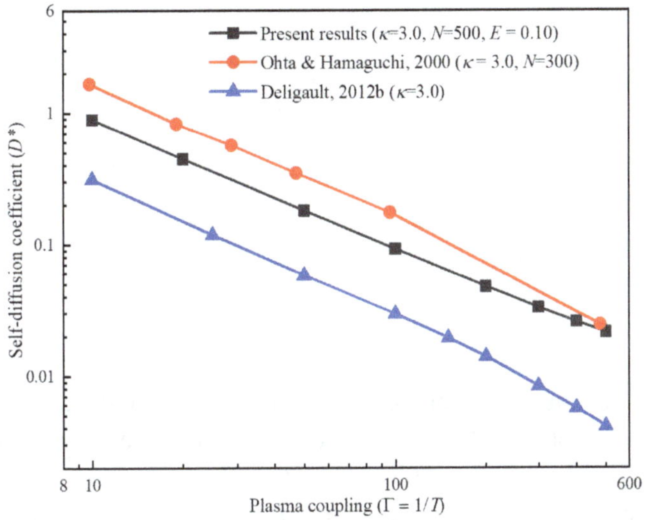

The NEMD outcomes of D^* for 3D SC-CDPs have good agreements with MD simulation results of strongly coupled dusty plasma and higher than OCPs and ionic mixtures. It is hypothesized that due to electric field charged dust particles are polarized and thus induced dipole-dipole coupling. Due to dipole moment, dust particles have extra interaction energy via external E strength (Ivlev et al., 2008). It is supposed that the externally applied electric field also changed the structure of SC-CDPs. Due to the external applied electric field, the charged dust particles are derived in the electric field's direction. According to the current state of our knowledge, no significant studies have been done to understand the elementary process of DPs under the static electric field's influence on the diffusion motion. The simulation of the dust particles in a moderate electric field can lead to a linear response. The drift velocity of the dust particle also increased with increasing E strength. The electric field also caused breaking the caged region of the dust particles in the direction to force field but in all directions. That effect caused an increase in the diffusion coefficient. With increasing κ, the D^* gradually increased as was expected.

Figure 4. Effect of external applied electric field on self-diffusion coefficient in three-dimensional low temperature dusty plasmas obtained through non-equilibrium molecular dynamics simulation for $\kappa = 1.4$, $N = 500$ *(a)* $\Gamma = 50$, *(b)* $\Gamma = 200$

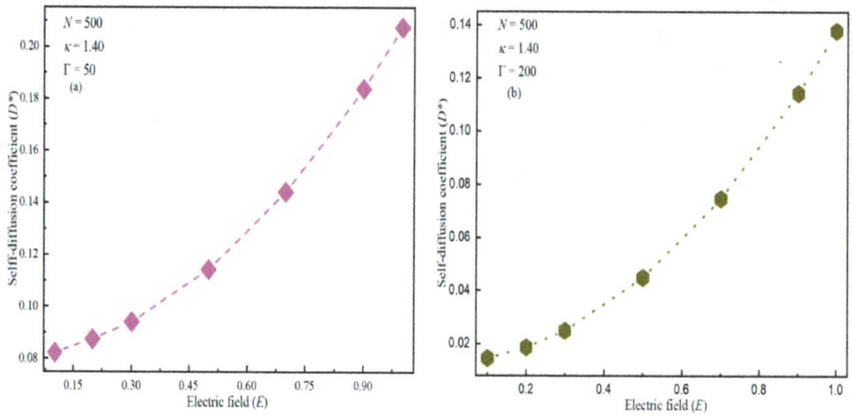

The primary task of the present research is to investigate the effects of variations of external E on D^* for LT-DPs. The NEMD data of plasma D^* against the external applied electric field with $N = 500$, $\kappa = 1.4$, and two different values of plasma coupling are shown in Figure 4; (a) $\Gamma = 50$ and (b) $\Gamma = 200$, respectively. The obtained plasma D^* at equilibrium strength of external applied electric field ($E = 0.10$) was compared and discussed in Figure 1 for $\kappa = 1.4$. It is observed from two panels of figure 4 that the electric field significantly affects D^* in LT-DPs. The result shows that various electric fields ($E = 0.1, 1.0$) can affect the interactions and substantially change SC-DPs structure. The low, increasing trend in D^* is noted at the low strength of the electric field (E = 0.1-0.5). It is also stated that at the higher strength of the electric field (E = 0.6-1.0), the increasing trend of D^* is prominent. The higher strength of E has higher power; therefore, the electric field has a more massive effect on the D^*. This increasing behavior of D^* values shows that strongly coupled dusty plasma in liquids states the same as conventional electrorheological fluids and ionic liquids under the electric field influence. It is observed that the D^* values increased under the applied strength of the external electric field 250% for the case of $\kappa = 1.4$ and $\Gamma = 50$. In cold-liquid state nearly melting temperature (κ = 1.4, Γ=200), the D^* increased almost 944%.

Figure 5. Effect of external applied electric field on self-diffusion coefficient in three-dimensional low temperature dusty plasmas obtained through non-equilibrium molecular dynamics simulation for κ =2, N = 500 (a) Γ = 50, (b) Γ = 400

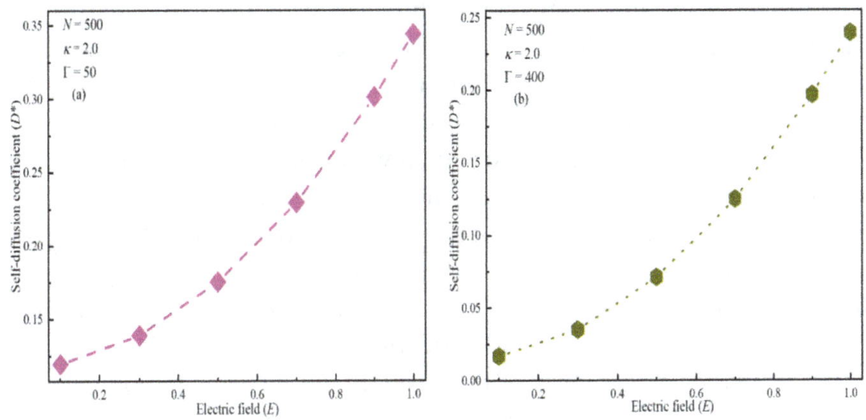

Figure 6. Effect of external applied electric field on self-diffusion coefficient in three-dimensional low temperature dusty plasmas obtained through non-equilibrium molecular dynamics simulation fo κ =3, N = 500 (a) Γ = 50, (b) Γ = 500

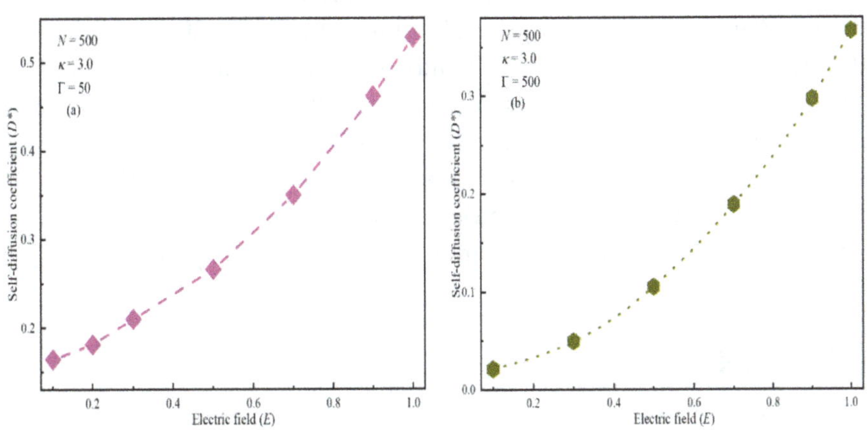

Figures 5 and 6 demonstrated the NEMD outcomes of plasma $D*$ as a function of the external electric field (E) for $\kappa = 2.0$ and 3, respectively. The effect of E on plasma $D*$ at low temperature with $N = 500$ is investigated. Two different regimes of LT-DPs for each screening strength for the same number of particles and external applied electric field strength are selected. Results are compared and discussed with previous MD data for low strength of $E = 0.10$ are given in Figures 2 and

3, respectively. It is observed that D^* increased with increasing external electric field and screening strength. The D^* decreases with increase plasma coupling and increases with increasing screening strength and electric field. It is noted that the increasing trend of D^* in LT-DPs under the influence of E is nearly the same as ionic liquids and electrorheological fluids. Obtained results show that the presented algorithm gives precise data. In the overall analysis, we can say that LT-DPs have electrorheological properties for the electric field with respective to diffusion coefficients. The effects of electric field on low system temperature are prominent, compared with high temperature, especially at higher electric field strength (= 0.5, 1.0). The results also demonstrated that the electric field effect on plasma D^* at high screened dust particles is more prominent than lower screening strength.

The obtained result from Figs. 4-6 suggests that the drift velocity of the dust particle increased with increasing electric field strength. The electric field also caused breaking the caged region of the dust particles. The main outcomes of this work D^* increase with increasing E. It was noted from the calculations that at $E =$ 1.0, the total energy not remained entirely conserved. The total energy error was observed in the low order of variation of the interacting charged dust particles. The validity and consistency of our MD code under Yukawa and electric field interaction were checked and verified. Our outcome gives more precise and reliable data for diffusion in strongly coupled dusty plasma. These data have a good agreement with the previous MD and experimental (Nunomura et al., 2006). These results indicate striking similarities of ionic liquids. It is hypothesized that charged dust particles are driven in the applied external electric field's direction, which caused an increase in the D^* values. The simulation outcomes are consistent with conventional electrorheological fluid (Kanu & Shaw, 1995) and ionic liquids (Clark et al., 2019). So what under the influence of the electric field, the LT-DPs show electrorheological behavior. It may be helpful to explore more properties of dusty plasma.

CONCLUSION

The Nonequilibrium molecular dynamics (NEMD) simulations have been used to investigate the external electric field's effect on the self-diffusion coefficient (D^*) in low-temperature dusty plasma. Results are obtained by employing the integral formula for the velocity autocorrelation function of the Green-Kubo relation. The D^* was equilibrated with previous numerical results of dusty plasma and one-component plasma at equilibrium strength of electric field ($E = 0.10$). By comparing this research work, our developed MD code gives the precise data for D^* in low temperature dusty plasma. The simulation outcomes show linear behavior as a function of the plasma coupling parameter under the constant electric field. It is observed that

D^* increased with increasing electric field strength in the low-temperature state of dusty plasma. The electric field mainly affects the D^* in fluids to the solid-like phase of dusty plasma for different screening strengths. It has been demonstrated that the presented numerical results for a given range of plasma parameters (Γ, κ) and electric field strength have accuracy and consistency. It is concluded that the developed NEMD simulations approach is excellent performance concerning the electric field as an additional force with the Yukawa force.

ACKNOWLEDGMENT

The authors would like to thank Dr. X. D. Zhang at the Network Information Center of Xi'an Jiaotong University for support of the HPC platform and also to the national center for physics (NCP) Islamabad allocation computational power to check and run the MD code.

REFERENCES

Adamovich, I., Baalrud, S. D., Bogaerts, A., Bruggeman, P. J., Cappelli, M., Colombo, V., Czarnetzki, U., Ebert, U., Eden, J. G., Favia, P., Graves, D. B., Hamaguchi, S., Hieftje, G., Hori, M., Kaganovich, I. D., Kortshagen, U., Kushner, M. J., Mason, N. J., Mazouffre, S., ... Vardelle, A. (2017). The 2017 Plasma Roadmap: Low temperature plasma science and technology. *Journal of Physics. D, Applied Physics*, *50*(32), 323001. doi:10.1088/1361-6463/aa76f5

Begum, M., & Das, N. (2016). Self-diffusion as a criterion for melting of dust crystal in the presence of magnetic field. *The European Physical Journal Plus*, *131*(2), 1–12. doi:10.1140/epjp/i2016-16046-2

Boufendi, L., & Bouchoule, A. (2002). Industrial developments of scientific insights in dusty plasmas. *Plasma Sources Science & Technology*, *11*(3A), A211–A218. doi:10.1088/0963-0252/11/3A/332

Chu, J. H., & Lin, I. (1994). Direct observation of Coulomb crystals and liquids in strongly coupled rf dusty plasmas. *Physical Review Letters*, *72*(25), 4009–4012. doi:10.1103/PhysRevLett.72.4009 PMID:10056356

Clark, R., Von Domaros, M., Mcintosh, A. J. S., Luzar, A., Kirchner, B., & Welton, T. (2019). Effect of an external electric field on the dynamics and intramolecular structures of ions in an ionic liquid. *The Journal of Chemical Physics*, *151*(16), 164503. doi:10.1063/1.5129367 PMID:31675893

Daligault, J. (2012a). Diffusion in ionic mixtures across coupling regimes. *Physical Review Letters, 108*(22), 1–5. doi:10.1103/PhysRevLett.108.225004 PMID:23003608

Daligault, J. (2012b). Practical model for the self-diffusion coefficient in Yukawa one-component plasmas. *Physical Review E: Statistical, Nonlinear, and Soft Matter Physics, 86*(4), 1–2. doi:10.1103/PhysRevE.86.047401 PMID:23214710

Dzhumagulova, K. N., Ramazanov, T. S., & Masheeva, R. U. (2012). Velocity Autocorrelation Functions and Diffusion Coefficient of Dusty Component in Complex Plasmas. *Contributions to Plasma Physics, 52*(3), 182–185. doi:10.1002/ctpp.201100070

Hollingsworth, S. A., & Dror, R. O. (2018). Molecular Dynamics Simulation for All. *Neuron, 99*(6), 1129–1143. doi:10.1016/j.neuron.2018.08.011 PMID:30236283

Ivlev, A. V., Morfill, G. E., Thomas, H. M., Räth, C., Joyce, G., Huber, P., Kompaneets, R., Fortov, V. E., Lipaev, A. M., Molotkov, V. I., Reiter, T., Turin, M., & Vinogradov, P. (2008). First observation of electrorheological plasmas. *Physical Review Letters, 100*(9), 1–4. doi:10.1103/PhysRevLett.100.095003 PMID:18352717

Ivlev, A. V., Thoma, M. H., Räth, C., Joyce, G., & Morfill, G. E. (2011). Complex plasmas in external fields: The role of non-hamiltonian interactions. *Physical Review Letters, 106*(15), 1–4. doi:10.1103/PhysRevLett.106.155001 PMID:21568565

Kanu, R. C., & Shaw, M. T. (1995). Effects of DC and AC Electric Fields on the Response of Electrorheological Fluids Comprising Cylindrical PBZT Particles. *Progress in Electrorheology*, 303–323.

Laroussi, M. (2020). Cold Plasma in Medicine and Healthcare: The New Frontier in Low Temperature Plasma Applications. *Frontiers in Physics, 8*, 1–7. doi:10.3389/fphy.2020.00074

Nunomura, S., Samsonov, D., Zhdanov, S., & Morfill, G. (2006). Self-diffusion in a liquid complex plasma. *Physical Review Letters, 96*(1), 13–16. doi:10.1103/PhysRevLett.96.015003 PMID:16486468

Ohta, H., & Hamaguchi, S. (2000). Molecular dynamics evaluation of self-diffusion in Yukawa systems. *Physics of Plasmas, 7*(11), 4506–4514. doi:10.1063/1.1316084

Rosenfeld, Y. (1977). Relation between the transport coefficients and the internal entropy of simple systems. *Physical Review A, 15*(6), 2545–2549. doi:10.1103/PhysRevA.15.2545

Schwabe, M., Khrapak, S. A., Zhdanov, S. K., Pustylnik, M. Y., Räth, C., Fink, M., Kretschmer, M., Lipaev, A. M., Molotkov, V. I., Schmitz, A. S., Thoma, M. H., Usachev, A. D., Zobnin, A. V., Padalka, G. I., Fortov, V. E., Petrov, O. F., & Thomas, H. M. (2020). Slowing of acoustic waves in electrorheological and string-fluid complex plasmas. *New Journal of Physics*, *22*(8), 083079. doi:10.1088/1367-2630/aba91b

Shahzad, A., Aslam, A., & He, M.-G. (2014). Equilibrium molecular dynamics simulation of shear viscosity of two-dimensional complex (dusty) plasmas. *Radiation Effects and Defects in Solids*, *169*(11), 931–941. doi:10.1080/10420150.2014.968852

Shahzad, A., & He, M. G. (2012a). Thermal Conductivity of Three-Dimensional Yukawa Liquids (Dusty Plasmas). *Contributions to Plasma Physics*, *52*(8), 667–675. doi:10.1002/ctpp.201200002

Shahzad, A., & He, M.-G. (2012b). Thermodynamic characteristics of dusty plasma studied by using molecular dynamics simulation. *Plasma Science & Technology*, *14*(9), 771–777. doi:10.1088/1009-0630/14/9/01

Shahzad, A., & He, M.-G. (2012c). Shear viscosity and diffusion motion of two-dimensional dusty plasma liquids. *Physica Scripta*, *86*(1), 015502. doi:10.1088/0031-8949/86/01/015502

Shahzad, A., & He, M.-G. (2015a). Nonlinear Effects of Dusty Plasmas using Homogenous Nonequilibrium Molecular Dynamics Simulations. *Journal of Applied Nonlinear Dynamics*, *4*(3), 259–265. doi:10.5890/JAND.2015.09.006

Shahzad, A., & He, M.-G. (2015b). Calculations of thermal conductivity of complex (dusty) plasmas using homogenous nonequilibrium molecular simulations. *Radiation Effects and Defects in Solids*, *170*(9), 758–770. doi:10.1080/10420150.2015.1108316

Shahzad, A., & He, M.-G. (2015c). Numerical experiment of thermal conductivity in two-dimensional Yukawa liquids. *Physics of Plasmas*, *22*(12), 123707. doi:10.1063/1.4938275

Shahzad, A., He, M.-G., Haider, S. I., & Feng, Y. (2017). Studies of force field effects on thermal conductivity of complex plasmas. *Physics of Plasmas*, *24*(9), 093701. doi:10.1063/1.4993992

Shahzad, A., He, M.-G., & He, K. (2013). Diffusion motion of two-dimensional weakly coupled complex (dusty) plasmas. *Physica Scripta*, *87*(3), 035501. doi:10.1088/0031-8949/87/03/035501

Shahzad, A., He, M.-G., & Shakoori, M. A. (2017). Thermal Transport and Non-Newtonian behaviors of3D Complex Liquids Using Molecular Simulations. *2017 14th International Bhurban Conference on Applied Sciences and Technology (IBCAST)*, 472–474.

Shahzad, A., Khan, M. Q., Shakoori, M. A., He, M.-G., & Feng, Y. (2020). Thermal Conductivity of Dusty Plasmas through Molecular Dynamics Simulations. *Thermophysical Properties of Complex Materials*, 13.

Shahzad, A., Shakoori, M. A., & He, M.-G. (2020). Wave Spectra in Dusty Plasmas of Nuclear Fusion Devices. *Fusion Energy*, 13. https://www.intechopen.com/books/advanced-biometric-technologies/liveness-detection-in-biometrics

Shahzad, A., Shakoori, M. A., He, M.-G., & Bashir, S. (2018). *Sound Waves in Complex (Dusty) Plasmas*. In *Computational and Experimental Studies of Acoustic Waves*. Intech Open. doi:10.5772/intechopen.71203

Shahzad, A., Shakoori, M. A., He, M.-G., & Yan, F. (2020). *Numerical Approach to Dynamical Structure Factor of Dusty Plasmas*. Intechopen. doi:10.5772/intechopen.78334

Shahzad, A., Shakoori, M. A., He, M.-G., & Yang, F. (2019). Dynamical structure factor of complex plasmas for varying wave vectors. *Physics of Plasmas*, *26*(2), 023704. doi:10.1063/1.5056261

Spreiter, Q., & Walter, M. (1999). Classical molecular dynamics simulation with the velocity verlet algorithm at strong external magnetic fields. *Journal of Computational Physics*, *152*(1), 102–119. doi:10.1006/jcph.1999.6237

Sukhinin, G. I., Fedoseev, A. V., Khokhlov, R. O., & Suslov, S. Y. U. (2012). Plasma Polarization Around Dust Particle in an External Electric Field. *Contributions to Plasma Physics*, *52*(1), 62–65. doi:10.1002/ctpp.201100063

Sukhinin, G. I., Fedoseev, A. V., Salnikov, M. V., Rostom, A., Vasiliev, M. M., & Petrov, O. F. (2017). Plasma anisotropy around a dust particle placed in an external electric field. *Physical Review. E*, *95*(6), 1–8. doi:10.1103/PhysRevE.95.063207 PMID:28709289

Vaulina, O. S., Nefedov, A. P., Petrov, O. F., & Fortov, V. E. (2001). Transport properties of macroparticles in dust plasma induced by solar radiation. *Journal of Experimental and Theoretical Physics*, *92*(6), 979–985. doi:10.1134/1.1385637

Chapter 5
Synthesis of Carbon Nanotubes via Plasma Arc Discharge Method

Adeel Aabir
University of Agriculture, Faisalabad, Pakistan

Muhammad Yasin Naz
University of Agriculture, Faisalabad, Pakistan

Shazia Shukrullah
University of Agriculture, Faisalabad, Pakistan

ABSTRACT

CNTs are the element that exists with predominant physio-chemical properties, which have been extensive researched today. These properties make carbon nanotubes (CNTs) valuable in a wide potential range of applications. The production of high-quality carbon nano-tubes (CNTs) via different precursors has been reported for many years. The arc discharge is a pristine technique to form CNTs with a high-quality yield. This technique has been elucidated for a long time, but the growth condition and mechanism of affected synthesized parameters and coorelation between synthesized parameters and nucleation of carbon have not been explored. In this chapter, the authors present the factors affecting temperature, geometry, grain size, electrodes, pressure, catalyst, arc current, power supply, and growth mechanism of CNTs. The variation in parameters has been elicited along with challenges and gaps.

DOI: 10.4018/978-1-7998-8398-2.ch005

1. INTRODUCTION

Carbon nanotubes possess extensive chemical, mechanical, optical, and thermal properties listed in Table 1. They are broadly classified as SWCNTs, MWCNTs, and DWCNTs. Moreover, SWCNTs are divided into three kinds, namely arm-chair, zigzag, and chiral carbon nanotubes. CNTs can show metallic or semiconducting behavior, depending on the nature of the structure or chirality. While armchair carbon nanotubes are always metallic with indices m=n=3q, others can be metallic or semiconductors. The indices n and m identify the electronic structure and q be the integer. In arm-chair carbon nanotubes, there is no band-gap between the valence and the conduction band but in the case of zig-zag and chiral carbon nanotubes, a narrow bandgap exists which is the nature of the semiconductor materials.

Researchers devised different ways to synthesize CNTs via various precursors. The most popular synthesis methods are arc discharge, laser ablation, and chemical vapor deposition (Chau et al., 2020). Hydro-thermal, ball milling and electrolysis have also been used to synthesize CNTs. CNTs were synthesized by Iijima via the arc discharge method (Bahgat et al., 2011). While the literature has no more comprehensive research on the formation mechanism of CNTs. There is a need of an elevated correlation between parameters and growth conditions of nano-tubes. Yoshinori et al. (Yoshinori, 2010) presented the chronological feature of CNTs under hydrogen atmosphere. Tessonnier (Tessonnier & Su, 2011) revealed the growth and nucleation process during production. Journet et al. (Journet et al., 2012) explained medium and low-temperature routes to synthesized CNTs. In this literature, the growth mechanism of CNTs formation has been extensively explained.

2. SYNTHESIZED TECHNIQUE, PARAMETERS AND THEIR EFFECTS ON YIELD

2.1 Arc Discharge Setup

The arc discharge is an old technique to generate plasma using electric current, firstly it was used for the production of CNTs by Iijima. A model of the arc discharge chamber is presented in Figure 1. It consisted of two electrodes, anode and cathode having positive or negative charges respectively. The anode was filled-up with the precursor of carbon and catalyst, the cathode electrode was typically a graphite rod and the chamber was usually filled with a gas. An intermediate gap of 1 mm to 2 mm was maintained to generate arc discharge between the electrodes whenever a power supply of alternating or direct current (AC, DC) was applied. For a steady arc discharge, a uniform supply of current was maintained which was supportive to modify the gap, other else fluctuating arc discharge affected the quality of the product. This un-fluctuated current generates high-temperature plasma of 4000 K to 6000 K, which filled the carbon precursor inside the anode, vapor concentration of carbon in the gas phase flow towards the cathode and cools down because of the temperature gradient. After few minutes, the discharge arc was stopped and soots of CNTs were collected from the wall of the chamber. The soot was purified and investigated for morphology.

Table 1. Properties of SWCNTs and MWCNTs

				References
Mechanical properties	Young's Modulus	SWCNTs	~ 1 TPa	(Zhang, 2012)
		MWCNTs	~ 1-1.2 TPa	(Moradi et al., 2012)
	Tensile Strength	SWCNTs	~ 60 GPa	(Su et al., 2013)
		MWCNTs	~ 0.15 TPa	
Thermal properties	Thermal conductivity	SWCNTs	~ 1750-5800 W/mk	(Zhao et al., 2014)
		MWCNTs	>3000 W/mk	(Kim et al., 2012)
Electronic properties	Bandgap	In SWCNTs when n-m is divisible by 3	0 eV, Metallic	(Cai et al., 2012)
		When n-m is not divisible by 3	0.4-2 eV, Semiconductor	(Kia & Bonabi, 2012)
		In MWCNTs	~ 0 eV, Non-semiconductor	
Electrical Properties	Resistivity	SWCNTs and MWCNTs	10^{-6} Ωm	(Fang et al., 2013)
	Maximum current density	SWCNTs and MWCNTs	10^{7}-10^{9} A.cm^{-2}	(Fang et al., 2013; Su et al., 2013)
	Quantized conductance	SWCNTs and MWCNTs	12.9 kΩ^{-1}	(Chau et al., 2020; Su et al., 2012)

For the continuous deposition of carbon vapors, the anode was moved back to maintain the gap of 1 mm between the electrodes. The high temperature caused the breakdown of gas inside the chamber. For the formation of plasma, the gas was ionized into plasma. In the plasma emission photon was a cause of glow, which was due to the collusion between electrons and ions. The electrons, which were ejected from the cathode, strike at the end of the anode with very high speed and filled the precursor of carbon at the center. The high temperature converts the precursor of carbon into vapors and then the vicious cluster of carbon changes, vapors of carbon converted into liquid carbon. The gradient temperature affected the atmosphere for the deposition of liquid carbon crystallized on the cathode of the grey outer shell. These grey deposits were graphene cylinders which were named as CNTs (Arora & Sharma, 2014; Berry et al., 2014). But the actual mechanism of growth was arguable due to the different assumptions of CNTs growth, vapors phase, liquid phase, solid phase, and crystal phase.

Figure 1. The basic setup of the arc discharge method.

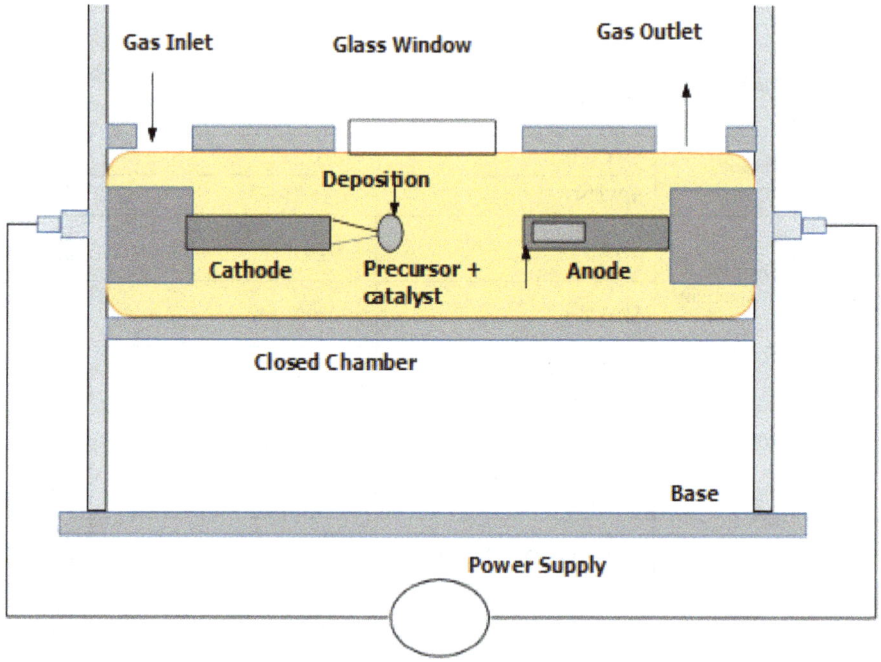

2.1 Vapors Phase Growth

The condensed vapors of carbon in MWCNTs are dominant to form CNTs in this model. The vapors phase model distributes the vapors of carbon into two groups, depending on their velocity. One group contains isotropic distribution of velocity and the other group contains a higher velocity than the first group due to the acceleration between the electrodes. CNTs were based on three steps named termination, tube growth, and seed formation. The vapors of carbon possess isotropic distribution at the beginning to form nanoparticles. With an increase in the current, the vapors of carbon that were directed to the first group become seeds. Whenever the applied current reaches a steady value, the ions of carbon start perpendicular flow towards the surface of the cathode.

2.1 Liquid Phase Growth

The amorphous carbon beads inside the nanotubes form when liquid droplets solidify. When an electric arc is used to heat a carbon precursor, the electrons interacting the surface induce localized heating, which causes the surface to liquefy. The anode's liquid droplets then expelled and float towards the cathode. At the cathode surface, these globules of liquid carbon tend to cool. Cooling moves from the outer to the inner layers, resulting in multilayered tubular structures.

2.1 Solid-Phase Growth

Amorphous carbon beads are found within the nanotubes that form when liquid droplets solidify. When an electric arc is used to heat a carbon precursor, the electrons bombarding the surface produce localized heating, causing the surface to liquefy. The liquid droplets from the anode are ejected and drift towards the cathode. These globules of liquid carbon tend to cool near the cathode surface. Multilayered tubular formations arise from cooling that travels from the outside to the inter layers.

3. NANO GROWTH IN ARC DISCHARGE

In the production of arc discharge, power supply plays a decisive role. Here we explain the effect of power supply, current, voltage, current, and frequency on the synthesis of nanotubes. In Figure 3, graphical representation shows various power supplies in the arc discharge method.

3.1. Classification of Power Supplies

In plasma formation, the power supply plays a significant role. Figure 2(a) shows electrons emission from cathode to anode with high velocity, whenever at the end of electrodes DC voltage was applied. The diameter of the anode was usually smaller than the diameter of the cathode because of the current density difference. The precursor of carbon-filled inside the anode was observed by the high-temperature gradient to form CNTs. While in AC arc discharge, no deposition was observed due to the periodic behavior across the electrodes. Due to this thermal effect, vapors of carbon escape of the plasma to form CNTs at the chamber wall, as shown in Figure

2(b). In Figure 2(c), pulsed arc method shows the formation of CNTs with a short range of milli-seconds width. The deposition takes place on the cathode surface whenever the pulse strikes to vaporize the anode surface.

In case of direct current supply, ionized gas flow towards cathode surface and hinder the carbon deposition. To remove gas ions, it was too much difficult to remove the negative charge on the cathode surface. This problem abstains from AC supply or pulsed power supply. But the drawback of AC power supply is the formation of CNTs at the positive end of the cathode surface. The pulsed arc method was most suitable in the formation of CNTs. Some variation was observed in both discharges by using different kinds of power supplies. The continuous arc discharge was not as energetic as pulsed discharge. At the ejection's time, the electrons have higher energies and take very short intervals of time (milliseconds) to approach isotropic distribution. Due to the continuous bombardment at the end of the anode, pulses of shorter width and increased energy of electrons result in high yield (He et al., 2020; Scalese et al., 2010; Shukrullah et al., 2015; Takekoshi et al., 2012; Zhao et al., 2013; Zhao et al., 2012).

3.2 Arc Current Effect

Arc current plays a significant role in the arc discharged method. The formation of carbon nanotubes affects the yield, size, and quality of synthesized nanotubes. It enables the emission of electrons from the cathode that moves towards the anode with high velocity. The emitter electrons hit on the anode surface and sputter more the precursor of carbon. The applied current generates high temperature due to which the precursor of carbon flows towards anode to form carbon vapors and formation of further nanotubes. A small amount of current known as chopping current also exist, which enables the production of plasma. To ablate the anode, the constant current of 50 to 100 A is maintained during this synthesis process. Nishizaka et al. (Nishizaka et al., 2011) found that in SWCNTs formation optical current density was about 250 to 270 Acm^2. In the graphical representation, it is clear that most of the experimental work has been performed in the range of 50 A to 100 A.

It was found that below 30 A of current results in low carbon yields. In the formation of carbon nanotubes, current anomalies were remarkable. The higher yield of CNTs obtained at a current of 100 A. Researchers found a lower yield of CNTs at low-level current (2 A to 10 A) (Su, 2010). However, the high-level amount of current (700 A to 900 A) was also used in the formation of CNTs. Arc current variation develops an interesting debate in the formation of CNTs. In this situation, we need a well-defined correlation between the formation of carbon nanotubes and the current level.

Figure 2. Effect of DC power supply, AC power supply, and pulsed arc supply on growth of CNTs.

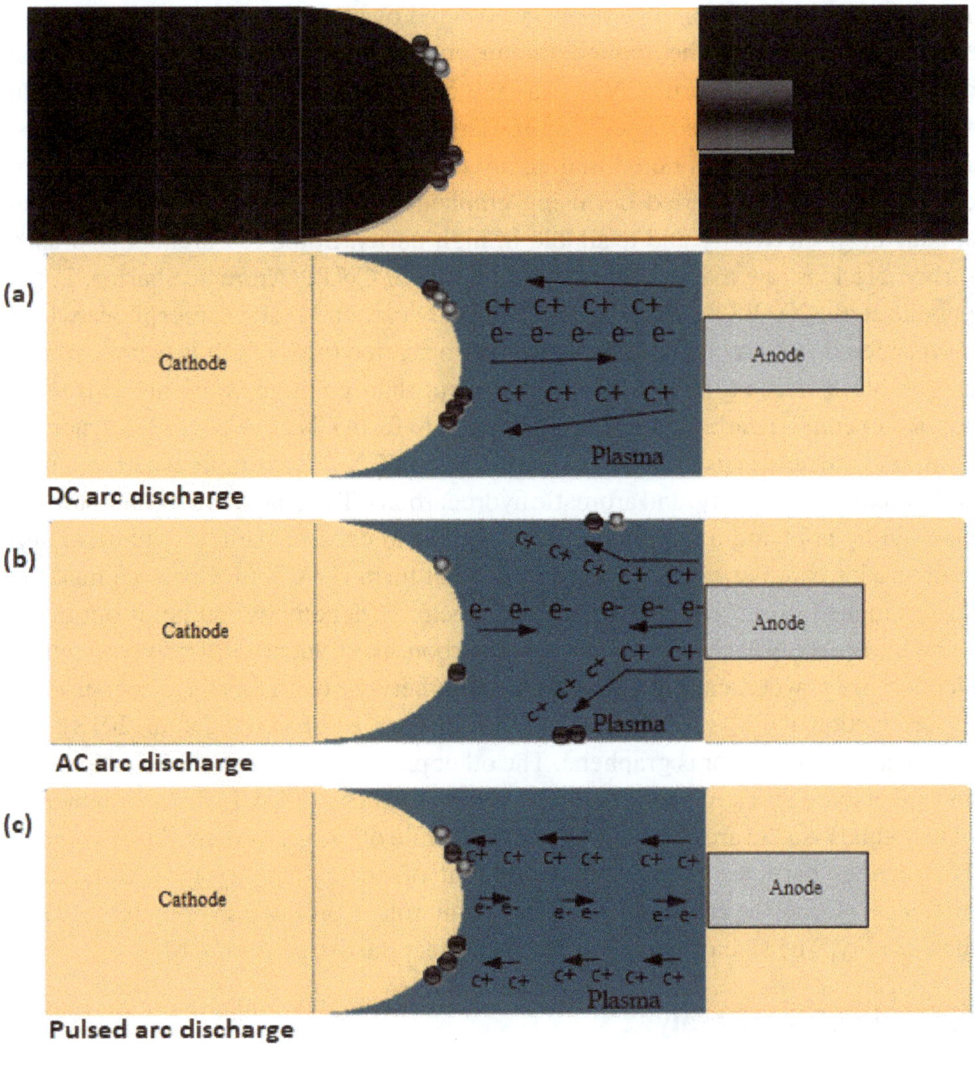

It was found that in the arc discharge method, frequency of the power supply influences the quality of deposition. The stronger growth at anode using a low range frequency of (50Hz) is possible, however a higher range of frequency (400 Hz) increases the rate of production.

3.3 Precursor of Carbon in Arc Discharge Method

The production of carbon nanotubes is executed by carbon precursors by using an eternal power source. The applied arc current evaporates the precursor of carbon into carbon vapors to form CNTs. The pie chart in Figure 3 shows the graphical representation of published papers using different power supplies, different precursors, and atmospheres. It was noted that an arc discharge method for the formation of carbon nanotubes is carried out using graphene precursor, which is an excellent conductor of electricity and available in high purity. Some researchers also used carbon black as raw material for the production of CNTs (Arora & Sharma, 2014; Łabędź et al., 2009; Qiu et al., 2010). Naturally carbon black is amorphous which is considered as a potential precursor in the formation of CNTs. It was noticed that carbon black needs a large amount of current than graphite when arc current is applied to convert carbon black into graphene to form CNTs of a tabular structure.

Some researchers used coal in the production of CNTs as a highly reactive source and a mixture of aliphatic and aromatic hydrocarbons. The coal was used as a starting material by applying a large amount of current to decompose the polymeric aryl structure into aromatic and alkyne species that form DWCNTs. Coal is a material that contained sulfur, which affects the diameter of nanotubes and helps out in the growth of DWCNTs. Various sources of carbon like toluene, fullerene, polyvinyl alcohol, tire powder, carbon nanofibers, and other hydrocarbons have been used in the production of CNTs by the method of arc discharge (Ghaleb & Jaafar, 2018). But the most used precursor is graphene. The other precursors need more investigations and understating for strong and better results on synthesis of CNTs. In the production of nanotubes via the arc discharge method, the role of the grain size is less explored. The smaller particles require small amount of energy to vaporize. Solid-state properties (Cohesive energy) play an important role in production of nanotubes and their quality. The size of particles can also affect the diameter of CNTs.

3.4 Role of the Catalyst

Generically, a metal catalyst is used in the arc discharge method for the production of well-defined carbon nanotubes. The catalyst is used in the form of powdered, which is filled in both sides of the electrodes along with precursor of carbon. In the formation of carbon nanotubes, the used catalyst should have high rate of evaporation and low activation temperature. The atoms of metal vaporize with the precursor of carbon and metal particles of carbon vapors agglomerate at the walls of the reactor. A possible reason for this behavior was the melted metal and vapors of carbon which move towards plasma, that's why particles of metal do not stick on the surface of the cathode. They move away towards the wall of the chamber due to their momentum and form SWCNTs. The suggested mechanism of growth of CNTs is illustrated in Figure4.

The metal catalyst favors the growth of SWCNTs rather than MWCNTs. The cathode changes the temperature and blocks the growth of MWCNTs. The commonly used catalysts are Fe (iron), Ni (Nickel), and Co (carbon monoxide) with Cr, S, and Y as a promoter. In the presence of these metal catalysts, single-walled carbon nanotubes were observed, whereas MWCNTs form in the absence of these catalysts. It was an interesting observation that MWCNTs were obtained by using a binary catalyst. No MWCNTs were obtained without the use of binary catalysts. Among all the metals, Nickel (Ni) and Fe (Iron) were the most used catalysts in the formation of better CNTs. Iron (Fe) is normally added with sulfur for better growth of SWCNTs. The main drawback of Fe (iron) was its oxides (Fe_2O_3), which do not behave as a metal catalyst and slow down the growth of SWCNTs (Zhong et al., 2018). The Fe (iron) support promotes the length of CNTs. In Fe metal catalyst addition of sulfur develops a shell at the surface of the cathode due to difference of melting points. These shells were helpful in the growth of DWCNTs. A constant distribution diameter of CNTs was observed by using the catalyst of FeS rather than Ni/Co alloy.

Figure 3. The percentage of published papers on arc discharge method to form CNTs using various power supplies, carbon precursors and different atmospheres.

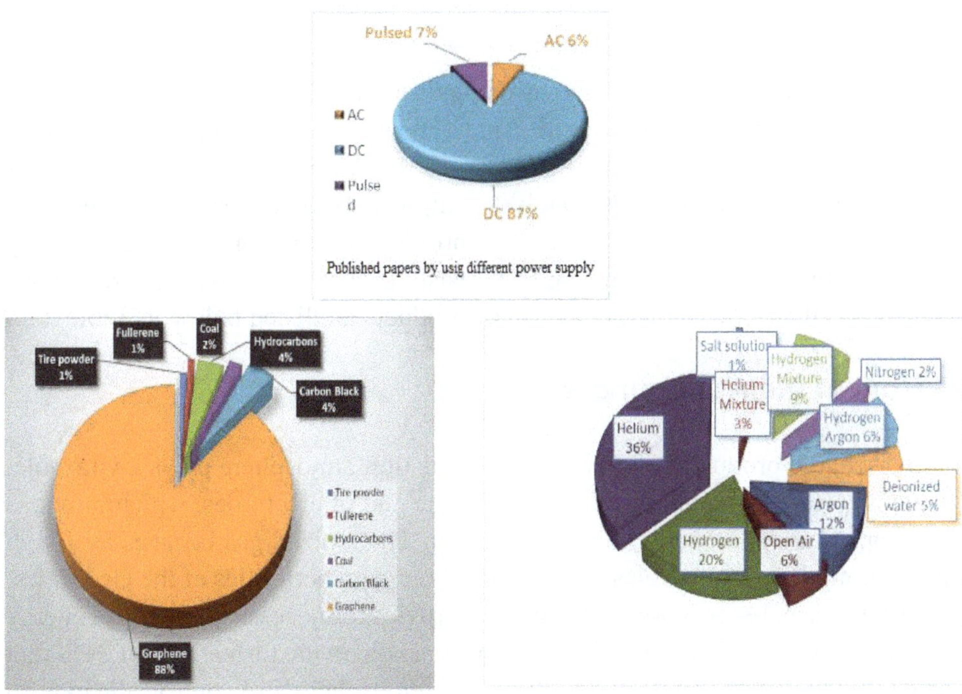

Figure 4. Illustration of role of catalyst in arc discharge growth of carbon nanotubes.

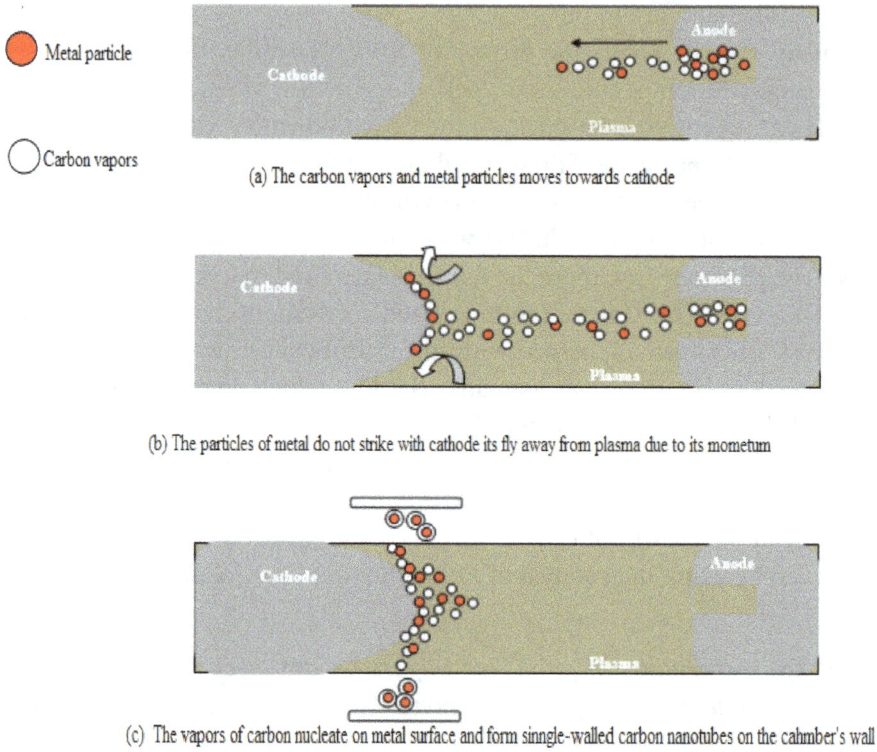

(a) The carbon vapors and metal particles moves towards cathode

(b) The particles of metal do not strike with cathode its fly away from plasma due to its mometum

(c) The vapors of carbon nucleate on metal surface and form sinngle-walled carbon nanotubes on the cahmber's wall

The nanotube's size is controlled by using different size catalysts. The vapors of carbon move with metal particles to form nanotubes. In the distribution of nanotubes diameter, the size of particles plays a vital role. The metal particle concentration affects the production of nanotubes and catalyst compositions also perform a significant role in the tube growth process. The concentration of iron cannot be exceeded by 10% in nanotubes production since it may resist the flow of carbon vapors. Various catalysts and their effects on the growth of products are listed below in Table 2.

3.5 Role of the Atmosphere

In arc discharge production of CNTs, the formation atmosphere plays a vital role. Plasma is formed by applying arc current at the end of electrodes, which provides high temperature and energy to ionize the gas. The ionized gas behaves as a high conductor medium and enables the transfer of mass on both sides of the electrodes. Due to this thermal energy, ions in the plasma provide the necessary amount of energy to the carbon vapors. The annealing effect is essential in the formation of CNTs that is provided through thermal conductivity. The atmosphere also controls the plasma

Table 2. Effect of pure and mixed catalysts on growth of carbon nanotubes.

Base catalyst	Mixture used	Comments
Ni	Ni	Promote SWCNTs growth
	Ni-Y	Alone Y is unable to synthesize CNTs
		Addition with Ni of Y reduced the yield of CNTs
		Y promote SWCNTs growth
	Ni-Mo	Mo doesn't affect the formation of CNTs
	Ni-Mo-Fe	Mo-Fe promote CNTs formation
	Ni-Co	SWCNTs formation
	Ni-Co+KCl	DWCNTs formation
	Ni-Cr	MWCNTs formation
	Ni-Ho	Increase quality and yield of CNTs
	Ni-Co-Fe-S	SWCNTs diameter increased
Fe	Fe	Affected the diameter and yield of MWCNTs
	Fe-FeS	Increased yield and growth of DWCNTs
	Fe- Fe_2O_3	Retard CNTs growth
	Fe- Fe_3O_4	SWCNTs yield increase
	Fe-W	The tungsten increases the yield of SWCNTs and reduced the diameter
	Fe-$Fe(C_3H_5)_2$	Improve SWCNTs yield
Ca	Ca	Production of small diameter then Ni in CNTs formation
Rh-pt	Rh-pt	CNTs yield increases
Co	Co	Production of DWCNTs
		Mean diameter increase of CNTs
	Co-CoS	Sulphur addition promote the production of CNTs

temperature that depends on the electrical conductivity and ionization potential of the gas. The thermal conductivity of hydrogen is high than other gases, which is a well-order quencher in the growth of CNTs. Su et al. (Su et al., 2011) investigated the CO effect and impurities that were present in the SWCNTs. It was observed that in the formation of SWCNTs, lower concentrations of CO are preferred over high concentrations. CNTs can also be produced by a liquid atmosphere rather gaseous environment, the reason for this preference is the formation of floating CNTs on the liquid surface. The NaCl solution is also used to promote the electrical conductivity and cooling ability of nanotubes which is a low-cost solution than the liquid helium and nitrogen (Yousef et al., 2013). The dissociation of NaCl into Na^{+1} and Cl^{-1} ions provides well-defined ionic conductivity. In the better yield of CNTs, Na^{+1}ion was a fundamental parameter but Na^{+1} ions resist the movement of carbon vapors and effect the yield of CNTs. While other solutions like H_3PO_4, LiCl, and beer froth have also been used in the arc discharge as the atmosphere. The effect of different environmental son arc discharge growth of CNTs is shown in Table 3.

Table 3. Effect of different atmospheres on yield of carbon nanotubes in arc discharge method.

Environment	Types of nanotubes	Effect on yield	Effect on nanotubes	Remarks
H_2	MWCNTs	Crystalline perfection.	Better than He, provide twice CNTs ratio,	More effective.
H_2-He			Provide CNTs with a small diameter.	
H_2-Ar		H_2 addition increased the yield.		
He	SWCNTs	Yield increases.	Synthesized CNTs diameter is smaller than He.	Uniform deposition.
Ar			Produced CNTs with a smaller diameter as compared to He.	
N_2	SWCNTs and MWCNTs	More yield due to lower pressure.	With the increase in N_2 diameter of MWCNTs decrease (Takekoshi et al., 2012).	Incorporation of N_2 atoms for closure of CNTs.
H_2-N_2			By a mixture of both gasses diameter distribution of diameter can be controlled.	
Open-air		Better yield than He.		
CO			Promote selective diameter growth.	
Kr	SWCNTs	Better yield than Ar.		
NaCl Solution			Promote short nanotubes in solution.	

3.6. Pressure Effect

In the method of arc discharged, the gas pressure controls the energy available for the gas molecules. The range of suitable gas pressure is found to be 300 torrs to 700 torrs. It was observed that in unstable plasma, lower level of yield has been obtained than high pressure, which provides a large number of precipitates. For better production of carbon nanotubes, an efficient pressure of nitrogen was 350 Torr. In the case of open-air,45 to 90 torr was the optimal pressure for CNTs production in arc discharge. The pressure in the range of 40-300 torr yields in less amount of CNTs. For helium gas, some researchers found the optimum range of pressure from 500 torr to 700 torr. The pressure range from 100-500 torr decreases the yield of nanotubes. Zhao et al. (Zhao et al., 2014) observed that1520 torr pressure of helium provides a 2.5 g/h yield. The pressure of hydrogen was found to be 500 torrs. It was also observed that yield of nanotubes has been increased from 100-700 torr, but beyond 700 torrs, it was found to be decreased.

3.7. Role of Temperature

The temperature is given to sublime the precursor for onward decomposition it provides thermal flux of vapors to form carbon ions. In the formation of CNTs, temperature helps to nucleate the vapors of carbon at the end of the cathode. The size and quality of carbon nanotubes were critically affected by temperature variations. In arc discharge, stable temperature for nucleation and growth was controlled by an electric arc. Increasing current density also increases the temperature. Plasma temperature can be regulated through thermal atmospheric properties. Higher temperatures (3600-3800 K) are used for hydrogen plasma while argon plasma is generated in the plasma temperature range of 2200-2400 K) The temperature across both the electrodes depends on the diameter of the electrodes. Smaller diameter results in higher temperature and current density, which are suitable for efficient decomposition of carbon precursor. The large diameter of cathode result in small current density, which facilitates the nucleation of carbon nanotubes. The anode diameter is generally selected smaller than the diameter of the cathode.

The electrode geometry also affects the quality and quantity of nanotube. As discussed above, the diameter of the cathode is kept larger than the anode The electrode's size affects the distribution of plasma temperature and also the synthesized product. However, the use of identical electrodes does not assist to obtain plasma at the surface of the cathode. The effect of identical and dissimilar electrodes was reported by few groups (Cai et al., 2012). The surface of cathode topography plays a significant role in the formation of nanotubes. The diameter of the anode should be less than 10 mm, since small anode diameter promotes the yield of nanotubes (Kia & Bonabi, 2012).

3.9. Yield and Size of Nanotubes

The concern of arc discharge was to obtain a better yield of CNTs. The yield of nanotubes depends on composition, the shape of electrodes, the effect of the atmosphere, size of particles, catalyst type, and arc current. The product of nanotubes also depends on the rate of cooling which also depends on temperature gradient, thermal conductivity, arc current, and the size of the cathode. The typical yield that runs few minutes was a few milli-gram which makes this method less preferable rather than chemical vapors deposition. However, the researcher has been trying to make this method efficient by controlling the experimental parameters. Few studies of setup modification have been discussed already. Few reports on variation in pressure values, the diameter of electrodes, composition of the catalyst, current and atmosphere have been proposed to optimize. The possible nanotubes' yield that can be obtained in arc discharge varies from 20 to 50 mg/min. However, some attempts of CNTs production by continuous and semi-continuous discharge have been reported as 6.5 g/h and 2 g/h. However, the effect of discharge parameters on diameter and size of nanotubes is not yet fully understood. The effect of experimental parameters on yield and size of nanotubes has been listed in Table 4.

Table 4. Effect of experimental parameters on yield and size of nanotubes.

Parameters	Effect on size	Effect on yield
Catalyst	S and CO increase the diameter of SWCNTs, while tungsten decreased the diameter. The Smaller diameter of SWCNTs is observed using Ca-Ni catalyst.	By optimizing catalyst composition 4 g/h yield is obtained. Yield increases by catalyst.
Temperature	Increasing temperature reduced diameter. Increasing temperature increase the diameter of MWCNTs.	
Electrodes	Promote purity of SWCNTs. Increase the diameter of the cathode from 0.8 to 2 cm. Length of SWCNTs increase from 1.2 to 1.8 μm.	The smaller diameter of the anode increases the yield. Better were yield found at the angle of 90^0.
Atmosphere	Small CNTs were observed in NaCl solution. Changing of the atmosphere fromH_2–Ar to H_2–He, the smaller diameter of CNTs observed. In a mixture of He-Ar, Ar affect the diameter of SWCNTs. By changing inert gas, the diameter of SWCNTs changes. The diameter of SWCNTs is smaller than He in the air With a large grain size, a smaller diameter of CNTs has been found.	
Current	Increasing current increases the diameter.	Yield increase. Yield is maximum at 8 Ampere
Setup modification	Diameter at 0G larger than 1G as 3.4 times.	Rotated electrodes increase the yield.
Pressure	SWCNTs were obtained at a pressure of 4 Kp.	Pressure increase (250 to 400 Torr), yield increase. Below than 100 torrs of pressure, the low-level yield obtained.

4. CONCLUSION

In the arc discharge method of synthesis of carbon nanotubes, the growth mechanism, the effect of parameters, precursors and gases is elucidated in this chapter. Applied current is the most important parameter in the arc discharge method to form high quality carbon nanotubes. The effect of temperature on growth and nucleation of CNTs is also not fully understood. The geometry of electrodes is an imperative parameter in the production of CNTs which requires more exploration to understand the role of shape of electrodes in the production of yield. Supply of DC power has been normally employed while other power supplies are not reported very well. The formation of CNTs via arc discharge involves graphite while other precursors such as carbon coal and carbon black have not been reported well. Some mathematical and theoretical investigations are also needed to develop a better arc discharge geometry.

REFERENCES

Arora, N., & Sharma, N. N. (2014). Arc discharge synthesis of carbon nanotubes: Comprehensive review. *Diamond and Related Materials*, *50*, 135–150. doi:10.1016/j.diamond.2014.10.001

Bahgat, M., Farghali, A. A., El Rouby, W. M. A., & Khedr, M. H. (2011). Synthesis andmodification of multi-walled carbon nano-tubes (MWCNTs) for water treatment applications. *Journal of Analytical and Applied Pyrolysis*, *92*(2), 307–313. doi:10.1016/j.jaap.2011.07.002

Berry, T. D., Filley, T. R., & Blanchette, R. A. (2014). Oxidative enzymatic response of white-rot fungi to single-walled carbon nanotubes. *Environmental Pollution*, *193*, 197–204. doi:10.1016/j.envpol.2014.06.013 PMID:25047356

Cai, X., Cong, H., & Liu, C. (2012). Synthesis of vertically-aligned carbon nanotubes without a catalyst by hydrogen arc discharge. *Carbon*, *50*(8), 2726–2730. doi:10.1016/j.carbon.2012.02.031

Chau, M. Q., Hoang, A. T., Truong, T. T., & Nguyen, X. P. (2020). Endless story about the alarming reality of plastic waste in Vietnam. *Energy Sources. Part A, Recovery, Utilization, and Environmental Effects*, 1–9. doi:10.1080/15567036.2020.1802535

Fang, L., Sheng, L., An, K., Yu, L., Ren, W., Ando, Y., & Zhao, X. (2013). Effect of adding W to Fe catalyst on the synthesis of SWCNTs by arc discharge. *Physica E, Low-Dimensional Systems and Nanostructures*, *50*, 116–121. doi:10.1016/j.physe.2013.03.005

Ghaleb, Z. A. A., & Jaafar, M. (2018). Carbon Nanotube-Metal Oxide Hybrid Nanocomposites Synthesis and Applications. *1-Dimensional Metal Oxide Nanostructures*, 281-312.

He, G., Duan, Y., & Pang, H. (2020). Microwave Absorption of Crystalline Fe/MnO@C Nanocapsules Embedded in Amorphous Carbon. *Nano-Micro Letters*, *12*, 1-16.

Journet, C., Picher, M., & Jourdain, V. (2012). Carbon nanotube synthesis: From large-scale production to atom-by-atom growth. *Nanotechnology*, *23*(14), 142001. doi:10.1088/0957-4484/23/14/142001 PMID:22433510

Kia, K. K., & Bonabi, F. (2012). Electric field induced needle-pulsed arc discharge carbon nanotube production apparatus: Circuitry and mechanical design. *The Review of Scientific Instruments*, *83*(12), 123907. doi:10.1063/1.4772575 PMID:23278005

Kim, Y. A., Muramatsu, H., Hayashi, T., & Endo, M. (2012). Catalytic metal-free formation of multi-walled carbon nanotubes in atmospheric arc discharge. *Carbon*, *50*(12), 4588–4595. doi:10.1016/j.carbon.2012.05.044

Łabędź, O., Lange, H., Huczko, A., Borysiuk, J., Szybowicz, M., & Bystrzejewski, M. (2009). Influence of carbon structure of the anode on the synthesis of single-walled carbon nanotubes in a carbon arc plasma. *Carbon*, *47*(12), 2847–2854. doi:10.1016/j.carbon.2009.06.031

Moradi, O., Yari, M., Zare, K., Mirza, B., & Najafi, F. (2012). Carbon nanotubes: A review of chemistry principles and reactions. *Fullerenes, Nanotubes, and Carbon Nanostructures*, *20*(2), 138–151. doi:10.1080/1536383X.2010.533312

Nishizaka, H., Namura, M., Motomiya, K., Ogawa, Y., Udagawa, Y., Tohji, K., & Sato, Y. (2011). Influence of carbon structure of the anode on the production of graphite in single-walled carbon nanotube soot synthesized by arc discharge using a Fe–Ni–S catalyst. *Carbon*, *49*(11), 3607–3614. doi:10.1016/j.carbon.2011.04.063

Qiu, J., Chen, G., Li, Z., & Zhao, Z. (2010). Preparation of double-walled carbon nanotubes from fullerene waste soot by arc-discharge. *Carbon*, *48*(4), 1312–1315. doi:10.1016/j.carbon.2009.01.036

Scalese, S., Scuderi, V., Bagiante, S., Gibilisco, S., Faraci, G., & Privitera, V. (2010). Order and disorder of carbon deposit produced by arc discharge in liquid nitrogen. *Journal of Applied Physics*, *108*(6), 064305. doi:10.1063/1.3475726

Shukrullah, S., Mohamed, N. M., Naz, M. Y., & Shaharun, M. S. (2015). Synthesis and Characterization of Multiwalled Carbon Nanotubes Using Ferrocene and Aluminum Oxide/Iron Nitrate Catalysts. *Analytical Letters*, *48*(2), 340–348. doi:10.1080/00032719.2014.938342

Su, J. S. (2010). Direct route to grow CNTs by micro-electrodischarge machining without catalyst in open air. *Materials and Manufacturing Processes*, *25*(12), 1432–1436. doi:10.1080/10426914.2010.500339

Su, Y., Yang, Z., Wei, H., Kong, E. S. W., & Zhang, Y. (2011). Synthesis of single-walled carbon nanotubes with selective diameter distributions using DC arc discharge under CO mixed atmosphere. *Applied Surface Science*, *257*(7), 3123–3127. doi:10.1016/j.apsusc.2010.10.127

Su, Y., Zhang, Y., Wei, H., Qian, B., Yang, Z., & Zhang, Y. (2012). Length-controlled synthesis of single-walled carbon nanotubes by arc discharge with variable cathode diameters. *Physica E, Low-Dimensional Systems and Nanostructures*, *44*(7-8), 1548–1551. doi:10.1016/j.physe.2012.03.025

Su, Y., Zhou, P., Zhao, J., Yang, Z., & Zhang, Y. (2013). Large-scale synthesis of few-walled carbon nanotubes by DC arc discharge in low-pressure flowing air. *Materials Research Bulletin*, *48*(9), 3232–3235. doi:10.1016/j.materresbull.2013.04.092

Takekoshi, K., Kizu, T., Aikawa, S., & Nishikawa, E. (2012). One-step synthesis of metalencapsulated carbon nanotubes by pulsed arc discharge in water. *e-Journal of Surface Science and Nanotechnology, 10*, 414-416.

Tessonnier, J. P., & Su, D. S. (2011). Recent progress on the growth mechanism of carbon nanotubes: A review. *ChemSusChem*, *4*(7), 824–847. doi:10.1002/cssc.201100175 PMID:21732543

Yoshinori, A. (2010). Carbon nanotube: Story in ando laboratory. *Journal of Siberian Federal University. Mathematics and Physics*, *3*(1), 3–22.

Yousef, S., Khattab, A., Osman, T. A., & Zaki, M. (2013). Effects of increasing electrodes on CNTs yield synthesized by using arc-discharge technique. *Journal of Nanomaterials*. doi:10.1155/2013/392126

Zhang, Y. (2012). Synthesis of few-walled carbon nanotube–Rh nanoparticles by arc discharge: Effect of selective oxidation. *Materials Characterization*, *68*, 102–109. doi:10.1016/j.matchar.2012.03.013

Zhao, J., Su, Y., Yang, Z., Wei, L., Wang, Y., & Zhang, Y. (2013). Arc synthesis of doublewalled carbon nanotubes in low pressure air and their superior field emission properties. *Carbon*, *58*, 92–98. doi:10.1016/j.carbon.2013.02.036

Zhao, J., Zhang, J., Su, Y., Yang, Z., Wei, L., & Zhang, Y. (2012). Synthesis of straight multiwalled carbon nanotubes by arc discharge in air and their field emission properties. *Journal of Materials Science*, *47*(18), 6535–6541. doi:10.100710853-012-6583-z

Zhao, T., Hou, C., Zhang, H., Zhu, R., She, S., Wang, J., & Wei, B. (2014). Electromagnetic wave absorbing properties of amorphous carbon nanotubes. *Scientific Reports*, *4*(1), 1–7. PMID:25007783

Zhong, B., Wang, C. J., Wen, G. W., Yu, Y. L., & Xia, L. (2018). Facile fabrication of boron and nitrogen co-doped carbon@ $Fe_2O_3/Fe_3C/Fe$ nanoparticle decorated carbon nanotubes three-dimensional structure with excellent microwave absorption properties. *Composites. Part B, Engineering*, *132*, 141–150. doi:10.1016/j.compositesb.2017.09.001

Chapter 6
Foodborne Pathogen Inactivation by Cold Plasma Reactive Species

Linda Agun
Universiti Teknologi Malaysia, Malaysia

Chang Shu Ting
Universiti Tunku Abdul Rahman, Malaysia

Norizah Redzuan
Universiti Teknologi Malaysia, Malaysia

Santhana Krishnan
Universiti Teknologi Malaysia, Malaysia

Siti Sarah Safaai
Universiti Teknologi Malaysia, Malaysia

Zarita Zakaria
Universiti Teknologi Malaysia, Malaysia

Muhamad Nor Firdaus Zainal
Universiti Teknologi Malaysia, Malaysia

Mohd Fadthul Ikmal Misnal
Universiti Teknologi Malaysia, Malaysia

Norhayati Ahmad
Universiti Teknologi Malaysia, Malaysia

DOI: 10.4018/978-1-7998-8398-2.ch006

ABSTRACT

Cold plasma is implemented in the food industry for protecting the agricultural product from foodborne pathogens. In this case, dielectric barrier discharge cold plasma pen (DBD-CP) was applied to study its efficiency in inactivation of bacterial on oyster mushroom. The surface of the fresh oyster mushroom was treated with 5 kV of AC voltage with variable of treatment times (0-4 min). Data showed sufficient energy by DBD-CPP has inactivated the existence of bacterial on the oyster mushroom surface with undetectable of bacteria colony. The reactive species generated by cold plasma undoubtedly irreversibly damage the deoxyribonucleic acid, ribonucleic acid, and enzymes of gram bacterial, which eventually causes cell death. Above all, an understanding of the microorganism cell structure, the food surface types, and roughness is an essential in manipulating cold plasma processing parameters to achieve the maximum rate of microbial inactivation.

INTRODUCTION

As the awareness of food safety and quality increases among consumers, demand for fresher, safer, and higher quality foods is increasing (Ji et al., 2018). High production and consumption of fresh crops, "ready to eat" such as vegetables and fruits, have resulted in the rise of agricultural associated foodborne outbreaks worldwide (Castro-Ibáñez et al., 2017). A wide range of fresh fruits and vegetables were implicated with foodborne diseases (FBD) due to the consumption of raw and minimally processed foods that do not meet the food safety and hygiene standard practices in a food supply chain (Shah et al., 2019). FBD caused either by infectious agents or their toxins that enter the body through the ingestion of food (Bayliss, 2012; Bastien et al., 2019). Viruses causing FBD attack cells of the digestive tract and propagate inside them subsequently they attack other cells of the digestive tract or enter other organs such as the liver or central nervous system and cause disease (Vasickova et al., 2005; O'Shea et al., 2019).

Several techniques were introduced to inactivate microorganisms during postharvest processing which includes the use of aqueous chemical sanitizers such as chlorine (Gabriel et al., 2016; Ramos et al., 2020) and organic acids (Sabillón et al., 2020), gaseous sanitizers (Rodríguez et al., 2017), non-thermal processing techniques such as high hydrostatic pressure, ozone and gamma irradiation, nanotechnology and electro-spraying of water (Atungulu et al., 2012; Neetoo & Chen, 2014). However, the use of these techniques have resulted in lower consumer acceptance, reduction in the nutritional values, affected the quality attributes of foods, increase the acid concentration and their reaction with organic matter can produce various toxic disinfection by-products (Ramos et al., 2020; Giaouris et al., 2018). Some postharvest handling techniques are not safe for consumers and environment, for example, quarantine crops in Japan are fumigated with methyl bromide for pesticide control although the fumigant is an ozone-depleting substance (Sakudo et al., 2017). In this respect, safe and effective alternative technology is needed to protect crops from pathogens without using biocides.

New intervention technology, cold plasma (CP) was reported to achieve more desirable inactivation of different microbial species without comprising food product quality (Jiang et al., 2017; Charoux et al., 2020). Cold plasma is an ionized gas characterized by active particles such as electrons, ions, free radicals and atoms that is produced by applying energy to a gas or a mixed system with gas (Han et al., 2019; Zhao et al., 2020). Severe bombardment by CP reactive species provokes surface lesions of the microorganism living cells (Chizoba Ekezie et al., 2017). The chemical species of plasma responsible for microorganism inactivation are primarily active radicals, which also tend to react with food components via physical and chemical routes (Misra, 2016). The advantage of cold plasma treatment over conventional methods includes non-thermal, non-toxic, in-package, rare side-effect since cold plasma could be performed at atmospheric conditions (Zhao et al., 2020).

The unique characteristic of CP is the temperature required during the treatment is very close to the ambient temperature which is suitable for the processing of temperature-sensitive products such as fresh-cut fruits and vegetables (Hati et al., 2018). Short treatment time owing to high current spike reduced the thermal effects due to a quick drop in temperature that indicates the suitability of CP application for foodborne pathogen inactivation process (Mandal et al., 2018).

There is limited information in the literature on the processing parameter of cold plasma with foodborne pathogen for inactivation or reducing of foodborne pathogens on fresh food and vegetables. Consequently, understanding of pathogen characteristics is important to apply CP technology effectively as the difference in the initial levels of the attached bacterial populations can affect the bacterial sensitivity towards the CP treatment based on the bacterial cell membrane characteristics (Ziuzina et al., 2014; Los et al., 2017). The cold plasma mechanism of reactive species on foodborne pathogen inactivation also included. This review critically investigates the usage of cold plasma reactive species and its processing parameter towards destructions the foodborne pathogen on fresh foods. The mechanism on the inactivation of food borne pathogen, its influencing factors is explained in depth. Finally, the cold plasma voltage current and exposure time efficacy of the treatment was gathered together with foodborne pathogen inactivation count.

Significance of Cold Plasma Reactive Species

Microbial decontamination denotes the decomposition or removal of microorganisms. The reduction of pathogenic bacteria in fresh agricultural products using cold plasma (CP) technology is a relatively new approach aiming to improve microbiological safety while maintaining the sensory attributes and an extended shelf-life of the treated foods (Chizoba Ekezie et al., 2017; Agun et al., 2021; Redzuan et al., 2020). Several experimental investigations have been conducted thus far on known applications while others explored new pathways for exploitation of non-thermal plasma for the food industry. Notwithstanding, limitations of plasma may include decline in the firmness of fruit and vegetables, discoloration, and heightened acidity (Chizoba Ekezie et al., 2017).

Generating plasmas by using noble gases is more beneficial as it prevents the oxidative degradation of food on exposure to oxygen. However, this can increase the cost of treatment. Nevertheless, the most economical operations in food systems employ atmospheric air (Mandal et al., 2018). Atmospheric cold plasma (ACP) technology is a novel approach dealing with the inactivation of microorganisms to improve the food safety, along with the preservation of the organoleptic properties (Charoux et al., 2017). The technology takes its advantages from its use of reduced water quantity, low processing temperatures, and low cost. ACP generated in ambient air (Prasad et al., 2017). When plasma is generated at room pressure and temperature, it is termed ACP (Barba et al., 2017). It is a weakly ionised gas which can be applied at a temperature of around 20–40 °C and is, therefore, unlikely to cause thermal injury to the surrounding tissue (Becker et al., 2019). ACP provides inactivation effects mainly by the generation of cell-lethal reactive species (Han et al., 2016) (Han et al., 2016) such as photons, electrons, positive and negative ions, free radicals, and excited molecules such as reactive oxygen species (ROS) and reactive nitrogen species (RNS). These charged species and molecules independently or synergistically are expected to play a significant role in inactivation or killing of microbes (Ziuzina et al., 2020; Gök et al., 2019).

Generally, the types of reactive species of photons, electrons, positive and negative ions, free radicals, and excited molecules formed comprise RONS (reactive oxygen and nitrogen species). Most of reactive species such as excited O_2 and N_2, as well as free radicals (NO and OH) leads to etching and eventually diffuse macromolecules within the cells (Mandal et al., 2018; Laroussi et al., 2003; Fernández et al., 2012; Ohta, 2016; Cheng-yang et al., 2019; Mahdavian Mehr et al., 2020). The plasma activated atoms and ions causes molecular 'sandblasting' and can break down organic contaminants (Pankaj et al., 2013).

MICROBIOLOGICAL DIFFERENCES OF FOODBORNE PATHOGENS ON COLD PLASMA EFFECTS

Foodborne pathogens are the major targets of several industrial food processes since they cause foodborne outbreaks and epidemics and health risk to the consumers (Khelissa et al., 2017; Lafarga et al., 2019). The commonly recognised foodborne diseases (FBD) are *listeriosis*, diarrhoea illness, *haemolytic uremic syndrome* (HUS), *giardiasis*, and *salmonellosis*, etc. (Kljujev et al., 2018). These pathogens have been isolated from soil, sewage sludge, vegetation, and water therefore, has the predominant foodborne pathogens to contaminate the food surfaces detected in fresh vegetables and fruits meat, fish, eggs, etc., that are often associated with FBD (Ukuku et al., 2019). Microorganisms are categorised into Gram (+) and Gram (-) groups by their cell wall structures using Gram staining method. Figure 1 shows the comparison of morphology between *E. coli* and *B. subtilis* cells of Gram (-) and Gram (+) bacteria, respectively. The surface of Gram (+) bacteria were observed to be smoother than that of the Gram (-) bacteria (Laroussi et al., 2003; Liao et al., 2017). Bacteria with a single membrane are known as Gram (+) as the bacteria rich and thick with peptidoglycan layers (~15–80 nm) providing them with higher strength and rigidity (Hati et al., 2018; Shahriari et al., 2018) such as *listeria monocytogenes, B. subtilis* and *staphylococcus aureus*. However, in Gram (-) bacteria cell, the cytoplasmic region is restricted by thin peptidoglycan layer (~2 nm) overlaid by an asymmetrical phospholipids bilayer and lipopolysaccharides containing some proteins (Han et al., 2016), such as *salmonella spp*., *campylobacter spp*. and E. *coli*.

According to Hati et al. (2018), the cytoplasmic membrane protects a cell from surrounding environment hence the disturbance of microorganism cytoplasmic membrane is a strategy to inactivate foodborne pathogens by physical or chemical techniques (Hati et al., 2018). Nishime et al., (2017) compared the plasma antimicrobial efficiency of the Gram (-) and Gram (+) bacteria. Results showed that the Gram (+) bacterium *E. faecalis* and Gram (-) bacterium *P. aeruginosa* was the most susceptible species to plasma effect (Nishime et al., 2017). Ukuku et al. (2019) observed the differences between gram (-) *salmonella* and *citrobacter freundii* and Gram (+) bacterium *L. monocytogenes* during plasma treatment. Gram (-) outer membrane react with plasma causes damage on the bacterial cell wall and membrane, resulting in cell death. However, for Gram (+) plasma components need to penetrate the bacterial cell wall and membrane before they can interact with intracellular bio-functional molecules or organelles and inactivate bacteria (Ukuku et al., 2019).

Figure 1. Scanning electron micrographs of E. coli and B. subtilis cells (Laroussi et al., 2003)

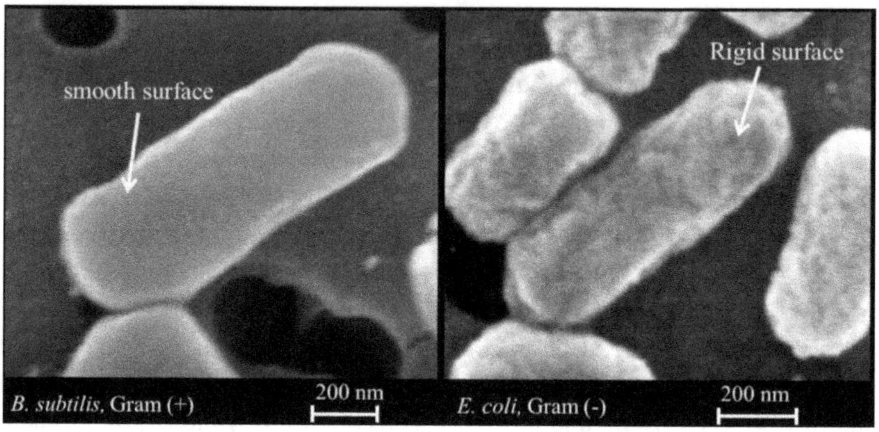

The efficiency of plasma treatment is largely influenced by the surface topography of the sample. Cold atmospheric pressure plasma treatment causes a continuous bombardment of the bacterial cell with reactive plasma components, causing an erosion of the cell atom by atom through etching (Hertwig et al., 2017). CP, the fourth state of matter is a valuable alternative for food decontamination due to its lethal microbiological power, up to a 5-log reduction in microorganism count (Ziuzina et al., 2014; Tanino et al., 2019; Sen et al., 2019; Min et al., 2018; Min et al., 2016; Ha et al., 2016; Ziuzina et al., 2020) and reaching in some cases below detection limit of 1.0 log10 CFU/ml (Ziuzina et al., 2014; Patange et al., 2018; Yeon et al., 2017) as listed in Table 1. Huang et al (2020) shows DBD plasma treatments are more effective against Gram (-) negative bacteria than Gram (+) bacteria when shows difference damage in physical due to different cell layers of peptidoglycan. Peptidoglycan consists of a complex network of repeating disaccharide backbones that are cross-linked by a *pentaglycine peptide* chain and provides the structural framework for the cell wall.

The findings in the review provide beneficial information for developing cold plasma treatment to inactivate and reducing the FBD thereby improve microbial safety and quality of the "ready to eat" foods. Various FBD occurrences globally have motivated the food industry to apply strict and robust hygienic protocols to avoid food contamination. Markedly, the review must significantly studied on effects of plasma reactive species on bacterial Gram (+) cell due to more prone to plasma exposure in comparison to other microbes. Focused on compromising safety and nutritional parameters of the processed foods also need to included.

Table 1. Bacterial inactivation by cold plasma treatment for agricultural produces

Agricultural crops	Plasma Treatment Parameter	Results	References
Black paper	4 min, 30 kHz, 21 W, a flow rate of 0.3L/min, humidify gas rotating plasma reactor	Inactivation of the *Aspergillus* spores on the whole black peppers. A decrease in the survival ratio of three orders (3 log) was achieved.	Tanino et al., 2019
Hazelnuts	30 min, 100 W, low-pressure air plasma	*Aspergillus* spp. spores were 7.8–7.9 cfu/g in hazelnuts. Plasma treatment with reductions of up to 5.6 log cfu/g	Sen et al., 2019
Grape tomatoes	3 min, 1.1A, 35 kV DBD reactor, atmospheric air	Reduction of *Salmonella* from 0.9±0.2 log/CFU to 3,.3±0.5 log/CFU Reduced the number of total mesophilic aerobic bacteria and yeast and molds in the double-layer configuration by 1.3 ± 0.3 and 1.5 ± 0.2 log CFU/tomato, respectively	Min et al., 2018
Romaine Lettuce	1-10 min, 30–60kV, 1–3A current, DBD reactor	Microbial reduction: 1.1±0.3 log CFU/g (*E. coli* O157: H7), 0.5±0.3 log CFU/g (*Salmonella*), 1.0±0.5 log CFU/g (*L. monocytogenes*) & 1.3±0.2 1 log PFU/g (*Tulane* virus)	Min et al., 2016
Brown rice	5-20 min, atm. pressure, 250 W power input, DBD reactor	Reduce from 4.29 CFU/g to 1.30 log CFU/g (*B. cereus*), 4.23 log CFU/g to 1.29 log CFU/g (*B. subtilis*) & 5.26 log CFU/g to 2.96 log CFU/g (*E. coli* O157:H7).	Ha et al., 2016
Strawberries Spinach	2.5 min, 100 kV, atmospheric air, DBD pilot-scale reactor 2.5 min, 100 kV, atmospheric air, DBD pilot-scale reactor	Reductions 2.0 log10 CFU/ml reductions for *E. coli* Reductions1.3 log10 CFU/ml reductions for *Listeria innocua* Reductions 2.2 log10 CFU/ml *for E. coli* Reductions 1.7 log10 CFU/ml for *L. innocua*	Ziuzina et al., 2020 Ziuzina et al., 2020
Strawberry Cherry tomatoes	ACP, 70kV, DBD reactor 120 s & 300 s 10 s 60 s 120 s	L. *monocytogenes* average reductions of approximately 4.2 ±0.5 log10 CFU/sample were recorded. Populations of *Salmonella* were reduced by 1.7±0.1 and 3.8 ±0.4 log10 CFU/sample Significantly different reductions in *E.coli* of 3.5±0.7 log10 CFU/sample *Salmonella* populations on tomato were undetectable Reduced populations of E. *coli* to undetectable levels L. *monocytogenes* populations were reduced to levels below detection limits.	Ziuzina et al., 2014 Ziuzina et al., 2014
Lettuce	1-5 min, 60–80 kV, atmospheric cold plasma	*S. enterica, L. monocytogenes and P. fluorescen* were reduced from 8.06, 7.29 and 7.50 $\log_{10}$ CFU/ml (0 s) to non-detachable (below detection limit of 1.0 $\log_{10}$ CFU/ml) after treatment.	Patange et al., 2018
Mandarin	10 min, 900W, nitrogen-cold plasma treatment	*P. italicum* inhibition on the mandarin surface without noticeable alteration of surface. Unaffected the carbon dioxide generation, weight loss, and surface colour of fruit and the soluble solid contents, titratable acidity, maturity index, pH, and ascorbic acid concentration of flesh. 84% reduction in disease incidence.	Yeon et al., 2017

COLD PLASMA REACTIVE SPECIES MECHANISM FOR BACTERIAL CELL INACTIVATION

Over Plasma was defined as ionized gases by an American chemist and physicist, Irving Langmuir for the first time in 1929 (Sanlisoy et al., 2017). Generally cold plasma (CP) is an ionized gas containing varieties of highly reactive species like electrons, negative ions, positive ions, free radicals, excited or non-excited atoms and photons (Mandal et al., 2018). During generation of CP, most of the applied energy is focused onto electrons instead of heating the entire gas stream. So, gas molecules remain at around the ambient temperature collectively. The working gas for CP generation is one of the most important variables that manipulate the functional effects of plasma treatments (Pankaj & Kevin, 2017). These gases contribute in collision reactions with electrons resulting in the formation of various chemically active species (Mir et al., 2019). Plasma treatment through the charged particles generated will be a great platform for inactivation and decontamination of pathogenic fungi, thus consequently stopping the formation of mycotoxin (Khamsen et al., 2016). When the positive and negative ions combine with each other, the energy is released as visible and UV light (Zainal et al., 2015), although UV is known to have the ability to inhibit bacterial replication, plasma-radiated UV is not a key mechanism in the process of killing or deactivating microorganisms because of its ultrathin penetration depth (Khamsen et al., 2016).

Review by Patange et al. (2018) suggested that ACP treatment can generate excellent sources of reactive oxygen and nitrogen species (RONS). The RONS mediated damage can result in alteration of intracellular functions leading to oxidation of free fatty acids, lipids, and protein components. However, wider variety of aquatic organisms needs to employ for better understanding and complete toxicity evaluation of long-term effects. The etching activity (Figure 2) leads to irreversible damage to deoxyribonucleic acid (DNA), ribonucleic acid (RNA) and enzymes resulting in cell lysis or cell death (Patange et al., 2018; Misra & Jo, 2017). Nevertheless, plasma reactive species only penetrate for surface decontamination and do not cover biofilms protected by thick living tissues or growing in body cavities (Zhao et al., 2020; Alkawareek et al., 2012). The plasma bombardment by electrons, ions and radicals can cause dissociation of chemicals bonds, cell wall erosion and surface lesion stimulation via an etching process. Plasma etching is based on the interaction of relative energetic ions and activated species with the molecules of the substrate. The accumulation of charged particles at the external surface of the membrane leads to electrostatic stress resulting in cell wall rupture (Pignata et al., 2017).

ROS are the major bactericidal agents which causes damage to DNA and proteins in microbial cells (Xu et al., 2016) without comprising food product quality. Types of ROS include ozone (O_3), atomic oxygen (O), excited oxygen (O_2), singlet oxygen

($^1O_2$) and superoxide anion (O_2-). The O_2 is the most popular gas investigated to date where the gas produces O, OH, OOH and a mixture of other radicals that can contribute to the sterilisation effects. Among these, the OH radical is often considered the most efficient for microbial inactivation despite its extremely short lifetime (Shintani et al., 2010; Phan et al., 2018). According to Mahnot et al. (2019), the use of gases like He, He/O_2, Ar, Ar/O_2, N2/ O_2, $N_2/O_2/CO_2$ and air have shown to be effective in microbial inactivation using cold plasma. Apart from DNA and protein alterations, OH, O, O_3 and H_2O_2 could break the structural bond, for instance, C-O, C-N bond in cell wall component, the peptidoglycan leading to cell wall destruction (Misra, 2016) in CP treated samples. The H_2O_2 is the main bactericidal agent which damages iron-sulphur and mononuclear iron enzymes (Mandal et al., 2018). During plasma treatment, generated ROS can react with both lipopolysaccharide and peptidoglycan, thus breaking the molecular structure by disintegrating C-O, C-N, and C-C bonds (Han et al., 2016).

Figure 2. A schematic diagram of plasma bombardment by reactive species of CP for microorganism inactivation (Patange et al., 2018; Misra & Jo, 2017)

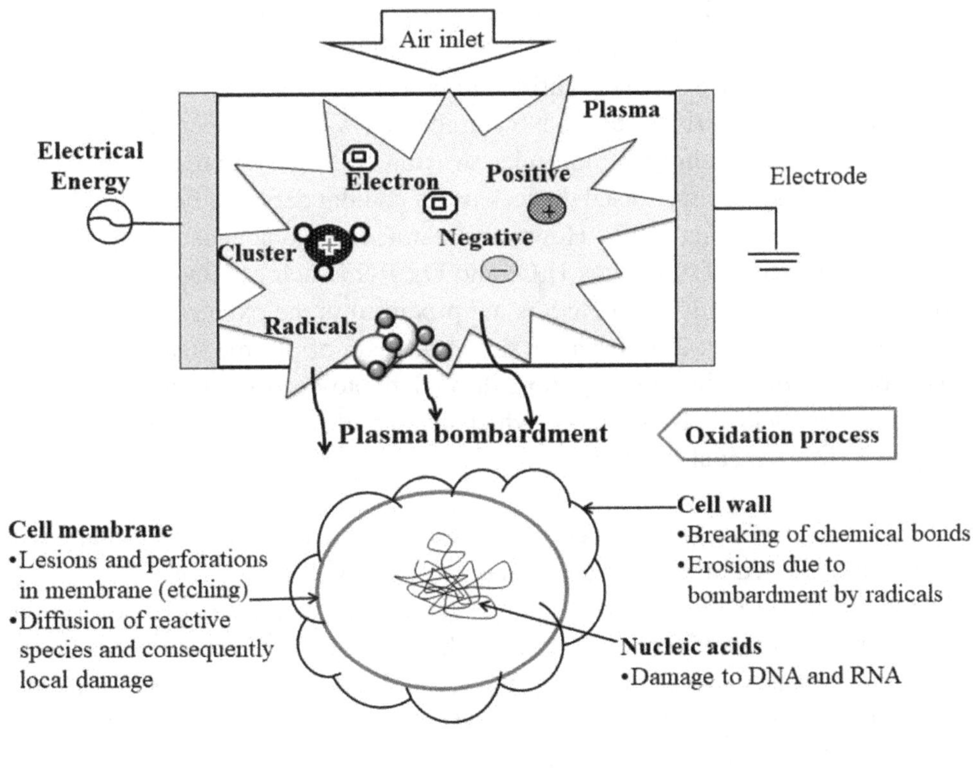

However, in indirect treatment, the inactivation effect of the air plasma treated liquid is better than that of the oxygen plasma treated water mainly due to formation of nitrogen-based reactive species in the liquid. Complete disinfection (7 log reduction) is achieved after 15 min O_2 and 25 min air plasma treatment (Shen et al., 2019). Although the potential in the antimicrobial function of reactive oxygen species (ROS) is considered to be the prime plasma-borne disinfectant, the effectiveness that bioactive species formed in plasma for Gram (+) still open for investigating. The plasma parameter and ability to get through the porous peptidoglycan layer and allows small particles must be further studied.

RNS is another reactive species of plasma besides ROS. The RNS comprises species such as atomic nitrogen (N), excited nitrogen (N_2), nitric oxide (NO•). For N_2 species, the active species had no or insufficient ability to shrink bacterial endospores, suggesting a different mechanism of action compared to oxygen (O_2) that could be due to difficulties in ionising the N_2. Chemically, N_2 is a molecule with a triple bond where the dissociation energy is high (approximately 9.91 eV) that explains its resistance to ionisation and it is not known which active species are present to cause the sterilisation effects observed with N2 gas plasma. The NO and nitrogen dioxide NO_2, have a role in microorganism inactivation by damaging the chemical constituents like proteins, lipids, and nucleic acids (Pignata et al., 2017). Apart from RNS such as N, N_2, NO, and NO_2, nitrous acid (HNO_2) and nitric acid (HNO_3) are well known for their antimicrobial efficacy as well (Schnabel et al., 2012; Gilmore et al., 2018). The RONS can interact with both lipopolysaccharide and peptidoglycan, resulting in molecular structure damage via breaking of COO, CON, and COC bonds that caused the leakage of cellular debris of bacteria (Huang et al., 2020; Prasad et al., 2017). However, treatment of an aqueous solution with air plasma produces ROS such as H_2O_2 and O_3, RNS such as NO_2-, NO_3-, and ONOOH, and acidification, but RNS is not produced in the oxygen plasma (Phan et al., 2018; Shen et al., 2019). The basic mechanism of plasma treatment on the erosion of microbial cells, atom by atom, through photo-desorption and etching are as explained by (Schnabel et al., 2012) which involves three phases and is illustrated in Figure 3 (Moisan et al., 2002):

1. First phase: Microorganism genetic material destruction by UV irradiation, a process that requires a sufficient amount of DNA strands lesions.
2. Second phase: Through intrinsic photo-desorption, photon-induced desorption that results from UV photons breaks chemical bonds in the microorganism surface, causing volatile compounds formation from atoms intrinsic to the microorganism cell. The volatile by-products of this non-equilibrium chemistry are small molecules such as CO and CHx.

3. Third phase: Through etching, where the reactive species adsorption on the microorganism subsequently undergoes chemical reactions to form volatile compounds referred to as spontaneous etching. The reactive species can be atomic (O) and molecular radicals (O_3), and excited molecules in a metastable state ($^1O_2$ singlet). These reactive species that formed under thermodynamic equilibrium conditions yield small molecules such as CO_2, H_2O that are the final products of the oxidation process 3 (Moisan et al., 2002).

According to Rossi et al. (2008), sterilisation efficiency is more related to O atom density than UV radiation intensity, i.e. sterilisation is not related to DNA damage as in UV sterilisation but more likely to the etching of spores (Rossi et al. 2008). Oxidative damage of membrane lipids affects their ability to regulate mass transport across the cell (Hertwig et al., 2018). The precise mechanisms through which cold plasma affects antigenicity require further elaboration. There was a strong relationship between either spark or glow plasma process duration and efficacy for allergenicity attenuation, which may also be affected by concentration. The results achieved point to application for reducing the allergenicity of food allergen residues in food processing environments (Ng et al., 2021). Significant increases in the generation of RONS were directly related to increase in plasma processing parameters such as treatment time, voltage/power (Prasad et al., 2017), frequency and gas/gas mixture (Chizoba Ekezie et al., 2017). Plasma processing parameters

Figure 3. Survival curve characterising plasma sterilisation during the three phases (Moisan et al., 2002).

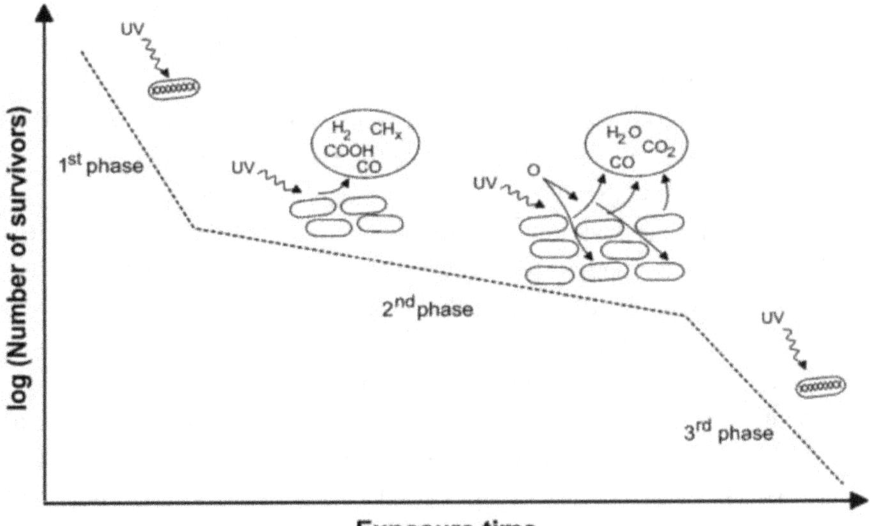

used during food treatments were reported to influence the antimicrobial efficacy of CP against microbiota population and foodborne pathogens (Khelissa et al., 2017; Norhayati et al., 2016).

PROCESSING PARAMETERS OF COLD PLASMA FOR MICROBIAL DESTRUCTIONS

Cold plasma (CP), a chemical-free process is an emerging non-thermal processing technology for inactivation of various foodborne microorganisms found at different food processing stages (Mandal et al., 2018; Ukuku et al., 2019; Sakudo & Onodera, 2019). The reactive species of plasma responsible for microbial inactivation was affected by CP processing parameters (Liao et al., 2017). The processing parameters involved in microbial inactivation such as applied voltage, discharge frequency, target cell properties, treatment time, medium surface, and the flow of the carrier gas influence the CP generation and thus the efficacy of the process (Ziuzina et al., 2020; Norhayati et al., 2016; Sakudo & Onodera, 2019; Lu et al., 2014). Longer plasma exposure times, higher discharge voltages, and regulated volume flow rate for CP inactivation process led to increase of microbial inactivation rates (Wang et al., 2018). According to Prasad et al., (2017), irreversible inactivation of E. coli was significantly achieved the highest log reduction of 6 log CFU mL^{-1} on tomato stored in a sealed bag with extended of 15 min ACP treatment time and increased voltage up to 60 kV. In another study, plasma treatment for 10, 60 and 120 s resulted in a reduction in *salmonella, E. coli* and *L. monocytogenes* counts on tomato to undetectable levels from initial populations of 3.1, 6.3 and 6.7 log CFU per gram respectively. However, treatment times of up to 300 s were required to attain substantial reductions on strawberry surfaces (Ziuzina et al., 2014). Inline, Shah et al (2019) reported the difference in microbial load on kale leaves between plasma treatment times 120, 240, and 300 s was statistically insignificant. Levels of *E. coli* O157:H7 on the kale leaves were reduced below the detection limit of 5.5×103 CFU/ mL after plasma treatment for 300 s with no significant change in color values.

A higher bacterial inactivation occurred when apples surfaces treated with cold plasma for 40 s was immediately followed with the antimicrobial treatment for 180 s or 3600 s. These results suggest that a 40 s pre-treatment with cold plasma reduces sanitizer treatment time from 1 h to 3 min to achieve approximately 3.5–4 log inactivations of *L. monocytogenes* on apples. However, the food hygienic without any detrimental effect on nutritional values as well as sensory properties of the apples must be done due to longer exposure treatment time. Yet, long time of treatment was defect the nutritional of the foods. The food fat content and oxidized to smaller chain fatty acids, keto acids and aldehyde might be damaging (Hati et al., 2018). Nonetheless, Charoux et al., (2020) demonstrated a limitation impact on food quality as evidenced by no significant reductions in the phenolic content of black pepper grains post treatment (Charoux et al., 2020). In contra, Zhang et al., (2021) shows ACP short treatment time enabled pea protein denaturation at reduced temperature (70–90 °C) to improve its gelling properties and low protein concentration. The gel prepared by heating ACP treated pea protein at 80–90 °C for 30 min has the potential for the development of meat binders, fat replacers, and meat analogues in food industry. In agreement, According to Giannoglou et al., (2021), the strawberries with the ACP treated was enhancing aromatic profile of the fruits derived from the sensory evaluation. The enhancement of nutritional value of fruits in addition to the quality retention related with safety and consumer perceived traits. The ACP processing also led to a significant inactivation at approximately 61.4% of the initial load of TVC, resulting in the retention of the microbial load at lower levels during storage.

The strong antimicrobial effects of CP are also due to chemical and bioactive radicals generated during electrical discharge caused by microbe cell permeabilisation which can result in cellular components leakage and reactive species can affect the intracellular proteins and nucleic acids (Mandal et al., 2018) According to Misra (2016), the reactive species produced in plasma reacted with the amino acid in proteins which further caused structural changes in proteins and resulted in the destruction of the microbial spores. Mode of plasma exposure impacts decontamination efficacy. Reduction in inactivation efficacy of CP treatment both in liquid cereal-based media and inoculated rubber surface was noted for bacterial endospores as compared to vegetative cells (Wang et al., 2018). Usta et al., (2019) stated that the amount of ROS in the liquid phase elevated after extended exposure to CP and resulted in a greater bacterial zone of inhibition. Selcuk et al. (2008) stated that destruction rate of both Aspergillus spp. and Penicillum spp. were dependent not only on plasma gas and the duration of the exposure but also on the type and surface of the contaminated of grains and legumes seeds.

According to Zhang et al. (2017), the plasma field is generated when gas molecules are ionised with a highly energetic electron in a gas discharge system. The nature of reactive species produced for the initiation and propagation of plasma chemical reactions are significantly correlated to the type of carrier gas and hence, affects the performance of CP. Besides that, gas flow affects the gas velocity with which the active species are transported to the target cells, some short-time living species may not reach the sample as affected by flow rate. Besides transporting radicals and active species to the cell surface, the gas flow also contributes to cooling effect. Niemira (2012) showed air yielded greater log reductions than did ammonia or argon when the microorganisms were reduced by 6 logs within 30 seconds (air), 60 seconds (ammonia), and 180 seconds (argon). Nonetheless, the interaction of treatment efficacy with distance from plasma emitter head must consistent for better plasma reaction chemistry within the plume. Reineke et al. (2015) studied the impact of pure argon and admixtures with oxygen (0.135% v/v) and nitrogen (0.2% v/v.) on the inactivation of Bacillus spores of an atmospheric pressure plasma jet. The strain was inactivated by more than 2 log10 in 0.5 min for all tested process gases with the highest inactivation (3.3 log10) for the argon with an admixture of oxygen (0.135% v/v) and nitrogen (0.2% v/v).

Extensive DNA damage was related to the bacterial type and processing parameters. Higher voltage and extended treatment time were associated with greater inactivation efficacy (Ziuzina et al., 2020). Olatunde et al. (2019) also reported that increasing applied voltage to 80 kV showed higher bacterial inactivation to undetectable levels within a short period of treatment which was in agreement with findings by Patange et al., (2018) where increment in voltage level and the longer treatment time were found to gradually enhance the microbial inactivation rate in dairy effluent by ACP treatment]. This contributes to an increased number of active species that amplifies the effect on nucleic acids and DNA coupled with a loss of cell culturability. Therefore, understand the parameters that influence bacterial characteristics in order to find solution to overcome their activation is necessary. Reactive species, voltage and treatment time are such factors can influence the microbial inactivation rate by bacterial metabolism reduction or through the change in the permeability of the cell membrane.

On the other hand as far as quality characteristics are concerned, the treatment effect was very limited and mainly related to a slight increase of both dry matter content and translucent appearance during storage. In addition, a slight reduction of enzymatic activity was observed but this effect was dependent on the type of enzyme considered, and did not seem to have any positive reflection on related qualitative parameters. A reduction of metabolic heat was obtained along with an alteration of the respiratory pathway, indicating a stress response of the tissue to the treatment that should be further clarified (Tappi et al., 2016).

RESULTS AND DISCUSSION

In this section, the treatment of the oyster mushroom by using dielectric barrier discharge cold pen plasma gave us data regarding the efficiency of the cold plasma in inactivation of the bacterial. For the research design and procedures, plasma treatment was done onto the surface of the oyster mushrooms with four different treatment times (0 min, 2 min, 3 min and 4 min) of parameters. The setup of dielectric barrier discharge cold plasma pen reactor for mushroom treatment was developed. The voltage current applied for the pileus of oyster mushroom treatment was ~6kV. The atmospheric pressure gas was flow by using 5 standard liter per minute (SLM) of flow rate. Then, some of the microbiology techniques were applied such as isolation, serial dilution and spread plate on NA agar to obtain the bacteria colonies from the mix culture. In details, the mushroom sample which untreated and treated by plasma treatment was cut into 1.0 X 1.0 cm in square and grew in 10mL of sterile distilled water. Pure cultures are best obtained by streak plate method. In the streak plate method, a loopful of the inoculum is placed near the periphery of the plate with agar medium and spread or streaked on the upper portion of the plate with parallel overlapping strokes. The inoculum is streaked over other portions of the plate so that isolated colonies could be observed in the last streaked area. Next, the colony morphologies analysis were carried out to identify the reduction of bacteria from the given samples treated with different optimal parameters follows by Gram staining method.

Table 2 illustrated influences of the DBD CP pen when exposure on the mushroom with varies of treatment time. Data gathered showed increment of DBD CP pen treatment time was decreased the bacteria colonies on the mushroom surfaces. Control samples shows the colonies on the mushroom was too numerous to count. The 2 min of treatment time shows the colonies was decreased observably with 17 colonies forming only. However the plasma still has insufficient discharge to disturb the bacterial cell until the plasma treatment time was increased up to 3 min. The 3 min displays 10 numbers of colonies only while 4 min treatment shows the DBD CP pen efficiently inactivated to undetectable levels (0 colonies). The cells and internal bacteria damage occur by membrane destruction. Data gather was in line with Sakudo et al., 2019 whereas *Salmonella enterica serovar Abony* was completely inactivated after employing the 5 min of treatment regime. According to Ziuzina et al. (2014), atmospheric cold plasma represents a potential alternative to traditional methods for non-thermal decontamination of foods. Their studied shows an extended cold plasma treatment time was necessary to reduce bacterial populations attached on the more complex surface of strawberries to undetectable levels from initial populations. Patange et al. (2018) also demonstrated continuous cold plasma treatment was applied to synthetic dairy and meat effluents. Microbial inactivation showed treatment time dependence with significant reduction in microbial populations within 120 s, and to undetectable levels after 300 s.

Table 2. Bacteria colonies viable on the plates on day 7 for direct dilution (20^{-1}) with varies exposures times by cold plasma treatment

Treatment times (min)	**control (0)**	**2**	**3**	**4**
No of Colonies	Too numerous to count (TNTC)	17	10	0
Colonies types	1	3	4	0
CFU/ml	TNTC	3.4 X 10^4	2.0 X10^4	undetectable
Direct dilution (20^{-1})	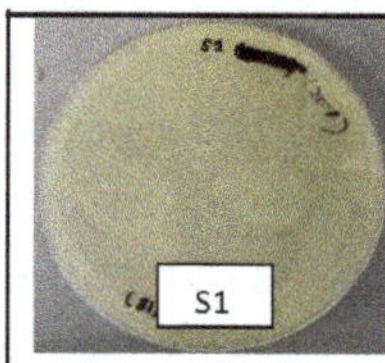	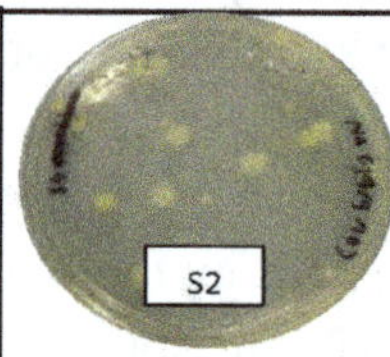	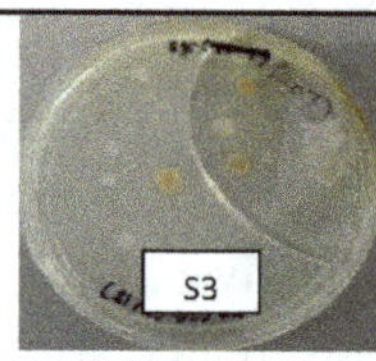	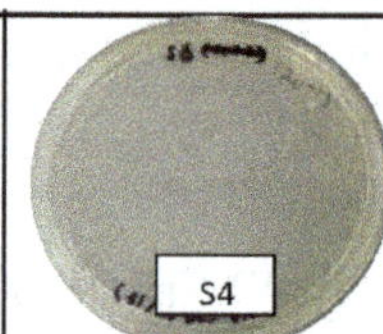

Table 3 and Table 4 showed the morphologies characterization, bacterial shapes and gram staining results for S1 and S3 samples of oyster mushroom treated up to 3 minutes by DBD cold plasma pen respectively. Each single colony with different appearance was picked up and inoculates on the agar plate by quadrant streak. The colony morphologies from different plates on S1 and S3 showed the different characteristics strain observed with naked eyes which is white, orange and yellow in colours. To be exact, colonies from the best plate was chosen to obtain pure culture. In order to obtain pure culture, few series of streaking method had been done. The S1 sample showed only one type of colony with shapes of *diplobacillus* whereas known as gram positive. The S3 strains monitored existence four types of colonies with different shapes identified as *appendaged, spiral* and *bacillus* from gram negative and *filamentous* bacteria from gram positive respectively. The data gathered illustrated that 3 min of DBD cold plasma treatment not has sufficient time to inactivate the positive and negative gram of bacterial compared to 4 min treatment. According to Tanino et al., 2019, by using humidified gases for inactivation treatment in the rotating plasma reactor also facilitated the inactivation of the *aspergillus* spores on the whole black peppers, and no viable spores of *aspergillus* sp. were detected after 4 min treatment. Nishime et al., (2017) was in line when their presented results reveal that the distance is an essential parameter for fungi inactivation especially for long treatment time. The Gram (+) bacterium presented the greatest susceptibility to plasma exposure compared to the others microorganisms same as gathered data as showed in Table 4.

Table 3. Gram staining results after isolation of control samples of oyster mushroom

Strain	Streak plate	Images/ Bacteria Shape
Control (S1)	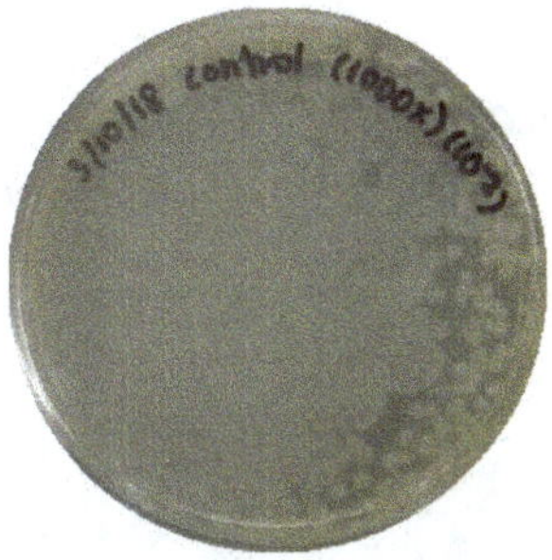	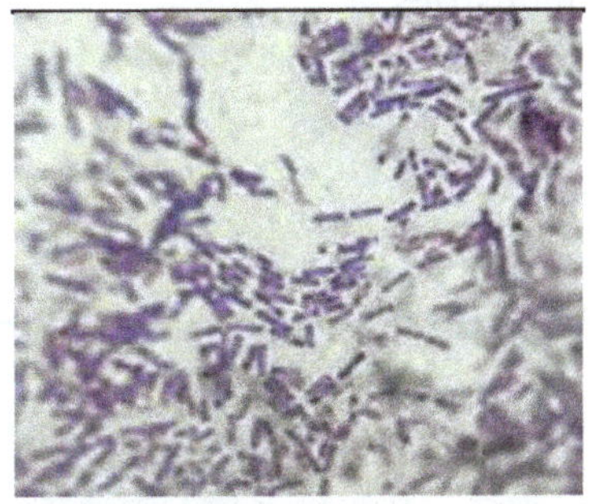 **Gram staining: Gram Positive** Shape: *Diplobacillus*

CONCLUSION

Overall, cold plasma (CP) is an advanced microbial decontamination technology which is promising to be applied in various food industries especially in postharvest processing to avoid foodborne diseases. This studied shows that dielectric barrier discharge (DBD) cold plasma pen treatment at 5 kV by atmospheric pressure for 4 minutes was effective in achieving undetected bacterial on oyster mushroom surface. With plasma mechanism, the etching activity causes irreversible damage to DNA, RNA and enzymes that eventually leads to cell lysis or microorganism cell death of the Gram (+) and Gram (-) bacteria due to their different outer membrane structures and sensitivities to CP reactive species. In addition, CP processing parameters such as voltage and duration treatment were reported to significantly influence the rate of the food pathogen inactivation. Above all, an understanding of the microorganism cell structure and the food surface types and roughness is an essential in manipulating CP processing parameters to achieve the maximum rate of microbial inactivation. Cold plasma has shown its potentiality as an antimicrobial treatment and has been tested on different food products, but the impact on product quality and metabolism is still scarcely known. Therefore, further studied and analysis must emphasis on:

- the optimized life duration of the vegetables and fruits after blasting by using optimized cold plasma processing parameters.
- the cold plasma optimized processing parameters on bacterial Gram (-) whereby this bacterial are more susceptible to cold plasma than Gram (+) bacteria attributes to improve microbiological of the treated foods.

- to be carried out to completely understanding the food safety issues and food properties associated with cold plasma technology in food processing industries.

Table 4. Screening, morphology characterization and gram staining of 3 minutes treated sample of oyster mushroom by DBD cold plasma pen treatment

Strain	Streak plate	Images/ Bacteria Shape
1		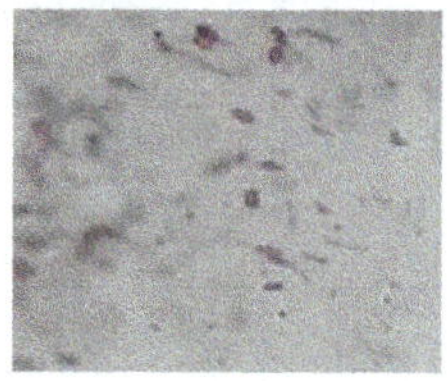 **Gram staining: Gram Negative** Shape: *Appendaged*
2	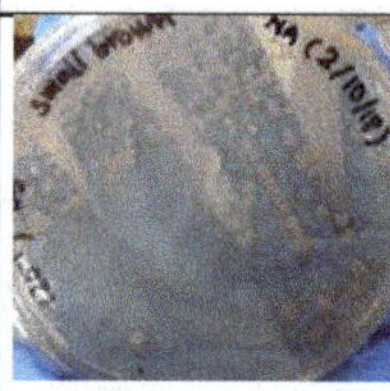	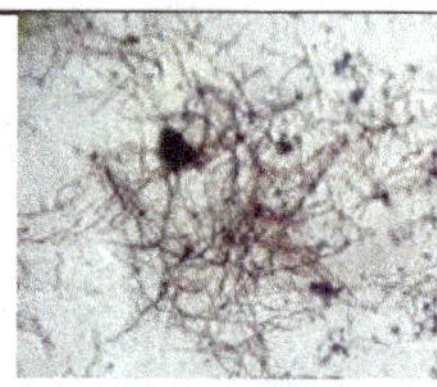 **Gram staining: Gram Positive** Shape: *Filamentous bacteria*
3	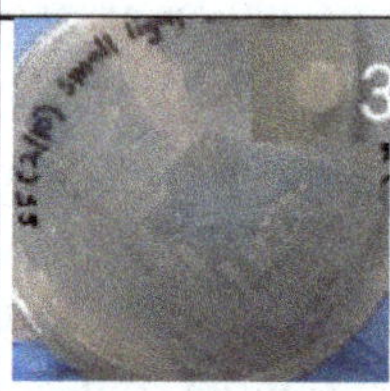	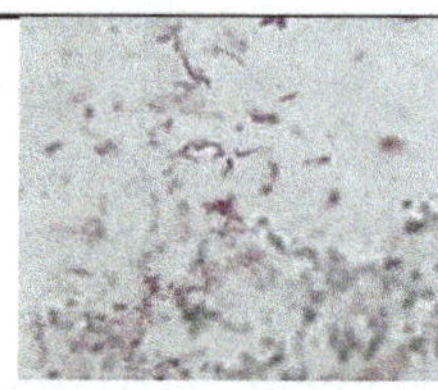 **Gram staining: Gram Negative** Shape: *Spiral*
4	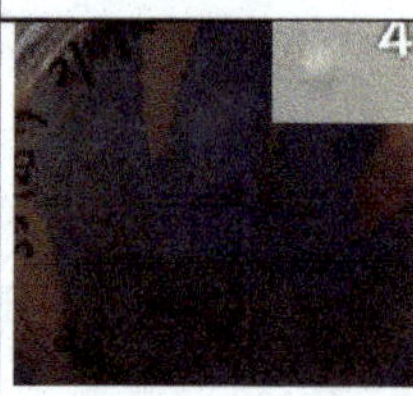	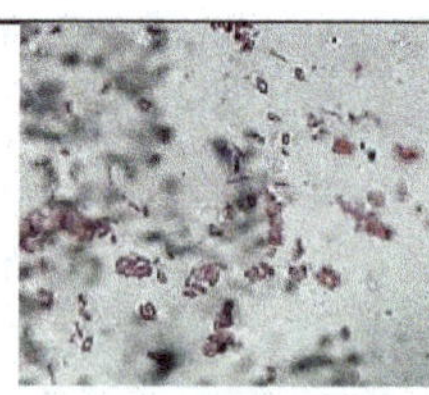 **Gram staining: Gram Negative** Shape: *Bacillus*

REFERENCES

Agun, L., Ahmad, N., Redzuan, N., Idirs, N. A. S., Taib, S. M., Zakaria, Z., & Ibrahim, R. K. R. (2021). Sterilization of oyster mushroom crop residue substrate by using cold plasma technology. *Materials Today: Proceedings*, *39*, 903–906. doi:10.1016/j.matpr.2020.03.584

Alkawareek, M. Y., Algwari, Q. T., Laverty, G., Gorman, S. P., Graham, W. G., O'Connell, D., & Gilmore, B. F. (2012). Eradication of Pseudomonas aeruginosa Biofilms by Atmospheric Pressure Non-Thermal Plasma. *PLoS One*, *7*(8), e44289. doi:10.1371/journal.pone.0044289 PMID:22952948

Atungulu, G. G., & Pan, Z. (2012). Microbial decontamination of nuts and spices. In Microbial Decontamination in the Food Industry: Novel Methods and Applications. Woodhead Publishing Series in Food Science, Technology and Nutrition. doi:10.1533/9780857095756.1.125

Barba, F. J., Koubaa, M., Prado-silva, L., Orlien, V., De Souza, A., & Ana, S. (2017). Mild processing applied to the inactivation of the main foodborne bacterial pathogens: A review. *Trends in Food Science & Technology*, *66*, 20–35. doi:10.1016/j.tifs.2017.05.011

Bastien, M., Vaniscotte, A., Combes, B., Umhang, G., Raton, V., Germain, E., Villena, I., Aubert, D., Boué, F., & Poulle, M. L. (2019). Identifying drivers of fox and cat faecal deposits in kitchen gardens in order to evaluate measures for reducing contamination of fresh fruit and vegetables. *Food and Waterborne Parasitology*, *14*, e00034. doi:10.1016/j.fawpar.2018.e00034 PMID:32095604

Bayliss, D. L. (2012). *Understanding the inactivation mechanism of foodborne pathogens using cold atmospheric plasma*. Academic Press.

Becker, S., Zimmermann, J. L., Baumeister, P., Brunner, T. F., Shimizu, T., Li, Y. F., ... Welz, C. (2019). Effects of cold atmospheric plasma (CAP) on bacteria and mucosa of the upper aerodigestive tract. *Auris, Nasus, Larynx*, *46*, 294–301.

Castro-Ibáñez, I., Gil, M. I., & Allende, A. (2017). Ready-to-eat vegetables: Current problems and potential solutions to reduce microbial risk in the production chain. *Lebensmittel-Wissenschaft + Technologie*, *85*, 284–292. doi:10.1016/j.lwt.2016.11.073

Charoux, C. M. G., Free, L., Hinds, L. M., Vijayaraghavan, R. K., Daniels, S., O'Donnell, C. P., & Tiwari, B. K. (2020). Effect of non-thermal plasma technology on microbial inactivation and total phenolic content of a model liquid food system and black pepper grains. *Lebensmittel-Wissenschaft + Technologie*, *118*, 108716. doi:10.1016/j.lwt.2019.108716

Cheng-yang, L., Zhang, N., Guan, B., Zhou, Z., & Mei, F. (2019). Reactive oxygen species are involved in cell death in wheat roots against powdery mildew. *Journal of Integrative Agriculture*, *18*(9), 1961–1970. doi:10.1016/S2095-3119(18)62092-1

Chizoba Ekezie, F.-G., Sun, D.-W., & Cheng, J.-H. (2017). A review on recent advances in cold plasma technology for the food industry: Current applications and future trends. *Trends in Food Science & Technology*, *69*, 46–58. doi:10.1016/j.tifs.2017.08.007

Fernández, A., & Thompson, A. (2012). The inactivation of Salmonella by cold atmospheric plasma treatment. *Food Research International*, *45*(2), 678–684. doi:10.1016/j.foodres.2011.04.009

Gabriel, A. A., Ugay, M. C. C. F., Siringan, M. A. T., Rosario, L. M. D., Tumlos, R. B., & Ramos, H. J. (2016). Atmospheric pressure plasma jet inactivation of Pseudomonas aeruginosa biofilms on stainless steel surfaces. *Innovative Food Science & Emerging Technologies*, *36*, 311–319. doi:10.1016/j.ifset.2016.07.015

Giannoglou, M., Xanthou, Z. M., Chanioti, S., Stergiou, P., Christopoulos, M., Dimitrakellis, P., Efthimiadou, A., Gogolides, E., & Katsaros, G. (2021). Effect of cold atmospheric plasma and pulsed electromagnetic fields on strawberry quality and shelf-life. *Innovative Food Science & Emerging Technologies*, *68*, 102631. doi:10.1016/j.ifset.2021.102631

Giaouris, E. E., & Simões, M. V. (2018). *Pathogenic Biofilm Formation in the Food Industry and Alternative Control Strategies.* In A. M. Holban & A. M. Grumezescu (Eds.), *Foodborne Diseases: Handbook of Food Bioengineering* (pp. 309–377). Academic Press.

Gilmore, B. F., Flynn, P. B., O'Brien, S., Hickok, N., Freeman, T., & Bourke, P. (2018). Cold Plasmas for Biofilm Control: Opportunities and Challenges. *Trends in Biotechnology*, *36*(6), 627–638. doi:10.1016/j.tibtech.2018.03.007 PMID:29729997

Gök, V., Aktop, S., Özkan, M., & Tomar, O. (2019). The effects of atmospheric cold plasma on inactivation of Listeria monocytogenes and Staphylococcus aureus and some quality characteristics of pastırma—A dry-cured beef product. *Innovative Food Science and Emerging Technologies, 56*(November), 102188.

Ha, K. L., Kim, H., Sik, K. W., Jo, C., Kim, J., Hun, S. K., ... Han, W. K. (2016). Evaluation of cold plasma treatments for improved microbial and physicochemical qualities of brown rice. *Lebensmittel-Wissenschaft + Technologie*, *73*, 442–447. doi:10.1016/j.lwt.2016.06.055

Han, L., Patil, S., Boehm, D., Milosavljević, V., Cullen, P. J., & Bourke, P. (2016). Mechanisms of inactivation by high-voltage atmospheric cold plasma differ for Escherichia coli and Staphylococcus aureus. *Applied and Environmental Microbiology*, *82*(2), 450–458. doi:10.1128/AEM.02660-15 PMID:26519396

Han, Y. X., Cheng, J. H., & Sun, D. W. (2019). Changes in activity, structure and morphology of horseradish peroxidase induced by cold plasma. *Food Chemistry*, *301*, 125240. doi:10.1016/j.foodchem.2019.125240 PMID:31387040

Hati, S., Patel, M., & Yadav, D. (2018). Food bioprocessing by non-thermal plasma technology. *Current Opinion in Food Science*, *19*, 85–91. doi:10.1016/j.cofs.2018.03.011

Hertwig, C., Leslie, A., Meneses, N., Reineke, K., Rauh, C., & Schlüter, O. (2017). Inactivation of Salmonella Enteritidis PT30 on the surface of unpeeled almonds by cold plasma. *Innovative Food Science & Emerging Technologies*, *44*, 242–248. doi:10.1016/j.ifset.2017.02.007

Hertwig, C., Meneses, N., & Mathys, A. (2018). Cold atmospheric pressure plasma and low energy electron beam as alternative nonthermal decontamination technologies for dry food surfaces: A review. *Trends in Food Science & Technology*, *77*, 131–142. doi:10.1016/j.tifs.2018.05.011

Huang, M., Zhuang, H., Zhao, J., Wang, J., Yan, W., & Zhang, J. (2020). Differences in cellular damages induced by dielectric barrier discharge plasma between Salmonella typhimurium and Staphylococcus aureus. *Bioelectrochemistry (Amsterdam, Netherlands)*, *132*, 107445. doi:10.1016/j.bioelechem.2019.107445 PMID:31918057

Ji, S. H., Ki, S. H., Ahn, J. H., Shin, J. H., Hong, E. J., Kim, Y. J., & Choi, E. H. (2018). Inactivation of Escherichia coli and Staphylococcus aureus on contaminated perilla leaves by Dielectric Barrier Discharge (DBD) plasma treatment. *Archives of Biochemistry and Biophysics*, *643*, 32–41. doi:10.1016/j.abb.2018.02.010 PMID:29454864

Jiang, Y., Sokorai, K., Pyrgiotakis, G., Demokritou, P., Li, X., Mukhopadhyay, S., Jin, T., & Fan, X. (2017). Cold plasma-activated hydrogen peroxide aerosol inactivates Escherichia coli O157: H7, Salmonella Typhimurium, and Listeria innocua and maintains quality of grape tomato, spinach and cantaloupe ☆. *International Journal of Food Microbiology*, *249*, 53–60. doi:10.1016/j.ijfoodmicro.2017.03.004 PMID:28319798

Khamsen, N., Onwimol, D., Teerakawanich, N., Dechanupaprittha, S., Kanokbannakorn, W., Hongesombut, K., & Srisonphan, S. (2016). Rice (Oryza sativa L.) Seed Sterilization and Germination Enhancement via Atmospheric Hybrid Nonthermal Discharge Plasma. *ACS Applied Materials & Interfaces*, *8*(30), 19268–19275. doi:10.1021/acsami.6b04555 PMID:27404121

Khelissa, S. O., Abdallah, M., Jama, C., Faille, C., & Chihib, N. E. (2017). Bacterial contamination and biofilm formation on abiotic surfaces and strategies to overcome their persistence. *Journal of Materials and Environmental Science*, *8*(9), 3326–3346.

Kljujev, I., Raicevic, V., Jovicic-Petrovic, J., Vujovic, B., Mirkovic, M., & Rothballer, M. (2018). Listeria monocytogenes – Danger for health safety vegetable production. *Microbial Pathogenesis*, *120*, 23–31. doi:10.1016/j.micpath.2018.04.034 PMID:29684542

Lafarga, T., Colás-Medà, P., Abadías, M., Aguiló-Aguayo, I., Bobo, G., & Viñas, I. (2019). Strategies to reduce microbial risk and improve quality of fresh and processed strawberries: A review. *Innovative Food Science and Emerging Technologies, 52*(May), 197–212.

Laroussi, M., Mendis, D. A., & Rosenberg, M. (2003). Plasma interaction with microbes. *New Journal of Physics*, *5*, 41. doi:10.1088/1367-2630/5/1/341

Liao, X., Liu, D., Xiang, Q., Ahn, J., Chen, S., Ye, X., & Ding, T. (2017). Inactivation mechanisms of non-thermal plasma on microbes: A review. *Food Control*, *75*, 83–91. doi:10.1016/j.foodcont.2016.12.021

Los, A., Ziuzina, D., Boehm, D., Cullen, P. J., & Bourke, P. (2017). The potential of atmospheric air cold plasma for control of bacterial contaminants relevant to cereal grain production. *Innovative Food Science and Emerging Technologies, 44*(2016), 36–45.

Lu, H., Patil, S., Keener, K. M., Cullen, P. J., & Bourke, P. (2014). Bacterial inactivation by high-voltage atmospheric cold plasma: Influence of process parameters and effects on cell leakage and DNA. *Journal of Applied Microbiology*, *116*(4), 784–794. doi:10.1111/jam.12426 PMID:24372804

Mahdavian Mehr, H., & Koocheki, A. (2020). Effect of atmospheric cold plasma on structure, interfacial and emulsifying properties of Grass pea (Lathyrus sativus L.) protein isolate. *Food Hydrocolloids*, *106*, 105899. doi:10.1016/j.foodhyd.2020.105899

Mahnot, N. K., Mahanta, C. L., Keener, K. M., & Misra, N. N. (2019). Strategy to achieve a 5-log Salmonella inactivation in tender coconut water using high voltage atmospheric cold plasma (HVACP). *Food Chemistry*, *284*, 303–311. doi:10.1016/j.foodchem.2019.01.084 PMID:30744862

Mandal, R., Singh, A., & Pratap Singh, A. (2018). Recent developments in cold plasma decontamination technology in the food industry. *Trends in Food Science & Technology*, *80*, 93–103. doi:10.1016/j.tifs.2018.07.014

Min, S. C., Hyeon, S., Niemira, B. A., Boyd, G., Sites, J. E., Fan, X., ... Jin, T. Z. (2018). In-package atmospheric cold plasma treatment of bulk grape tomatoes for microbiological safety and preservation ☆. *Food Research International*, *108*, 378–386. doi:10.1016/j.foodres.2018.03.033 PMID:29735070

Min, S. C., Hyeon, S., Niemira, B. A., Sites, J. E., Boyd, G., & Lacombe, A. (2016). Dielectric barrier discharge atmospheric cold plasma inhibits Escherichia coli O157: H7, Salmonella, Listeria monocytogenes, and Tulane virus in Romaine lettuce ☆. *International Journal of Food Microbiology*, *237*, 114–120. doi:10.1016/j.ijfoodmicro.2016.08.025 PMID:27562348

Mir, S. A., Siddiqui, M. W., Dar, B. N., Shah, M. A., Wani, M. H., Roohinejad, S., ... Ali, A. (2019). Promising applications of cold plasma for microbial safety, chemical decontamination and quality enhancement in fruits. *Journal of Applied Microbiology*, 1–13. doi:10.1111/jam.14541 PMID:31800143

Misra, N. N. (2016). Quality of Cold Plasma Treated Plant Foods. In Cold Plasma in Food and Agriculture: Fundamentals and Applications (pp. 253–271). Elsevier Inc. doi:10.1016/B978-0-12-801365-6.00010-X

Misra, N. N., & Jo, C. (2017). Applications of Cold Plasma Technology for Microbiological Safety in Meat Industry. *Trends in Food Science & Technology*, *64*, 74–86. doi:10.1016/j.tifs.2017.04.005

Moisan, M., Barbeau, J., Crevier, M., Pelletier, J., Philip, N., & Saoudi, B. (2002). Plasma sterilization. Methods and mechanisms. *Pure and Applied Chemistry*, *74*(3), 349–358. doi:10.1351/pac200274030349

Neetoo, H., & Chen, H. (2014). *Alternative Food Processing Technologies.* In S. Clark, S. Jung, & B. Lamsal (Eds.), *Food Processing: Principles and Applications* (pp. 137–169). John Wiley & Sons, Ltd. doi:10.1002/9781118846315.ch7

Ng, S. W., Lu, P., Rulikowska, A., Boehm, D., O'Neill, G., & Bourke, P. (2021). The effect of atmospheric cold plasma treatment on the antigenic properties of bovine milk casein and whey proteins. *Food Chemistry, 342*, 128283. doi:10.1016/j.foodchem.2020.128283 PMID:33067041

Niemira, B. A. (2012). Cold Plasma Reduction of Salmonella and Escherichia coli O157: H7 on Almonds Using Ambient Pressure Gases. *Journal of Food Science, 77*(3), 171–175. doi:10.1111/j.1750-3841.2011.02594.x PMID:22384964

Nishime, T. M. C., Borges, A. C., Koga-ito, C. Y., Machida, M., Hein, L. R. O., & Kostov, K. G. (2017). Non-thermal atmospheric pressure plasma jet applied to inactivation of different microorganisms. *Surface and Coatings Technology, 312*, 19–24. doi:10.1016/j.surfcoat.2016.07.076

Norhayati, M. N., Lee, B. K., Yap, S. S., Thong, K. L., & Yap, S. L. (2016). Cold Plasma Inactivation of Chronic Wound Bacteria. *Archives of Biochemistry and Biophysics*. PMID:27046340

O'Shea, H., Blacklaws, B. A., Collins, P. J., McKillen, J., & Fitzgerald, R. (2019). Viruses Associated With Foodborne Infections. *Reference Module in Life Sciences, 2*, 1–17.

Ohta, T. (2016). Plasma in Agriculture. In N. Misra, O. Schlüter, & P. Cullen (Eds.), *Cold Plasma in Food and Agriculture: Fundamentals and Applications* (pp. 205–222). Academic Press. doi:10.1016/B978-0-12-801365-6.00008-1

Olatunde, O. O., Benjakul, S., & Vongkamjan, K. (2019). High voltage cold atmospheric plasma: Antibacterial properties and its effect on quality of Asian sea bass slices. *Innovative Food Science & Emerging Technologies, 52*, 305–312. doi:10.1016/j.ifset.2019.01.011

Pankaj, S. K., Bueno-ferrer, C., Misra, N. N., Milosavljevic, C., Donnell, C. P. O., Bourke, P., ... Cullen, P. J. (2013). Applications of cold plasma technology in food packaging. *Trends in Food Science & Technology*, 1–13.

Pankaj, S. K., & Kevin, M. (2017). Cold Plasma Applications in Food Packaging. *Reference Module in Food Science*.

Patange, A., Boehm, D., Giltrap, M., Lu, P., Cullen, P. J., & Bourke, P. (2018). Assessment of the disinfection capacity and eco-toxicological impact of atmospheric cold plasma for treatment of food industry effluents. *The Science of the Total Environment, 631–632*, 298–307. doi:10.1016/j.scitotenv.2018.02.269 PMID:29525709

Phan, C. Q., Gangal, U., Granick, J. L., Kondeti, V. S. S. K., Hunter, R. C., Jablonowski, H., ... Bruggeman, P. J. (2018). Long-lived and short-lived reactive species produced by a cold atmospheric pressure plasma jet for the inactivation of Pseudomonas aeruginosa and Staphylococcus aureus. *Free Radical Biology & Medicine, 124*, 275–287. doi:10.1016/j.freeradbiomed.2018.05.083 PMID:29864482

Pignata, C., D'Angelo, D., Fea, E., & Gilli, G. (2017). A review on microbiological decontamination of fresh produce with nonthermal plasma. *Journal of Applied Microbiology, 122*(6), 1438–1455. doi:10.1111/jam.13412 PMID:28160353

Prasad, P., Mehta, D., Bansal, V., & Sangwan, R. S. (2017). Effect of atmospheric cold plasma (ACP) with its extended storage on the inactivation of Escherichia coli inoculated on tomato. *Food Research International, 102*, 402–408. doi:10.1016/j.foodres.2017.09.030 PMID:29195965

Ramos, B., Brandão, T. R. S., Teixeira, P., & Silva, C. L. M. (2020). Biopreservation approaches to reduce Listeria monocytogenes in fresh vegetables. *Journal of Food Microbiology, 85*, 103282. doi:10.1016/j.fm.2019.103282 PMID:31500713

Redzuan, N., Rabbi, M. R., Agun, L., Ahmad, N., Zakaria, K., Sirat, R. M., ... Yasak, R. (2020). Development of simultaneous substrate mixer and cold atmospheric plasma treatment for mushroom cultivation. In *AIP Conference Proceedings* (Vol. 2217, pp. 030125–1, 7). AIP Publishing.

Reineke, K., Langer, K., Hertwig, C., Ehlbeck, J., & Schlüter, O. (2015). The impact of different process gas compositions on the inactivation effect of an atmospheric pressure plasma jet on Bacillus spores. *Innovative Food Science & Emerging Technologies, 30*, 112–118. doi:10.1016/j.ifset.2015.03.019

Rodríguez, Ó., Gomes, W. F., Rodrigues, S., & Fernandes, F. A. N. (2017). Effect of indirect cold plasma treatment on cashew apple juice (Anacardium occidentale L.). *Lebensmittel-Wissenschaft + Technologie, 84*, 457–463. doi:10.1016/j.lwt.2017.06.010

Rossi, F., Kylian, O., & Hasiwa, M. (2008). Mechanisms of Sterilization and Decontamination of Surfaces by Low-Pressure Plasma. In R. D'Agostino, P. Favia, Y. Kawai, H. Ikegami, N. Sato, & F. Arefi-Khonsari (Eds.), Advanced Plasma Technology (pp. 319–340). Wiley-VCH Verlag GmbH & Co. KGaA.

Sabillón, L., Stratton, J., Rose, D., & Bianchini, A. (2020). Reduction in pathogenic load of wheat by tempering with saline organic acid solutions at different seasonal temperatures. *International Journal of Food Microbiology, 313*, 108381. doi:10.1016/j.ijfoodmicro.2019.108381 PMID:31670167

Sakudo, A. (2017). Recent advances in gas plasma technology for decontamination of food surfaces. In A. M. Grumezescu (Ed.), *Food Preservation: Nanotechnology in the Agri-Food Industry* (pp. 197–228). Acedemic Press. doi:10.1016/B978-0-12-804303-5.00006-7

Sakudo, A., Yagyu, Y., & Onodera, T. (2019). Disinfection and sterilization using plasma technology: Fundamentals and future perspectives for biological applications. *International Journal of Molecular Sciences*, *20*(20), 1–17. doi:10.3390/ijms20205216 PMID:31640211

Sanlisoy, A., & Carpinlioglu, M. O. (2017). A review on plasma gasification for solid waste disposal. *International Journal of Hydrogen Energy*, *42*(2), 1361–1365. doi:10.1016/j.ijhydene.2016.06.008

Schnabel, U., Niquet, R., Krohmann, U., Winter, J., Schlüter, O., Weltmann, K. D., & Ehlbeck, J. (2012). Decontamination of microbiologically contaminated specimen by direct and indirect plasma treatment. *Plasma Processes and Polymers*, *9*(6), 569–575. doi:10.1002/ppap.201100088

Selcuk, M., Oksuz, L., & Basaran, P. (2008). Decontamination of grains and legumes infected with Aspergillus spp. and Penicillum spp. by cold plasma treatment. *Bioresource Technology*, *99*(11), 5104–5109. doi:10.1016/j.biortech.2007.09.076 PMID:17993274

Sen, Y., Onal-Ulusoy, B., & Mutlu, M. (2019). Aspergillus decontamination in hazelnuts: Evaluation of atmospheric and low-pressure plasma technology. *Innovative Food Science & Emerging Technologies*, *54*, 235–242. doi:10.1016/j.ifset.2019.04.014

Shah, U., Ranieri, P., Zhou, Y., Schauer, C. L., Miller, V., Fridman, G., & Sekhon, J. K. (2019). Effects of cold plasma treatments on spot-inoculated Escherichia coli O157:H7 and quality of baby kale (Brassica oleracea) leaves. *Innovative Food Science & Emerging Technologies*, *57*, 102104. doi:10.1016/j.ifset.2018.12.010

Shahriari, S., Monajjemi, M., & Zare, K. (2018). Increasing the efficiency of some antibiotics on penetrating bacteria cell membrane. *Ukrainian Journal of Ecology*, *8*(1), 671–679. doi:10.15421/2018_265

Shen, J., Zhang, H., Xu, Z., Zhang, Z., Cheng, C., Ni, G., Lan, Y., Meng, Y., Xia, W., & Chu, P. K. (2019). Preferential production of reactive species and bactericidal efficacy of gas-liquid plasma discharge. *Chemical Engineering Journal*, *362*, 402–412. doi:10.1016/j.cej.2019.01.018

Shintani, H., Sakudo, A., Burke, P., & Mcdonnell, G. (2010). Gas plasma sterilization of microorganisms and mechanisms of action [Review]. *Experimental and Therapeutic Medicine*, *1*(5), 731–738. doi:10.3892/etm.2010.136 PMID:22993596

Tanino, T., Arisaka, T., Iguchi, Y., Matsui, M., & Ohshima, T. (2019). Inactivation of Aspergillus sp. spores on whole black peppers by nonthermal plasma and quality evaluation of the treated peppers. *Food Control*, *97*, 94–99. doi:10.1016/j.foodcont.2018.10.023

Tappi, S., Gozzi, G., Vannini, L., Berardinelli, A., Romani, S., Ragni, L., & Rocculi, P. (2016). Cold plasma treatment for fresh-cut melon stabilization. *Innovative Food Science & Emerging Technologies*, *33*, 225–233. doi:10.1016/j.ifset.2015.12.022

Ukuku, D. O., Niemira, B. A., & Ukanalis, J. (2019). LWT - Food Science and Technology Nisin-based antimircobial combination with cold plasma treatment inactivate Listeria monocytogenes on Granny Smith apples ☆. *Lebensmittel-Wissenschaft + Technologie*, *104*, 120–127. doi:10.1016/j.lwt.2018.12.049

Usta, Y. H., Çukur, E., Yıldırım, Ç., & Ercan, U. K. (2019). Design of a portable, battery-powered non-thermal atmospheric plasma device and characterization of its antibacterial efficacies. *Journal of Electrostatics*, *99*, 1–8. doi:10.1016/j.elstat.2019.03.002

Vasickova, P., Dvorska, L., Lorencova, A., & Pavlik, I. (2005). Viruses as a cause of foodborne diseases: A review of the literature. *Veterinarni Medicina*, *50*(3), 89–104. doi:10.17221/5601-VETMED

Wang, Y., Wang, T., Yuan, Y., Fan, Y., Guo, K., & Yue, T. (2018). Inactivation of yeast in apple juice using gas-phase surface discharge plasma treatment with a spray reactor. *Lebensmittel-Wissenschaft + Technologie*, *97*, 530–536. doi:10.1016/j.lwt.2018.07.049

Xu, Y., Tian, Y., Ma, R., Liu, Q., & Zhang, J. (2016). Effect of plasma activated water on the postharvest quality of button mushrooms, Agaricus bisporus. *Food Chemistry*, *197*, 436–444. doi:10.1016/j.foodchem.2015.10.144 PMID:26616972

Yeon, M., Jo, S., & Min, S. C. (2017). Mandarin preservation by microwave-powered cold plasma treatment. *Innovative Food Science & Emerging Technologies*, *39*, 25–32. doi:10.1016/j.ifset.2016.10.021

Zainal, M. N. F., Redzuan, N., & Misnal, M. F. I. (2015). Brief Review: Cold Plasma. *Jurnal Teknologi*, *10*, 57–61.

Zhang, H., Ma, D., Qiu, R., Tang, Y., & Du, C. (2017). Non-thermal plasma technology for organic contaminated soil remediation: A review. *Chemical Engineering Journal*, *313*, 157–170. doi:10.1016/j.cej.2016.12.067

Zhao, N., Ge, L., Huang, Y., Wang, Y., Wang, Y., Lai, H., Wang, Y., Zhu, Y., & Zhang, J. (2020). Impact of cold plasma processing on quality parameters of packaged fermented vegetable (radish paocai) in comparison with pasteurization processing: Insight into safety and storage stability of products. *Innovative Food Science & Emerging Technologies*, *60*, 102300. doi:10.1016/j.ifset.2020.102300

Ziuzina, D., Misra, N. N., Han, L., Cullen, P. J., Moiseev, T., Mosnier, J. P., Keener, K., Gaston, E., Vilaró, I., & Bourke, P. (2020). Investigation of a large gap cold plasma reactor for continuous in-package decontamination of fresh strawberries and spinach. *Innovative Food Science & Emerging Technologies*, *59*, 102229. doi:10.1016/j.ifset.2019.102229

Ziuzina, D., Patil, S., Cullen, P. J., Keener, K. M., & Bourke, P. (2014). Atmospheric cold plasma inactivation of Escherichia coli, Salmonella enterica serovar Typhimurium and Listeria monocytogenes inoculated on fresh produce. *Food Microbiology*, *42*, 109–116. doi:10.1016/j.fm.2014.02.007 PMID:24929725

Chapter 7
Atmosphere Non-Thermal Plasma for Seed Treatment

Siti Sarah Safaai
University Teknologi Malaysia, Malaysia

Linda Agun
Universiti Teknologi Malaysia, Malaysia

Norizah Redzuan
Universiti Teknologi Malaysia, Malaysia

Norhayati Ahmad
Universiti Teknologi Malaysia, Malaysia

ABSTRACT

Cold plasma has attracted lots of attention among researchers because it has a wide range of applications, such as the automotive industry, textile industry, microelectronics, packaging, biomedical technology, food preservation, and agricultural sectors. Scientists have shown a great interest in non-thermal plasma because of its advantages such as low temperature, scalable size, low operation cost, flexible operation, and high electron and reactive specie density. Also, non-thermal plasma can be operated at atmospheric pressure, which is an advantage in the agriculture industry rather than operating in a vacuum. Recently atmospheric cold plasma pressure was selected as one of the plasma technologies applied in the agricultural industry for treating the surface of the seed with environmentally friendly technology that produces no hazardous waste. DBD plasma is one of the cold plasma techniques, which can be easily triggered at atmospheric pressure and room temperature.

DOI: 10.4018/978-1-7998-8398-2.ch007

INTRODUCTION

Recently, non-thermal plasma has been widely used in biotechnology, medical treatment, food preservation, and agriculture. This is because non-thermal plasma does not cause hermal damage to living organisms, and it is also a green technology (Matra, 2016). Non-thermal plasma, which consisting of electrons, molecules, excited atoms, ionized gases, radicals, and a strong electric field considered as one of the physical methods for seed treatment (Lotfy,2017). A new approach is being proposed to assist germination and survival in the plasma treatment of seeds. Plasma-based treatment methods have been considered environmentally safe as used reactors were low energy consuming, the treatment process did not cause additional contamination, amounts of generated oxidants were low, and they did not retain in the ecosystem (Pawlat et al., 2018). It plays an important role in a wide range of plant developmental and physiological processes, including reducing the bacterial seed-bearing rate, altering seed coat structures, increasing seed coat permeability, and stimulating seed germination and seedling growth (Ling et al., 2014).

Based on preliminary investigations, it was verified that plasma pre-treatment of seeds of important agricultural crops is an effective tool for improvement of germination, shoot, and root growth because plasma has good fungicidal and bactericidal effects and increased water permeability by surface coat etching and stimulation of germination and seedlings growth (de Groot et al., 2018). Research on the active plasma-generated species such as ions, electrons, radicals, as well as heat and electromagnetic radiation and their effect on various materials, including plant tissues and microorganisms, produces promising results. Research on the effects of accelerated ions on plant tissue structures and even on microorganisms produces promising results (Pawlat et al., 2017). Many studies have been carried out using plasma treatment, and, surprisingly, most of them have shown positive results. For example, Sera et al. investigated the effect of plasma treatment on wheat and oat germination and early growth. They concluded from their studies that plasma treatment affected plant growth, improved footstalk on wheat seed, and accelerated rootlet generation on oat seeds (Park et al., 2018). Agun et al., (2021) studied the efficiency of DBD cold plasma pen treatment on the oyster mushroom bacterial decontamination. Their results show increments of exposure treatment times up to 3 min shows none growth of bacteria colonies. This because the bacteria cell wall was disrupt and destruction by the plasma bombardment. Thus able to extend the lifetime of the mushroom.

SIGNIFICANT OF RESEARCH

Seed treatments are usually done to protect it from pathogens which can damage the seed during the early seedling growth. If this step is neglected, it can simply cause plant diseases, insects, and pathogens to attack, the intervention of germination process and plant growth. However, some of the seed treatments involve chemical is very harmful because it might cause accidental poisoning, damage the seeds and dangerous for workers or farmers due to exposure of the chemical when treating huge quantities of seeds. The biggest concern in the agriculture industry today is the presence of microorganisms (bacteria and fungi) that will affect the crops. Recently in the agriculture industry, they have done several ways to control the reproduction of bacteria and fungi from crops. There are various ways to control the bacteria in the agriculture industry by the development of resistant varieties, using chemical fumigants, soil drainage, and specific tillage practices (Jiang et al., 2014).

A common way to keep the plant from microorganisms is by using the chemical. However, by using this method it will lead to the bad-side effect that will damage the leaves and stems area, also using chemical will cause environmental pollution. Therefore, we chose plasma technology in the agriculture industry because it does not harm living organisms and green technology. Then plasma technology gives good fungicidal and bactericidal effects, increased water absorption through the surface coat etching, and stimulation of germination of seedling growth (Joshi et al., 2018). There are also technologies like molecular breeding and genetic engineering that can cause gene mutations (Zhu et al., 2010). Therefore, we chose atmospheric plasma pressure as it only treats the surface of the seed and is also an environmentally friendly technology that produces no hazardous waste. Usually, the seed has a hard outer shell that makes it difficult for water absorption. After the plasma treatment, the seed will have small holes or poration to help water absorption and improve germination (seedling growth).

PLASMA CLASIFICATION

Plasma is the fourth state of matter which make up of neutral gas atoms/molecules, ions, electrons, and reactive species (Joshi et al., 2018). Plasma can be generated by supply a sufficient amount of energy to the gas to induce ionization. Plasma is classified based on electrons temperature into two groups that are non-thermal (cold) and thermal (hot) plasma. This group is classified based on their degree of ionization, atmospheric pressure, and temperature conditions (Mehmood et al.,2018).

Thermal Plasma

Electrons and heavy particles are at the same temperature and are in thermal equilibrium with each other. Thermal plasma is thermodynamic and can be found in larger numbers in the universe. It can be produced by electro-thermal and electromagnetic transmitters (Mehmood et al.,2018). Thermal plasma can be used in applications that require high temperature, such as casting processes in the metallurgical industry or in chemical synthesis processes (Lopex et al., 2019). The main drawback of the use of thermal plasma is represented by limited energy efficiency, high gas temperature, and high quenching effect of the chemical reaction (Attri et al., 2017).

Non-Thermal Plasma

The temperature of electrons for non-thermal plasma is much higher than the temperature of heavy particles. Non-thermal plasma is obtained under atmospheric pressure or reduced pressure (vacuum) and requires less power input. For these reasons, non-thermal plasma has been used due to the high selectivity and energy efficiency of plasma chemical reactions. Non-thermal plasma has the advantage where it can operate effectively at low temperatures and without quenching (Sahari, 2013). Due to its special properties, non-thermal plasma has been studied in various industrial and bioscience applications, for example, control the ozone production for water purification, pollution control, sterilization, food preservation, and agriculture (Matra,2016;Sahari, 2013). Non-thermal plasma can be generated by an electric discharge in a gas at lower pressure or using microwaves, corona discharge, dielectric barrier discharge, radio frequency plasma, and gliding arc discharge. In contrast, thermal plasma is generated at higher pressures and requires high-power inputs (Misra, et al., 2016).

PLASMA GENERATION

The electron must be separated from the neutral gas particles to generate plasma. The energy required for electric discharge is usually used to ionize the gas. Several methods of generating plasma are listed below:

Glow Discharge Plasma

Glow plasma discharge is a process produced by the high voltage between two electrodes in a gas-containing glass tube. According to Kumar et al., (2017), when the voltage reaches a certain value called the breakdown voltage to allow the gas near the electrodes to break down into positive and negative ions, the effect of this process is the production of the plasma. Sahari (2013) States that electrons from the cathode cannot withstand discharge without applying potential difference. In the presence of a voltage difference, the electrons at the cathode accelerated to the anode and collided with the gas atoms. This collision will contribute to the process of excitation and ionization (Sahari, 2013). The ionization process will produce new electrons and ions, which would induce the secondary emission of electrons to the cathode and give rise to a new ionization collision. Thus, the discharge in the plasma is self-sustaining plasma because of the process of electrons emission at the cathode and ionization in the plasma as shown in **Figure 1**. The excitation process gives rise to exciting species that decay to lower levels through the emission of light, and the process allows the plasma discharge to normally emit a characteristic glow (Kumar et al., 2017).

Figure 1. Schematic diagram of the basic plasma processes in a glow plasma discharge (Attri et al., 2017)

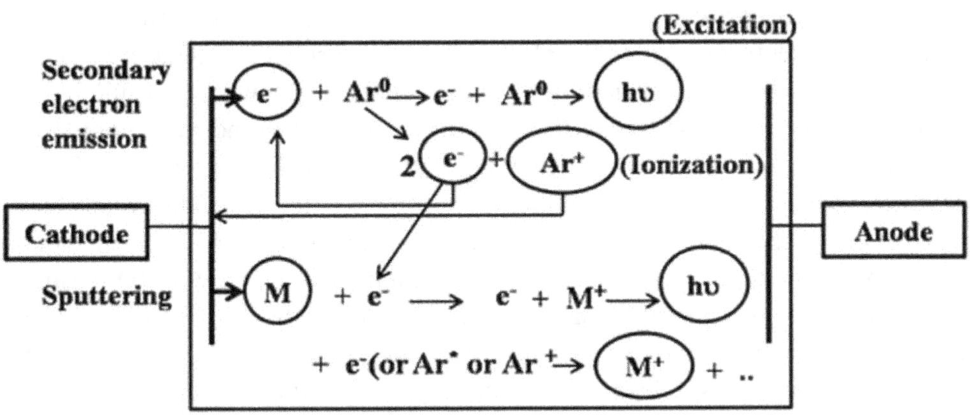

Dielectric Barrier Discharge

Dielectric barrier discharge (DBD) is the concept where the plasma discharges occurring between two electrodes, in which at least one electrode is covered by a thin layer of dielectric material (Feizollahi et al., 2021). The DBD plasma source is commonly used in laboratories to generate non-thermal plasma called silent discharge (Khan et al., 2020). DBD is called silent discharge because there is no spark which is accompanied by the generation of local shock waves and noise (Sahari, 2013). .

DBD plasma is one of the cold plasma techniques, which can be easily triggered at atmospheric pressure and room temperature. DBD plasma can generate ultraviolet rays, a strong electric field, high-energy electrons, and various active materials. No rare gas source or vacuum equipment is required in the DBD plasma process (Meng et al., 2017). The common planar DBD configuration is presented in **Figure 2**.

Figure 2.Common configuration of DBD system (Guo et al., 2015).

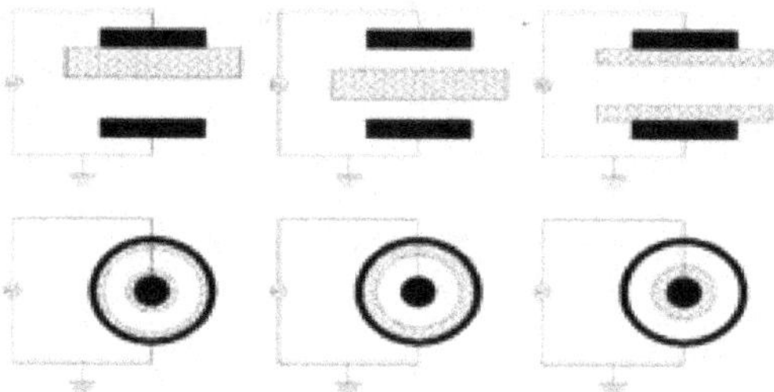

DBD can produce different patterns, which are filamentary discharge and glow discharge. Filamentary discharge is the formation of avalanches, and the development of streamers arises from separated local sites of the dielectric surface (Guo et al., 2015). These discharged cause uneven (inhomogeneous) plasma, which is not suitable for surface treatment. In some cases, where the uniformity and homogeneity of plasma effect are high required, a uniform glow DBD plasma has been developed (Guo et al., 2015). When a high potential difference is applied to the gas under atmospheric pressure or low pressure, glow DBD plasma is generated. Many research groups also reported the realization and application of atmospheric glow DBD plasmas. In particular, biological and medical applications using this kind of mild diffuse discharges have been extensively studied. Under low-pressure conditions, the generated plasma tends to be homogenous (**Figure 3**), which is suitable for many applications, such as surface treatment.

Figure 3. (A) Glow DBD plasma (B) Filamentary DBD plasma discharges(Lu et al., 2016)

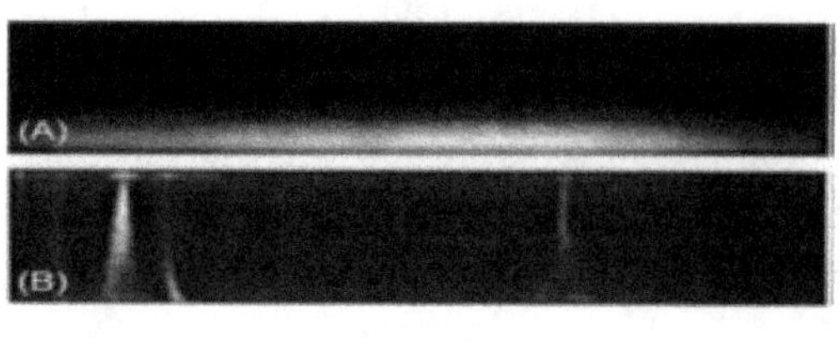

Plasma Jet

The plasma jet is often referred to as a plasma bullet due to its shape. The interest in plasma jets has gradually moved to atmospheric pressure non-thermal plasma jets (APNPJ) in the last ten years. In the biomedical field, APNPJ has been widely used, such as in bacterial inactivation, wound healing, and cancer treatment. Various configurations of electrodes have been developed to create APNPJ. These configurations are categorized by Lu et al. (2016) into four classifications: dielectric-free electrode jets, DBD jets, DBD-like jets, and single electrode jets as shown in Figure 4. A significant benefit of plasma jet is the ability to create certain chemical compounds that can penetrate the surface of the processed material (Baniya et al., 2020). Also, In comparison with the use of insecticides, the atmospheric plasmas obtained using He or He-air mixture as working gas does not cause undesirable effects such as residual toxicity, environmental pollution, and health hazards to humans and animals, and the atmospheric-pressure plasmas are extremely effective in inactivating plant pathogens and curing the infected plant leaves (Zhang et al., 2014).

Figure 4. Plasma jet system (Lu et al., 2016)

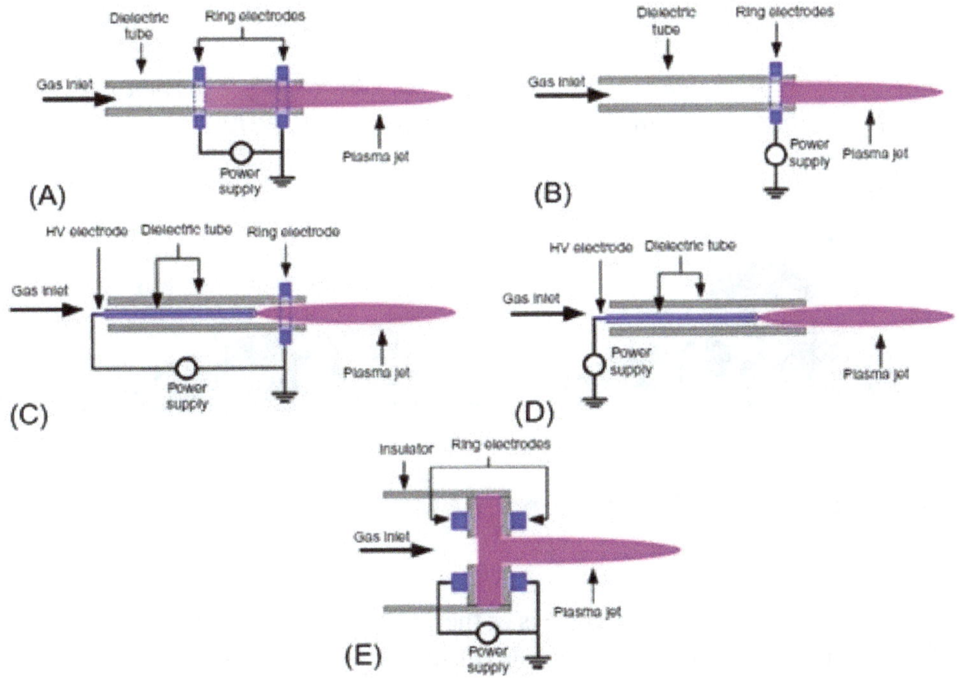

Seed Treatment

Plasma can interact with seed and change the seed surface characteristics. The plasma treatment can enhance the germination rate by removing microbial and increase water absorption on the seed surface. Research has been conducted to shows the impact of using dielectric barrier discharge plasma on wheat seed, in this research, they using various gas sources (oxygen, air, argon, and nitrogen) (Meng et al., 2017).. Another research focuses on the effect of cold atmospheric pressure plasma jet(CAPPJ) on seed germination and fenugreek seed growth by two-electrode argon CAPPJ system with and without an additional grounded electrode (Fadhlalmawla et al., 2019).

Contact Angel

Figure 5 shows the wetting of the seed surface before and after plasma treatment. Wetting for microscopically-scale rough seed surfaces can be described by the contact angle. The contact angle is the angle between the tangent to the liquid-air interface and the apparently solid surface as observed macroscopically (Bormashenko et al., 2012). The contact angle of wheat changes from 115±2° to zero, as shown in **Figure 5**. Therefore, the wetting properties have increased significantly, and wetting changes have led to a slight increase in water intake (Bormashenko et al., 2012)..

Figure 5. Change in contact angle of water on the wheat surface (a) before plasma treatment (b) after plasma treatment (Bormashenko et al., 2012).

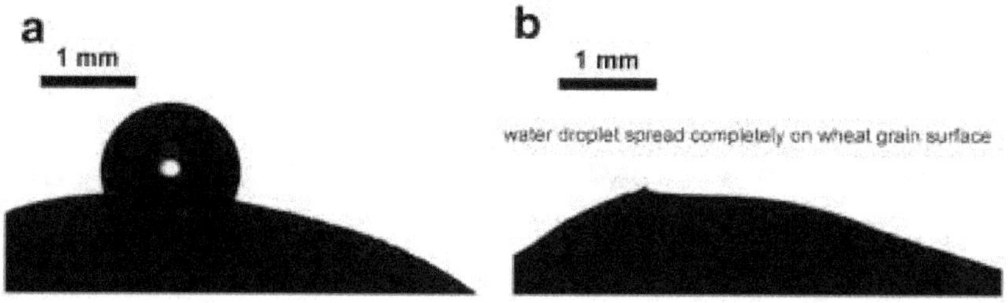

Scanning Electron Microscope (SEM)

SEM analysis is important for determining the scale, shape, and defect of the seed as it will help to improve the properties of the seed. The analysis measures the seed surface before it is treated and after it is treated by plasma to study the properties of the honeydew seed.

Furthermore, the imaging of the prepared sample can be magnified into different magnifications. Therefore SEM needs to compare the treated and untreated seed surfaces.

Figure 6 shows the picture by SEM image of A. thaliana seed; it shows the difference between treated and untreated seed coats. The images of SEM revealed a new and more rough appearance on the seed surface after plasma treatment. The change of appearance of the seed surface can be associated with a change in the absorption of water which may thus increase the rate of germination (Bafoil et al., 2018; Nantapan et al., 2019)

Figure 6. Scanning electron microscopy (SEM) images of A. thaliana seeds (A, C, E) are control and (B, D, F) are treated with plasma. Scale bars: 100μm A,B; 50μm C,D; 10μm E,F (Bafoil et al., 2018)

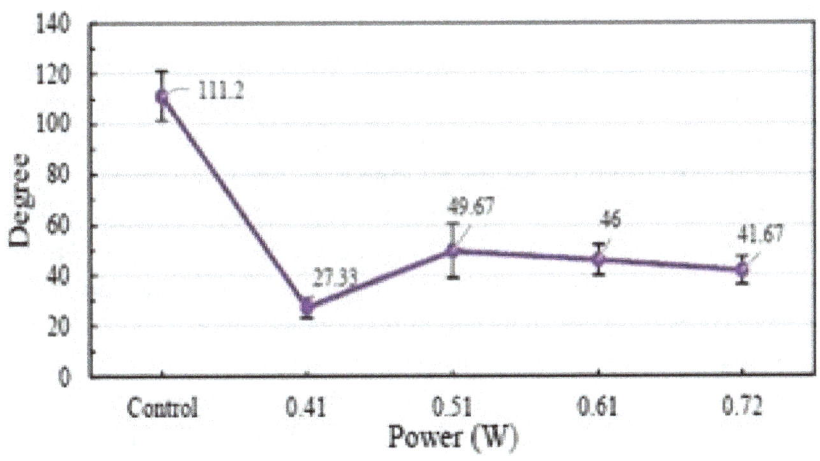

Germination

Then, another research about the impact of plasma treatment on the germination of radish seed. The experiment started by divided the radish seeds into four groups; each group was exposed to plasma treatment for 0, 2, 4, and 6 minutes. After plasma treatment, all radish seeds were soaked in the water for 2 hours. The seed was placed in the darkroom for cultivation. The growth rate of the seeds is determined by the length of the root. After five days of cultivation, the roots of the radish sprout grown straight first, and after seven days, the roots start to bend. This due to the stem substantially absorbs water, and the basement of the cultivation place was not suitable for holding the weight sprouts. From **Figure 7**, after seven days if we could observe that the root length and stem length of plasma treatment were rather longer than the other (Matra,2016).

Figure 7. Radish cultivation after plasma treatment for 5 and 7 days. (A) 2, (B) 4, (C) 6, (D) 0 minutes (Matra,2016).

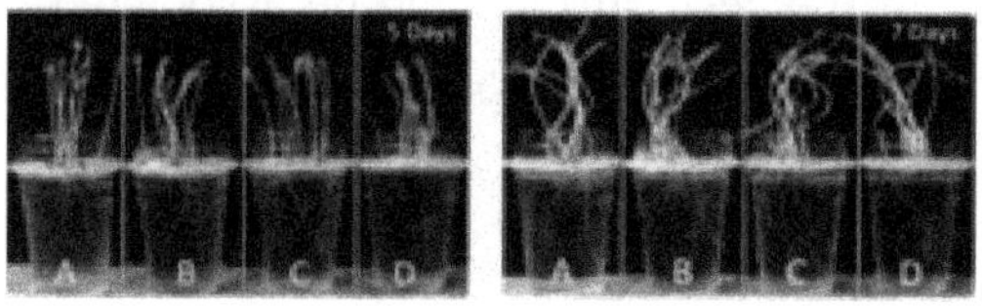

Sunflower Seed Treatment Use Plasma Jet

This study has shown the effects of atmospheric plasma jet treatment on the germination and enhancement growth of sunflower seeds. The sunflower seed was treated with argon plasma using the plasma technique. The argon plasma was operated by a high voltage DC pulse at 25 kHz. The dissipated plasma power was ranged from 0.41, 0.51, 0.61, and 0.72 watts, and the flow rate of argon gas was set at 4 liters per minute for 15 seconds. The physical properties, chemical properties, and growth rate were determined in the experiment. Concerning the physical properties, the morphology was checked by SEM via the contact angle and roughness. The chemical characteristics were analyzed by the contact angle. The growth rate was tested using four processes: the first is the water uptake, the second is the rate of germination count within five days, the third is the period of seeding within seven days, and the fourth is the dry weight of seeding in the oven for 24 hours at 80 ° C. The findings reveal that the contact angle of control is 41.08 degrees. The contact angle decreased after treatment with the plasma jet, and the wettability increased. The surface roughness increased with the use of increased plasma power. The best result was given by the water uptake of plasma with dissipated power at 0.41 Watts. Following plasma dissipated power at 0.41 and 0.61 watts, respectively, the germination count rates are 68% and 56%. When the power increased, the seeding length, as well as the dry weight, decreased seed quality*:* In this work, the power of the plasma was 0.41, 051, 0.61, and 0.72 watts by setting the exposure period to 15 seconds using the Argon gas plasma jet technique, which is the main factor for improving seed germination and plant growth or for controlling food surface microorganisms. The result shows that all the plasma treatments had a significant stimulatory effect on seed germination and vigor. As shown in **Figure 8** and **Figure 9**, the germination rate (GR%) increased when the exposure time increased.

Figure 8. Germination ate of sunflower seed treated/untreated by the plasma jet for five days (Nantapan et al., 2019)

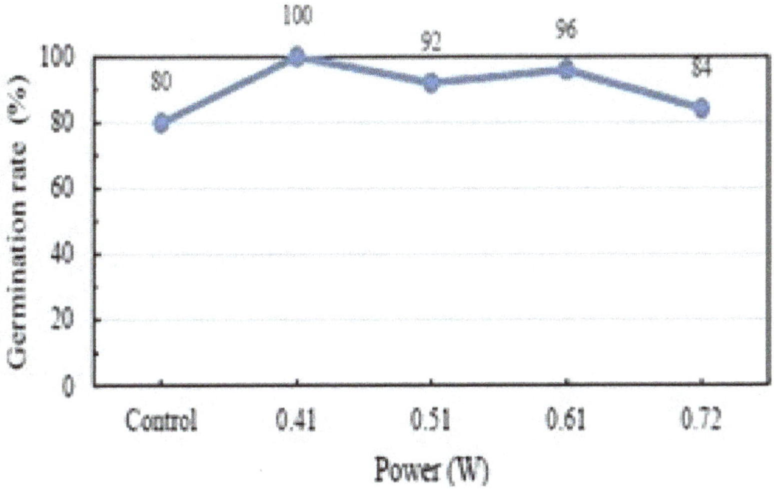

Figure 9. Sunflower seeds after treated/untreated by Argon gas plasma jet with 24 hrs. (a)Untreated, (b) 0.41 Watts, (c) 0.51 Watts, (d) 0.61 Watts and (e) 0.72 Watts (Nantapan et al., 2019)

The contact angle is the quality of its hydrophilic ability, which is relative to the germination seeds. When DI water was dropped into the 5 sample seed groups, the contact angle was measured. The apparent contact angle of the treated plasma was clearly smaller than the control, as well as the differential speed at which the coat absorbed the drop. The flat water on the surface indicates that the polar and nonpolar surface energy increased relative to the higher hydrophilicity properties, as shown in **Figure 10**.

Figure 10. Contact angle of water drop on the surface of the sunflower seed which treated/untreated plasma jet with power 0.41, 0.51, 0.61, and 0.72 watts (Nantapan et al., 2019)

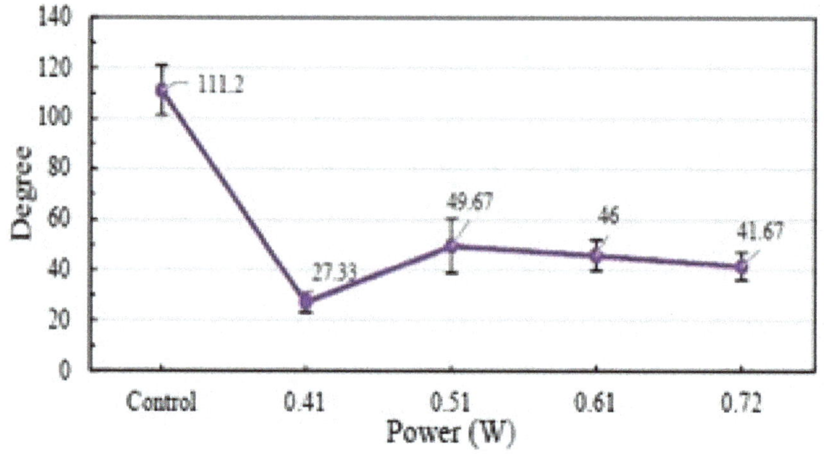

This study proved the effectiveness of an atmospheric pressure plasma jet on the sunflower seed operating at different power outputs. The 0.41 watts gave the best result in improving the sunflower seed germination and vigor, while the 0.51 watts showed the highest effect on water uptake.

Experimental Setup

The design of plasma DBD started from the observation of the existing product. The size of plasma DBD depends on its applications. The schematic diagram of the experiment set up for honeydew seed germination enhancement by plasma treatment is illustrated in **Figure 11**. The experiment was conducted in a room environment. We supply a high voltage (3.5 kV) into a gas port using air gas at a given frequency to generate the plasma. Air-gas flows into the air inlet from the multi-flow controller, which a multi-gas controller controls. Atmospheric pressure with a flow rate of 100 sccm by considering different time exposures (0, 120 s, 150 s, 180 s, 210 s, and 240 s) has flowed into the DBD plasma. The copper sheet was used as an electrode, while the Kapton tape was used as a dielectric barrier. The plasma DBD device is based on a double-ring structure with two copper electrodes covering the external glass tube. The top electrode was connected to a power supply, and the bottom electrode was grounded. The honeydew seed is placed between the plasma DBD at a different time to perform the seed treatment. After that, the seed will be planted to see the seed germination.

Figure 11. The schematic diagram of the DBD system for seeds treatments

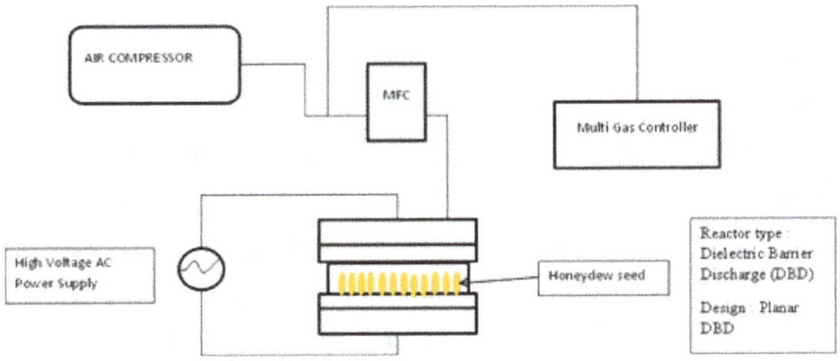

DBD Plasma Reactor Design

DBD plasma reactor is the main component in this research. This reactor consists of two plates, and the outer side of both plates was covered with Kapton tape, which acts as dielectric due to its wide range of temperature stability and electron isolation ability. It is used as a protection and insulation layer around the plates since it can sustain high temperatures throughout the plasma treatment process. In the upper plate, there is a copper sheet on the inner side of the plate which acts as an electrode to aid in plasma generation. The seeds were placed on the lower plate, which is within the discharged gap because plasma generation occurs in this gap: **Figure 12** shows a schematic diagram of the DBD plasma reactor.

Figure 12. Schematic diagram of DBD plasma

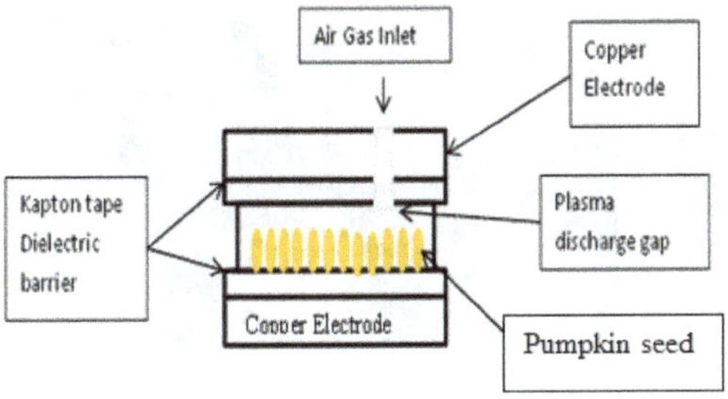

RESULTS AND DISCUSSIONS

Treatment Time

Figure 13 and **Figure 14** show the effects of cold plasma treatment within seven days with different treatment times. All the seeds were treated with a duration of 60 s, 90 s, 120 s, 150 s, and 180 s with a constant voltage of 3.5 kV and a fixed flow rate of 100 SSCM. All the seeds germinate at a steady increase from day 3 to day 7. On day 7, treated seeds of 120 s give the longest length of root and stem, and however untreated seeds give the shortest length of root and stem, which are 19.2 cm and 9.7 cm, respectively.

Figure 13. Effects of cold plasma treatment on the growth of root and stem for seven days at different treatment times

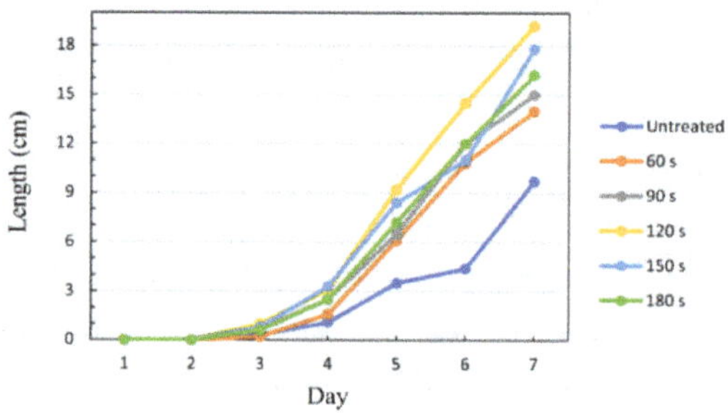

Figure 14. Pumpkin seed germination at different treatment times

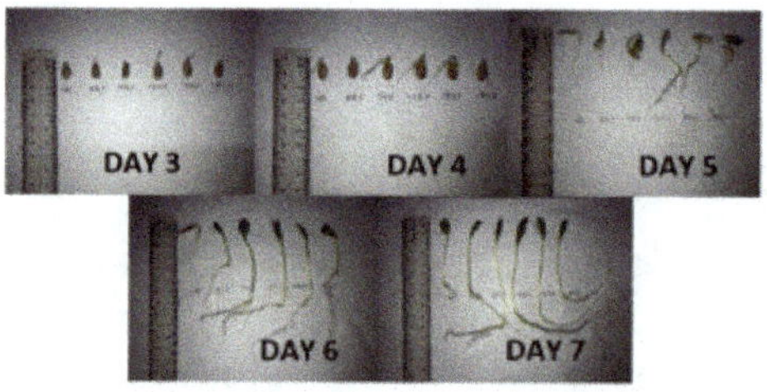

Surface Morphology

Three images of untreated and treated seeds for 150 s were taken with different magnification (100x, 500x, 1000x), as shown in **Table 1**. SEM image of untreated and treated seeds for 150 s. From the images of untreated seed, it can be seen there are some greyish spots on the surface suspected to be contaminants that can prevent water from penetrating the seed, but after 150 s of cold plasma treatment, some of the spots have been removed. Also, the seed surface appears to be rough, unlike before the treatment. The surface roughness helps to enhance the wettability of the seed. Besides, there are tiny holes created on the seed surface after the treatment. These tiny holes make the water easily seep into the seed surface.

Table 1. SEM image of untreated and treated seeds for 150 s

Magnification	100x	500x	1000x
Untreated	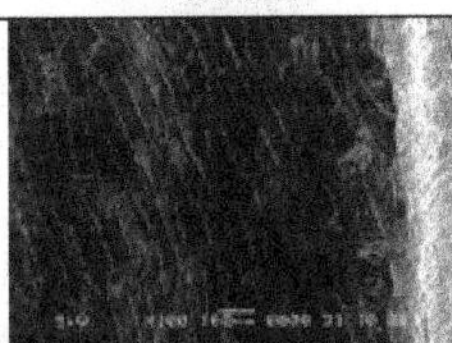	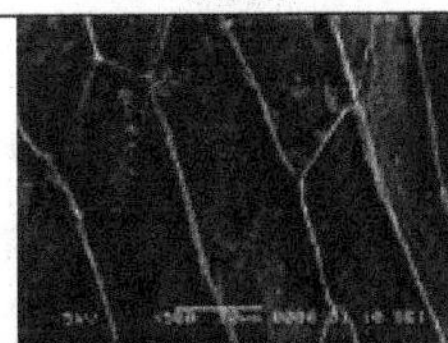	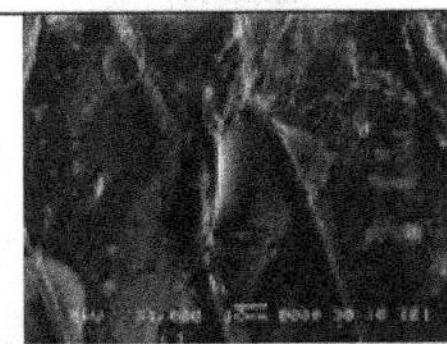
Treated	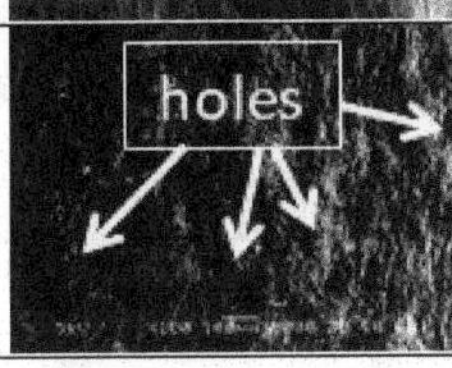		

CONCLUSION

The atmospheric pressure non-thermal plasma is a way to increase production in the agricultural sector by improving the germination rate of seeds. In this work, we carried out a seed treatment experiment using planar DBD discharge for pumpkin seeds. The optimization of the plasma reactor has been successfully determined using air as carrier gas with a flow rate of 100 SSCM and 3.5 kV of power voltage. Cold plasma treatment at different times proves that the best treatment time is 150 s. The surface morphology of the seeds has been successfully displayed on the surface of the untreated and treated seed. The presence of contaminants on the seed surface is removed after plasma treatment. After that, plasma treatment creates tiny holes or pores on the surface, which is helpful for water absorption. However, with the treatment time, the temperature of the seeds also increases and leads to a burn effect. The importance of this research is to explore the application of DBD discharge in the agriculture economy and helping to expand the potential of seed germination.

ACKNOWLEDGMENT

The authors are thankful to the Universiti Teknologi Malaysia (UTM) for the grant vote number 19J20.

REFERENCES

Agun, L., Sabirin, S., Ahmad, N., Zakaria, Z., Redzuan, N., & Ibrahim, R. R. (2021, April). The Efficiency of DBD Cold Plasma Pen Treatment on the Oyster Mushroom Bacterial Decontamination. *Journal of Physics: Conference Series*, *1892*(1), 012036. doi:10.1088/1742-6596/1892/1/012036

Attri, P., Arora, B., & Choi, E. H. (2017). Retraction: Utility of plasma: a new road from physics to chemistry. *RSC Advances*, *7*(26), 15735–15735. doi:10.1039/C7RA90034A

Bafoil, M., Jemmat, A., Martinez, Y., Merbahi, N., Eichwald, O., Dunand, C., & Yousfi, M. (2018). Effects of low temperature plasmas and plasma activated waters on Arabidopsis thaliana germination and growth. *PLoS One*, *13*(4), e0195512. doi:10.1371/journal.pone.0195512 PMID:29630641

Baniya, H. B., Guragain, R. P., Panta, G. P., Safaai, S. S., Dhungana, S., Qin, G., & Subedi, D. P. (2020). Investigation on parameters of atmospheric pressure plasma jet by electrical and optical methods. *Journal of Nepal Physical Society*, *6*(2), 50–56. doi:10.3126/jnphyssoc.v6i2.34857

Bormashenko, E., Grynyov, R., Bormashenko, Y., & Drori, E. (2012). Cold radiofrequency plasma treatment modifies wettability and germination speed of plant seeds. *Scientific Reports*, *2*(1), 1–8. doi:10.1038rep00741 PMID:23077725

de Groot, G. J., Hundt, A., Murphy, A. B., Bange, M. P., & Mai-Prochnow, A. (2018). Cold plasma treatment for cotton seed germination improvement. *Scientific Reports*, *8*(1), 1–10. doi:10.103841598-018-32692-9 PMID:30258075

Fadhlalmawla, S. A., Mohamed, A. A. H., Almarashi, J. Q., & Boutraa, T. (2019). The impact of cold atmospheric pressure plasma jet on seed germination and seedlings growth of fenugreek (Trigonella foenum-graecum). *Plasma Science & Technology*, *21*(10), 105503. doi:10.1088/2058-6272/ab2a3e

Feizollahi, E., Misra, N. N., & Roopesh, M. S. (2021). Factors influencing the antimicrobial efficacy of Dielectric Barrier Discharge (DBD) Atmospheric Cold Plasma (ACP) in food processing applications. *Critical Reviews in Food Science and Nutrition*, *61*(4), 666–689. doi:10.1080/10408398.2020.1743967 PMID:32208859

Guo, C. A., Tang, F., Chen, J., Wang, X., Zhang, S., & Zhang, X. (2015). Development of dielectric-barrier-discharge ionization. *Analytical and Bioanalytical Chemistry*, *407*(9), 2345–2364. doi:10.100700216-014-8281-y PMID:25510973

Jiang, J., Lu, Y., Li, J., Li, L., He, X., Shao, H., & Dong, Y. (2014). Effect of seed treatment by cold plasma on the resistance of tomato to Ralstonia solanacearum (bacterial wilt). *PLoS One*, *9*(5), e97753. doi:10.1371/journal.pone.0097753 PMID:24840508

Joshi, H., Kumar, R., & Meena, D. (2018). *Explosion of plasma technology in agriculture*. Academic Press.

Khan, T. M., Khan, S. U. D., Khan, S. U. D., Ahmad, A., & Abbasi, S. A. (2020). A new strategy of using dielectric barrier discharge plasma in tubular geometry for surface coating and extension to biomedical application. *The Review of Scientific Instruments*, *91*(7), 073902. doi:10.1063/5.0011890 PMID:32752834

Kumar, R., Singh, P., & Kumar, S. (2017). Plasma Treatment–A Tool to Improve Seed Quality–A Review. *Advances in Research*, *12*(4), 1–6. doi:10.9734/AIR/2017/36703

Ling, L., Jiafeng, J., Jiangang, L., Minchong, S., Xin, H., Hanliang, S., & Yuanhua, D. (2014). Effects of cold plasma treatment on seed germination and seedling growth of soybean. *Scientific Reports*, *4*(1), 1–7. PMID:25080862

López, M., Calvo, T., Prieto, M., Múgica-Vidal, R., Muro-Fraguas, I., Alba-Elías, F., & Alvarez-Ordóñez, A. (2019). A review on non-thermal atmospheric plasma for food preservation: Mode of action, determinants of effectiveness, and applications. *Frontiers in Microbiology*, *10*, 622. doi:10.3389/fmicb.2019.00622 PMID:31001215

Lotfy, K. (2017). Effects of cold atmospheric plasma jet treatment on the seed germination and enhancement growth of watermelon. *Ozean Journal of Applied Sciences*, *7*(12), 705–719. doi:10.4236/ojapps.2017.712050

Lu, P., Cullen, P. J., & Ostrikov, K. (2016). Atmospheric pressure nonthermal plasma sources. In *Cold plasma in food and agriculture* (pp. 83–116). Academic Press. doi:10.1016/B978-0-12-801365-6.00004-4

Matra, K. (2016). Non-thermal plasma for germination enhancement of radish seeds. *Procedia Computer Science*, *86*, 132–135. doi:10.1016/j.procs.2016.05.033

Mehmood, F., Kamal, T., & Ashraf, U. (2018). *Generation and applications of plasma (an academic review)*. Academic Press.

Meng, Y., Qu, G., Wang, T., Sun, Q., Liang, D., & Hu, S. (2017). Enhancement of germination and seedling growth of wheat seed using dielectric barrier discharge plasma with various gas sources. *Plasma Chemistry and Plasma Processing*, *37*(4), 1105–1119. doi:10.100711090-017-9799-5

Misra, N. N., Schlüter, O., & Cullen, P. J. (Eds.). (2016). *Cold plasma in food and agriculture: fundamentals and applications*. Academic Press.

Nantapan, J., Sarapirom, S., & Janpong, K. (2019, November). The effects of atmospheric plasma jet treatment to the germination and enhancement growth of sunflower seeds. *Journal of Physics: Conference Series*, *1380*(1), 012131. doi:10.1088/1742-6596/1380/1/012131

Park, Y., Oh, K. S., Oh, J., Seok, D. C., Kim, S. B., Yoo, S. J., & Lee, M. J. (2018). The biological effects of surface dielectric barrier discharge on seed germination and plant growth with barley. *Plasma Processes and Polymers*, *15*(2), 1600056. doi:10.1002/ppap.201600056

Pawłat, J., Starek, A., Sujak, A., Terebun, P., Kwiatkowski, M., Budzeń, M., & Andrejko, D. (2018). Effects of atmospheric pressure plasma jet operating with DBD on Lavatera thuringiaca L. seeds' germination. *PLoS One*, *13*(4), e0194349. doi:10.1371/journal.pone.0194349 PMID:29630623

Pawlat, J., Terebun, P., Kwiatkowski, M., Starek, A., Kiczorowski, P., Andrejko, D., & Kopacki, M. (2017, December). Effects of Helium-Air RF plasma jet on onion seeds' germination. In *2017 International Conference on Electromagnetic Devices and Processes in Environment Protection with Seminar Applications of Superconductors (ELMECO & AoS)* (pp. 1-4). IEEE. 10.1109/ELMECO.2017.8267743

Sahari, N. (2013). *Generation of Homogeneous Glow Discharge Using a Combination of Fine Wire Mesh Perforated Aluminium Alectrode* (Doctoral dissertation). Universiti Teknologi Malaysia.

Zhang, X., Liu, D., Zhou, R., Song, Y., Sun, Y., Zhang, Q., Niu, J., Fan, H., & Yang, S. Z. (2014). Atmospheric cold plasma jet for plant disease treatment. *Applied Physics Letters*, *104*(4), 043702. doi:10.1063/1.4863204

Zhu, X. G., Long, S. P., & Ort, D. R. (2010). Improving photosynthetic efficiency for greater yield. *Annual Review of Plant Biology*, *61*(1), 235–261. doi:10.1146/annurev-arplant-042809-112206 PMID:20192734

Chapter 8
Biosynthesis of Silver Nanoparticles for Study of Their Antimicrobial Effect on Plasma-Treated Textiles:
Silver Coating of Plasma-Treated Cotton Fabric

Shazia Shukrullah
University of Agriculture, Faisalabad, Pakistan

Muhammad Anwar
University of Agriculture, Faisalabad, Pakistan

Muhammad Yasin Naz
University of Agriculture, Faisalabad, Pakistan

Inzamam Ul Haq
University of Agriculture, Faisalabad, Pakistan

ABSTRACT

Dielectric barrier discharges (DBD) are the configurations for the production of electrical discharges using a dielectric medium between the metallic electrodes. Plasma treatment produces negative radicals, which increase the adhesion of fabric for nanoparticles. The plasma treatment made the fabric surface rougher because of the etching effect. UV-vis spectra of the Plasmon resonance band observed at 253-400 nm. X-ray diffraction results showed that AgNPs has a cubical structure and the average

DOI: 10.4018/978-1-7998-8398-2.ch008

crystalline size is 25 nm. SEM results determined that the morphology of the silver nanoparticles are flower shaped. The energy bandgap of AgNPs was observed at 2.59 eV. The silver nanoparticles were found to have enhanced antimicrobial properties and showed better zone of inhibition against isolated bacteria (Escherichia coli). DBD plasma treatment changed the chemical as well as physical properties of the cotton fabric. FTIR spectrum revealed that oxygen-containing groups, such as C-O, C=O, O-C-O, as well as O-C=O, increased on DBD treatment of cotton samples.

INTRODUCTION

The widely accepted finishing technique for textile is nanotechnology-based coatings. Nanostructured materials and nanoparticles are introduced on the surface of textiles for smart functionalization and surface modification. The benefits of nanoparticles depend on their surface area to volume ratio and size-dependent properties, which lead to their better characteristics as compared to bulk size particles. The natural fabric has well-known physical properties like softness, comfort, and better wettability (Hebeish et al., 2011). Promoting such inherent properties of natural fabrics with nanoparticles has fascinated the significant attention of manufacturers and scientists, which lead to the origination of garments and apparel in the past few years.

The coatings of nanoparticles have become important in recent times due to their ability to develop smart surfaces, acceptable cost and feasibility. In traditional coating formulations, it is necessary to combine various forms of nanoparticles to produce new materials with novel functionalities without serious changes to the existing machinery. Different types of nanoparticles are being used to impart unique properties to different products (Hebeish et al., 2011). Considering their enormous potential for practical applications, the production of multifunctional textile coatings is well-known as one of the most desirable areas for researchers. Researchers also tried to develop the features like electrical conductivity, antistatic strength, antimicrobial action, self-cleaning, flame retardant, UV safety, thermoregulation for numerous applications (Leroux et al., 2009; Aalipourmohammadi et al., 2019; Dong et al., 2019). Self-cleaning and conductive textiles are more important among different types of novel features due to their applicability in hospitals, underwear and army. Such functionalities can be accomplished by various care strategies and processes.

Properties of Coated Fabrics

Imparting novel and unique properties such as electrical conductivity, electromagnetic shielding, fire retardant, superhydrophobic, super hydrophilic, bactericidal activities, ultraviolet (UV) protection, and self-cleaning to natural fabrics has been initiated in the recent year (Leroux et al., 2009). Figure 1 illustrates some key applications and properties of silver nanoparticles coated fabrics. The major role of nanotechnology is to built materials and devices at the nano level. The analysis and applications of incredibly small structrues in nanoscience and nanotechnology can be used in all other areas of research, such as physics, chemistry, biology, materials science, and engineering. The surface area of nanomaterials is larger than the bulk materials. Nanomaterials are also labelled as environmentally friendly materials. Because of the small size, the physical and chemical characteristics of nanoparticles are totally different from their bulk counterparts.

Several methods of nanocoatings on textile have been deduced including immersion, spraying, in-situ synthesis, printing, padding, cross-linking, rolling and dyeing (Aalipourmohammadi et al., 2019). However, there are few important issues with these developed methods concerning the use of nanoparticles, including their shape and purity, size uniformity, toxicity, time consumption, durability on textile, health, safety as well as dispersibility in the colloid. Usually, nanoparticles aggregate in the form of microclusters due to electrostatic and van der Waals forces because of the lack of attraction between textiles and inorganic nanoparticles. Some fabrics like cotton and polyester are naturally hydrophobic, with less surface energy substances, which causes less durability of nanocoatings. The differences in surface energies between inorganic particles and organic fabrics cause repellency at their interfaces. Therefore, it is necessarily required to modify the surfaces of fabrics to enhance the surface energy without modifying their bulk properties before coating metal oxide nanoparticles (Dong et al., 2019; Kandregula et al., 2017).

Plasma Treatment of Fabrics

Various methods such as gamma irradiation, UV irradiation, chemical reactions, and plasma treatment have been developed to change the physical and chemical properties of textiles fabric. Plasma is the fourth fundamental form of matter also known as a moldable substance. Natural plasma only exists at extremely high temperatures or in a low-temperature vacuum; it does not react quickly and is extremely hot at temperatures above 20,000°C. The energy of hot plasma is so great that it vaporises everything in its path. The artificial plasma is produced by subjecting gas to the electromagnetic field or by heating, by applying a strong electromagnetic field the substance becomes more ionized and electrically conductive. Plasma treatment is a promising modification recently developed method, especially for its entire effects at the nanoscale. The best known plasmas are atmospheric pressure arc discharge, corona discharge, dielectric barrier discharge (DBD), capacitive discharge plasma, etc (Nalini and Vijayaraghavan, 2020). Recently, DBD plasmas have gained significant attention due to their environmentally friendly nature and cost-effectiveness. These plasmas offer different highly reactive active chemical species at ambient pressure which are necessary for surface processing. The ambient plasma pretreatment can generate polar groups such as carboxyl groups, acylamino groups, and hydroxide radicals that effectively improve the surface energies and wettability of fabric. Also, plasma increases the surface roughness of fabric by etching effect, which enhances the facilitation of attachment of nanoparticles (Moghimifar et al., 2014; Sinica et al., 2014).

Figure 1. Key properties of silver nanoparticles coated fabrics.

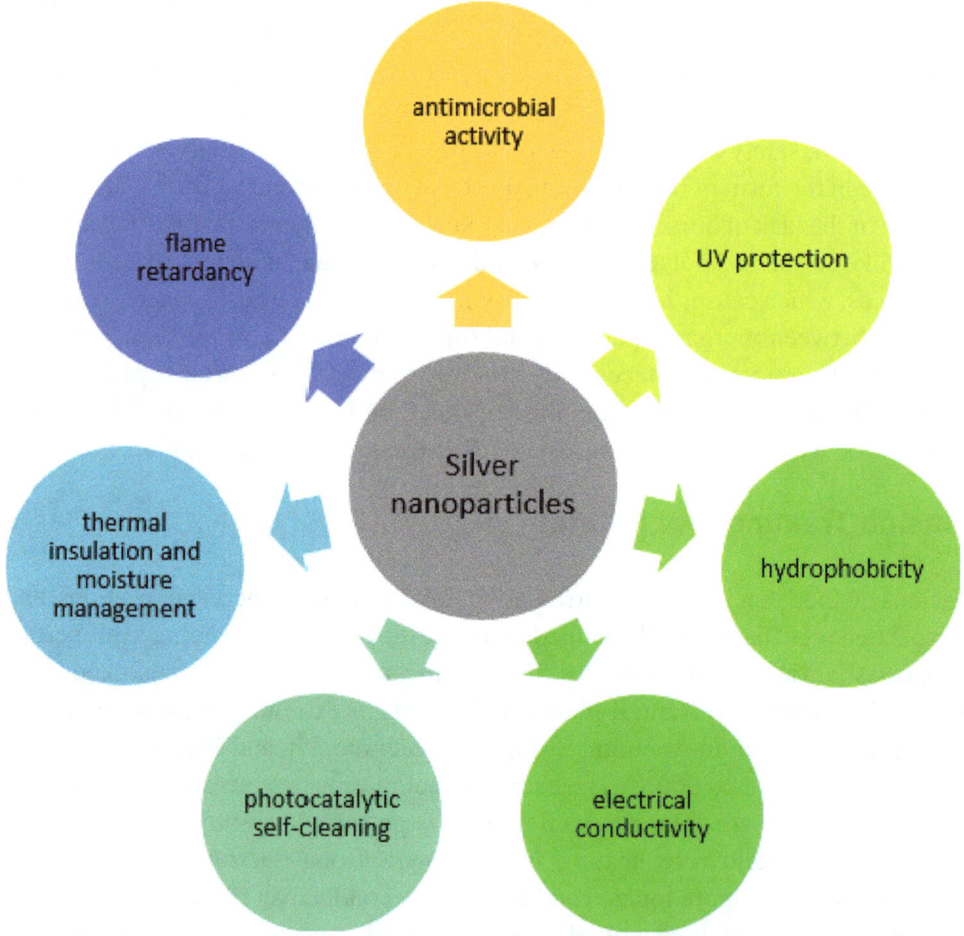

Methods of Nanocaotings

The chemical methods for the synthesis of nanoparticles require the use of toxic, highly superior, and hazardous chemicals that create a serious ecotoxicological concern. So, an eco-friendly, simple and less toxic green synthesis involving biodegradable materials such as plant extracts, enzymes, and microbes has gained much attention for the preparation of nanoparticles. Furthermore, a strategy based on plant extract reduces the chance of contamination and needs less time. The most prominent nanoparticles, silver, zinc oxides, carbon nanotubes (CNTs), SiO_2, nano clays, and TiO_2 are noteworthy. Silver based nanomaterials are vital biomedical agents that have been widely explored (Boddu et al., 2011). Silver has long been used as a disinfectant and is now employed in a variety of ways to treat infectious disorders. It has a broad-spectrum antibacterial activity and is low in toxicity to mammalian cells. The use of AgNPs to various textile fibres for imparting antimicrobial effects has recently become a key research focus in the textile industry (Xia et al., 2011). Cotton fabrics have been coated with metal and metal oxide nanoparticles to impart antibacterial characteristics, according to multiple publications. The biosynthesis of silver nanoparticles has been performed by different plant extracts like Azadirachta indica Citrus reticulate, Aloe vera, Muraya koenigii, and Curcuma longa. Azadirachta indica (Neem) is a succulent shrub, which has medicinal properties. Its leaves consist of phenolic, pectin, flavonoids, cellulose, polyphenols, hemicelluloses, anthraquinones, and acemannan. Neem leaf extract reduces silver salt to silver nitrate. Through aqueous neem, leaf excerpt decreases silver salt to silver nitrate this topped nanomaterials with neem extract existing the antimicrobial activity (Alemdar and Agaoglu, 2009; Dubey et al., 2010; Lee et al., 2010).

The above-mentioned problems such as toxicity, time consumption, expressivity, lack of binding ability between fabric and nanoparticles, long-term stability of nanoparticles, and generation of functional groups could be solved by biosynthesis and plasma pretreatment of fabric. In this chapter, the fabric samples were exposed to ionized gas DBD plasma having neutral and active species. The active species have modified the surface of the fabric by generating new active functional groups. Plasma treatment was considered as pre-treatment since the coating of functional nanoparticles was added afterward. The neem powder was used as a reducing and capping agent for the biosynthesis of silver nanoparticles to reduce the toxicity of particles. The sonochemical solution reduction method was used to synthesize and coat nanoparticles on the surface of the fabric. The ultrasonic method was consumed less time and create more stability and equal size distribution of nanoparticles on fabric surface (Lee et al., 2001; Selvi et al., 2011; Ali et al., 2018; Montazer et al., 2010).

MATERIALS AND METHODS

In this work, the silver nitrate ($AgNO_3$) with molecular weight (169.87 g/mol) was taken from the local market of Faisalabad. The 100% plain cotton fabric (200.34 g/m^2) with a density of 51 picks and 108 ends per inch was supplied by Masood Textile Mills Ltd. DBD plasma apparatus, used to treat the samples at standard atmospheric pressure and room temperature, is shown in Figure 2. Fresh neem leaves were collected from the University of Agriculture's botanical garden in Faisalabad, Pakistan. In this research work, commercially untreated and unbleached pure cotton fabric, DBD apparatus, Silver nitrate, power supply, beaker, distilled water, burette were used for the experiments.

Firstly, the cotton fabric samples were prepared before the plasma treatment to remove the impurities, proteins, and seizing particles. DBD is an electrical discharge in-between the two electrodes. The insulating dielectric barrier is used to separate the electrodes. A dielectric material is used to cover one out of both electrodes. It is a nonthermal discharge of plasma, produced by applying high voltage around the small gap where a dielectric layer stops the penetration of discharge into the arc. It has a high density of free electrons than corona discharge. The density of free electrons is 10^{10} electrons/cm^3. It produces asymmetrical and inconsistent discharge. Quartz glass is mostly used as a dielectric barrier in DBD. Gasses breakdown in the plane-parallel gap between insulated-electrodes. The discharge produces at atmospheric pressure, which leads to a large number of distinct breakdown channels known as micro-discharge or filament discharge. While a high voltage is supplied to the discharge gap, electrons sweeps to the saturation point. The desizing of cotton was done using enzyme (3-5 g/l), wetting agent (2 g/l), and sodium chloride (2g/l) in water. The water was boiled at 80ºC and the fabric was immersed in this boiling water (Frunza et al., 2018). The cotton fabric was washed and rinsed for one hour and then dried in a furnace at 70ºC. After completing the desizing, the fabric was cut into samples of dimensions 15×5 cm^2. The farbirc samples were treated by DBD plasma.

DBD apparatus consists of the variac, bulb, power supply, quartz tube, motor, roller, and copper rod. Variac is connected with power supply and light indicator. The output of the power supply connected to the roller and copper rod. The copper rod enclosed in a quartz tube acts as a dielectric. Plasma is generated between quartz tube and roller and applied on fabric that place on the roller attached with the motor. The cotton fabric was treated for 3 minutes on one side at 25 kV AC voltage. The cotton fabric sample was washed with non-detergent and acetone to remove the seizing particles. Then the sample was dried in the oven at 70ºC for 30 min. The sample was mounted on the cylindrical electrode of DBD which was rotated at 110 revolutions per min (rmp). The secondary power was applied on the two electrodes that break down the water and air molecule and produces plasma. The treatment time was kept for 3 minutes. The plasma was applied on both sides of the fabric. Finally, the sample was removed for further processing. The gap of electrodes was kept at 3 mm. The output power, current, voltage, and plasma treatment time were changed according to the experimental design (Anwar et al., 2021).

Figure 2. Schematic of DBD treatment apparatus.

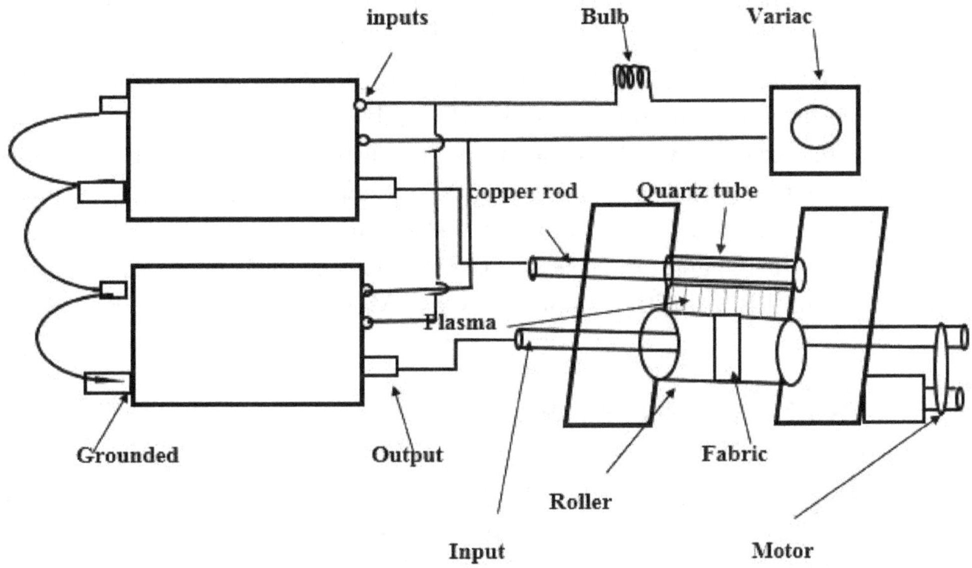

The neem leaves were cut into small pieces and then dried under sunlight for three days. After drying the leaves were grinded to make the powder using a mechanical grinder. Then we took 20 g of neem powder and mixed it with 100 ml methanol in the beaker which is also covered by aluminum foil. The solution was kept in a dark room for 24 hours. Thereafter, the solution was stirred for 15 min at a temperature of 70°C. The resulting extract was filtered by Whatman filter paper No.1 and the filtered extract was saved at room temperature for further experimentation. The solution of the required molarity was prepared by using 10 g of extract and distilled water in a 250 ml beaker. The measured quantity of silver nitrate was also added in the solution for formation of silver nanoparticles. The plasma pretreated sample was immersed in the prepared solution for 15 minutes and then the solution was kept in an ultrasonic bath. Initially, the ultrasonic bath was half-filled with distilled water and then switched on. Then the temperature of the bath was fixed at 65°C. The ultrasonic treatment time was also adjusted according to the experimental design. Ultrasonic waves with a frequency of 40 kHz created bubbles in the solution, which burst and transferred energy to it. In the presence of neem extract, the solution decreases to nanoparticles (Montazer et al., 2018; Nabi et al., 2020). The ultrasonic energy boosts not only the reaction rate but also the nanoparticle dispersion. After sonication, the samples were padded and cured by a pad dry cure machine. The padded samples were washed to remove the impurities.

The washed cotton was dried in an oven at 790ºC: this procedure was considered as one coating. Then each sample was coated seven times with silver nanoparticles (AgNPs). Now, the treated samples were ready for characterization and further applications. A change in the hue of the solution and the peak value of the surface plasmon resonance in the UV-Vis spectrum confirmed the creation of AgNPs. Scanning electron microscopy was used to determine the size and form of AgNPs. The crystalline structure was determined by X-ray diffraction studies. The reducing agents in neem powder and functional groups, introduced during plasma treatment of fabric, were checked by the Fourier transform infrared spectroscopy (FTIR). Using a disc diffusion approach, the antibacterial activity of AgNPs coated fabrics was tested against E. Coli strains (Gaminian and Montazer, 2015).

RESULTS AND DISCUSSION

The coated textile industry is not only opening new research avenues but also defining new textile functionalities, properties and performance. Textiles are being coated with metallic, nonmetallic and polymeric materials (Hebeish et al., 2011). Because of its unique physical and chemical qualities, low toxicity, and low cost, silver serves as a chemical agent in textile functionalization. Bulk silver has already been designated as a safe material by the US Food and Drug Administration (FDA). AgNPs' UV protection characteristics are due to their chemical stability in the presence of UV light. The suggested methods of UV protection involves refraction or scattering of UV rays at the surface, as well as absorption of UV radiations through high refractive index, which blocks direct and diffuse transmittance of UV rays through the fabric to reach the skin. Figure 3 illustrates UV protection mechanism of coated fabrics. The UV pretection efficiency is determined in terms of ultraviolet protection factor (UPF). According to EN 13758 standards and AS/NZ 4399 good protective properties are attained with a UPF value of 40-50 or greater and with UVA average transmittance smaller than 5%.

The efficiency of UV protection dependents on the particle size, shape, and concentration of the coating material. Results show that an increase in the concentration of AgNPs on fabric substrate increases UV protection factor. Also, the nanoparticles have higher UV protection factor than micro-sized nanoparticles because of their high specific area and refractive index. The suspension of nanoparticles with high agglomeration did not retain excellent UV protection properties, as agglomerated phenomena greatly inhabites the efficiency of UV blocking. A in situ synthesis of nanoparticles minimizes the particles' agglomeration. The small particle size in the presence of stabilizing and capping agents as well as reducing agents has been recommended to attain efficient UV protection. SEM images of samples, coated with AgNPs for 2 cycles, 4 cycles, and 6 cycles, are shown in Figure 4. The deposition of silver particles on the cloth surface was seen using scanning electron microscopy. The deposition of silver became more regular and dense as the coating thickness increased. The comparison of the untreated and the treated samples of cotton indicated that the untreated samples had a smooth surface than DBD plasma-treated surface (Pakdel et al., 2014). A strong interaction of DBD plasma with the surface of fabric is assumed due to presence of micropores on the fabric surface and moisture.

Figure 3. Schematic illustration of UV protection mechanism of coated textiles.

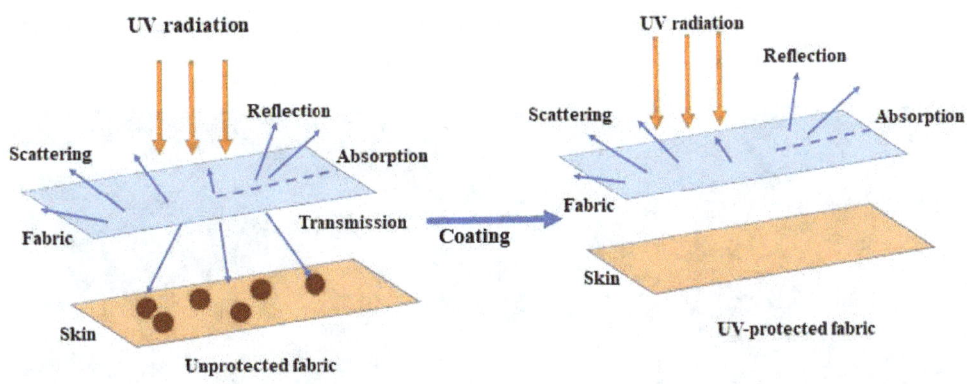

The structural properties of the prepared sample were examined through XRD analysis. Cu-Kα (λ = 1.540598 Å) source was used for production of XRD patterns. A typical XRD pattern of pure AgNPs is given in Figure 5. The 2θ values, FWHM, are d-spacing values wer extracted from XRD pattern, as shown in Table 1. The particle size was calculated by using the following formula: $D=K\lambda/(\beta \cos \theta)$. The structure of silver nanoparticles is a crystalline cubic (Palaskar et al., 2016). By using hkl and d-spacing values from XRD results, the lattice constant of all the prepared samples was calculated. XRD pattern showed the crystalline structure of silver nanoparticles of the dried biogenic. The distinct XRD peaks at scattering angle of 2θ values of 28°, 32°, 46°, and 77° could be indexed to the Bragg's reflections from the lattice planes (110), (111), (211) and (311), respectively. The unassigned peaks showed the presence of some biological moieties of neem leaves extract. By Scherrer's formula, the average crystallite size of silver was determined in the range of 25.1 nm (Qi et al., 2007). In general, the particular phase width of materials is proportional to the mean crystallite size of the material (Selvi et al., 2011).

The coated fabric was also tested for antibacterial activity. The bacterial culture in nutrient agar of 28 g was prepared and dissolved in 1000 ml of distilled water. The culture was incubated in the incubator for 24 hours at 37°C (Katewaraphorn and Aldred, 2016). The antimicrobial activity of AgNPs coated fabrics was tested against E.coli. Table 2 shows the zones of inhibition against E.coli (Katewaraphorn and Aldred, 2016). Figure 6 shows inhabitation zones after 2, 4 and 6 coating cycles. Zone of inhabitation increases with the coating cycles, which is attributed to the dense AgNPs coating in the fabric surface. The control showed 22 mm inhabitation zone. The coated fabric showed zone of inhabitation of 12 mm, 14 mm and 17 mm after 2, 4 and 6 coating cycles, respectively.

Figure 4. SEM micrograph of BDB plasma treated fabric for 3 min and coated with AgNPs for (a) 2 cycles, (b) 4 cycles, and (c) 6 cycles.

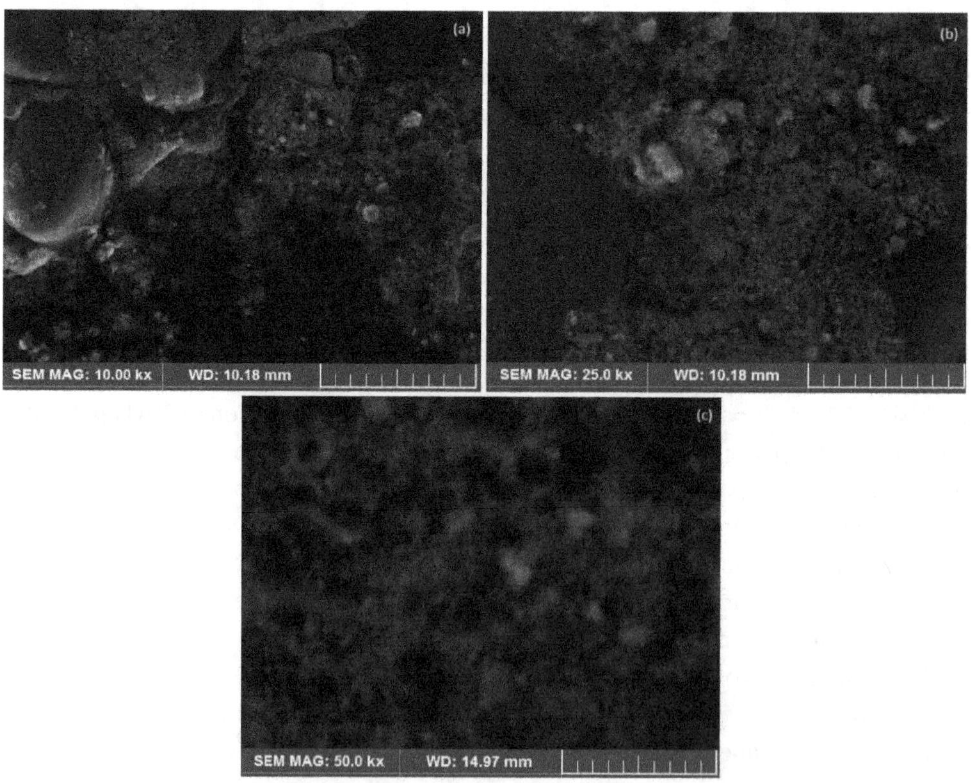

Figure 5. An XRD spectrum of silver nanoparticles.
Source: Anwar et al., 2021

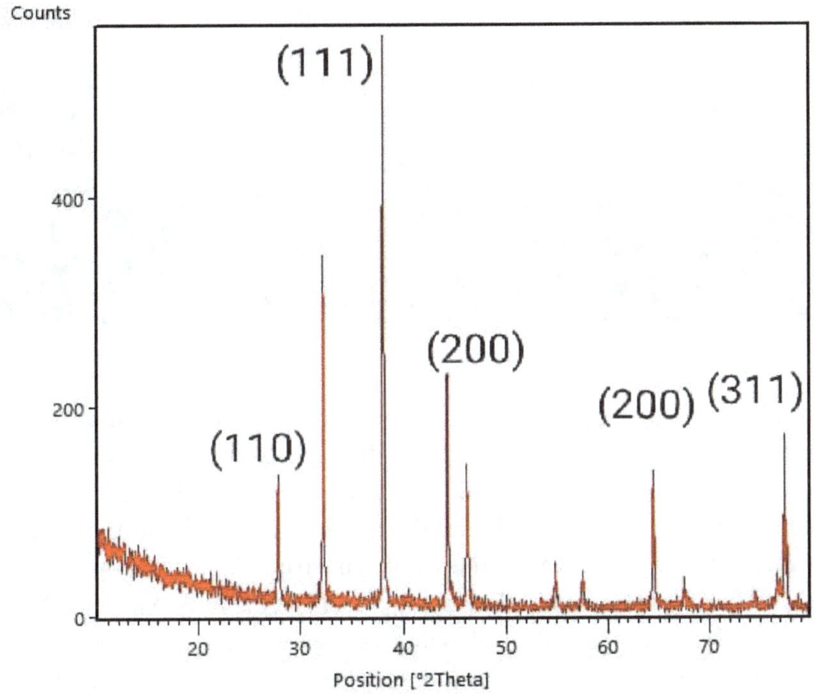

Table 1. Crystal structure data extracted from XRD pattern of AgNPs.

2Theta	Hkl	d-spacing	Lattice parameters a=b=c for cube	Volume $V=a^3$
28	1 1 0	3.1843	4.5032	91.3195
38	1 1 1	2.366	4.8408	113.436
47	2 0 0	1.9320	4.8291	112.61
66	2 0 0	1.4144	4.1039	69.1220
78	3 1 1	1.2242	4.0602	66.9307

Table 2. Zone of inhibition of coated fabric againt E.coli.

Coating cycles	Zone of inhabitation	Control Zone
2 cycles	12 mm	22 mm
4 cycles	14 mm	22 mm
6 cycles	17 mm	22 mm

Figure 6. Photographs of the antimicrobial zone of inhibition of AgNPs coate fabric after 2, 4, and 6 coating cycles.

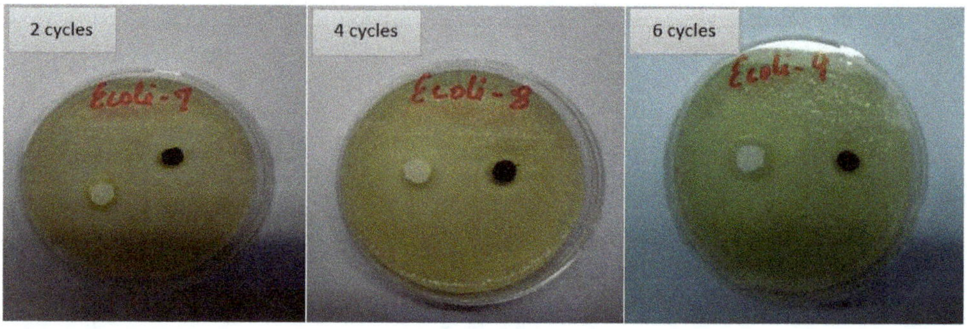

UV visible spectroscopy is a technique used for the optical properties of nanoparticles. Figure 7 shows the strong absorption spectrum of the green sysnthesized AgNPs (Anwar et al., 2021). Excitonic absorption peak was found at 300 nm that shows blue shift due to the small size of the nanoparticles. The supporting point is that only one peak was observed in the whole spectrum. These nanoparticles show a strong UV Vis absorption band that is not present in the bulk material (Naz et al., 2016; Vigneshwaran et al., 2006). This analysis shows that most particles were in nano size. The energy band gap of AgNPs was determined as 2.63 eV that shows that silver is a semiconducting material. With acceptable bandgap, the synthesized nanoparticles might have applications in advanced optoelectronic devices and sensors. They can also be used as a photocatalyst for degradation of dyes. The activation of the Surface Plasmon Resonance caused the brown colour to arise (Chinnammal and Arunkumar, 2014).

Figure 7. A typical UV plot and corresponding Taucplot of AgNPs.

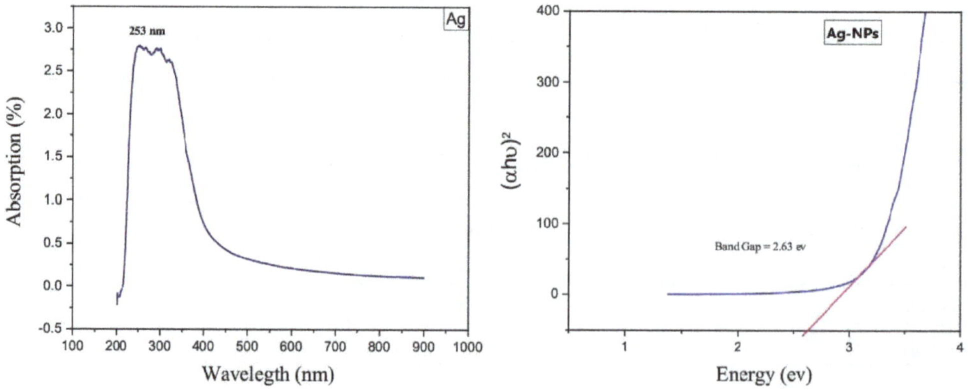

Wettability is the capability of any liquid like water to keep contact with solid surfaces. Wettability is measured through force balance between the cohesive and adhesive forces. Wetting deals with the three states of the matter i.e. solid, liquid, and gas. The contact angle and wettability are inverses of each other. The surface wetting of the coated samples was satisfactory and the liquid drop extended on the large part of the surface as the value of contact angle was below 90° (Vu et al., 2013). The wettability of treated and untreated cotton fabric was measured by a standard method that is BS 4554-1970 method. The wettability of DBD plasma-treated and untreated cotton samples is shown in Table 3. The absorbency of water drop relates to the wettability. From the table it is clear that DBD plasma treatment enhanced the wettability of cotton samples. The absorbence time of DBD plasma samples increased from 13 to 52 minutes. The water absorbance time of untreated samples was smaller than treated samples. As the voltage increased the wettability of DBD plasma-treated samples also increased (Shahidi et al., 2015).

Table 3. Wettability of cotton before and after DBD treatment.

The voltage across electrodes (kV)	Plasma treatment time (min)	Wetting time (mint)
Untreated Cotton	0	13
110	5	52
120	5	48
130	5	33
140	5	48

FTIR investigations were implemented on DBD plasma treated and AgNPs coated cotton fabric, as shown in Figure 8. FTIR spectrum confired the functional groups and biomolecules which were responsible for efficient stabilization and capping of AgNPs. The peaks at 980, 1030, 1400, 1570 and 3050 cm^{-1} for silver coated cotton fabric sample were observed. The peaks at 980 cm^{-1} and 1030 cm^{-1}(polysaccharide groups), 1400 cm^{-1} (nitrosamine), 1570 (diketones) and 3050 cm^{-1} (C-H and carboxyl acid) suggested the presence of terpenoids and flavanones adsorbed at the surface of cellulose fabric. FTIR spectrum of DBD trated fabric revealed oxygen-containing groups, such as C-O, C=O, O-C-O as well as O-C=O.

Figure 8. FTIR spectrum of AgNPs coated cotton fabric.

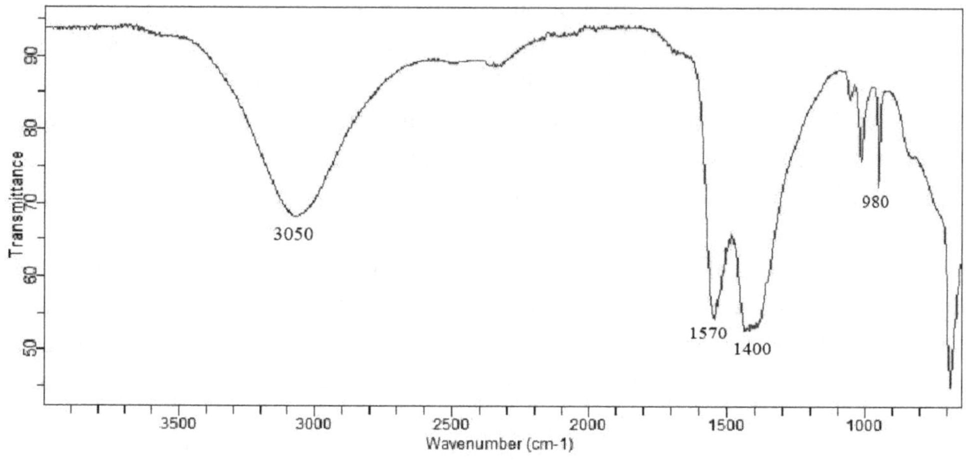

CONCLUSION

The cotton fabric was treated with DBD plasma in open air. The active species modified the surface of the fabric by generating new active functional groups. Plasma treatment was considered as pre-treatment since the coating of functional nanoparticles was added afterward. The sonochemical solution reduction method was used to synthesize and coat nanoparticles on the surface of the fabric. The ultrasonic method was consumed less time and create more stability and equal size distribution of nanoparticles on fabric surface. The plasma treatment made the fabric surface rougher because of the etching effect. UV-vis spectra of the Plasmon resonance band observed at 253-400 nm. X-ray diffraction results showed that AgNPs has a cubical structure and the average crystallite size is 25 nm. SEM results determined that the morphology of the silver nanoparticles is like flowers. The energy bandgap of AgNPs was observed at 2.59 eV. Silver nanoparticles were discovered to have improved antibacterial capabilities and a larger zone of inhibition against the pathogen Eschrichia coli..

REFERENCES

Aalipourmohammadi, M., Davodiroknabadi, A., & Nazari, A. (2019). Homogeneous Coatings of Titanium Dioxide Nanoparticles on Corona-Treated Cotton Fabric for Enhanced Self-Cleaning and Antibacterial Properties. *AUTEX Research Journal*, *3*(7), 797–799. doi:10.2478/aut-2019-0058

Alemdar, S., & Agaoglu, S. (2009). Investigation of in vitro antimicrobial activity of aloe vera juice. *Journal of Animal and Veterinary Advances*, *8*(1), 99–102.

Ali, A., Nguyen, N. H. A., Baheti, V., Ashraf, M., Militky, J., Mansoor, T., Noman, M. T., & Ahmad, S. (2018). Electrical conductivity and physiological comfort of silver coated cotton fabrics. *Journal of the Textile Institute*, *109*(5), 620–628. doi: 10.1080/00405000.2017.1362148

Anwar, M., Shukrullah, S., Haq, I. U., Saleem, M., AbdEl-Salam, N. M., Ibrahim, K. A., Mohamed, H. F., & Khan, Y. (2021). Ultrasonic Bioconversion of Silver Ions into Nanoparticles with Azadirachta indica Extract and Coating over Plasma-Functionalized Cotton Fabric. *ChemistrySelect*, *6*(8), 1920–1928. doi:10.1002lct.202004623

Boddu, S. R., Gutti, V. R., Ghosh, T. K., Tompson, R. V., & Loyalka, S. K. (2011). Gold, silver, and palladium nanoparticle/nano-agglomerate generation, collection, and characterization. *Journal of Nanoparticle Research*, *13*(12), 6591–6601. doi:10.100711051-011-0566-x

Chinnammal, S. K., & Arunkumar, K. (2014). Effect of plasma treatment on plain woven cotton fabric. *International Journal of Scientific Research*, *20*, 83–88.

Dong, P., Nie, X., Jin, Z., Huang, Z., Wang, X., & Zhang, X. (2019). Dual dielectric barrier discharge plasma treatments for synthesis of Ag–TiO_2 functionalized polypropylene fabrics. *Industrial & Engineering Chemistry Research*, *58*(19), 7734–7741. doi:10.1021/acs.iecr.9b00047

Dubey, S. P., Lahtinen, M., & Sillanpää, M. (2010). Tansy fruit mediated greener synthesis of silver and gold nanoparticles. *Process Biochemistry*, *45*(7), 1065–1071. doi:10.1016/j.procbio.2010.03.024

Frunza, L., Diamandescu, L., Zgura, I., Frunza, S., Ganea, C. P., Negrila, C. C., & Birzu, M. (2018). Photocatalytic activity of wool fabrics deposited at low temperature with ZnO or TiO_2 nanoparticles: Methylene blue degradation as a test reaction. *Catalysis Today*, *306*, 251–259. doi:10.1016/j.cattod.2017.02.044

Gaminian, H., & Montazer, M. (2015). Enhanced self-cleaning properties on polyester fabric under visible light through single-step synthesis of cuprous oxide doped nano-TiO_2. *Photochemistry and Photobiology*, *91*(5), 1078–1087. doi:10.1111/php.12478 PMID:26073930

Hebeish, A., El-Naggar, M. E., Fouda, M. M. G., Ramadan, M. A., Al-Deyab, S. S., & El-Rafie, M. H. (2011). Highly effective antibacterial textiles containing green synthesized silver nanoparticles. *Carbohydrate Polymers*, *86*(2), 936–940. doi:10.1016/j.carbpol.2011.05.048

Kandregula, W., Sarwar, Z., Abid, S., Munir, U., & Azeem, A. (2017). Aloe vera leaf gel extract for antibacterial and softness properties of cotton. *Journal of Textile Science & Engineering*, *7*(3), 1–6.

Katewaraphorn, J., & Aldred, A. K. (2016). A study of microcapsules containing Psidium guajava leaf extract for antibacterial agent on cotton fabric. *International Journal of Chemical Engineering and Applications*, *71*(1), 27–41. doi:10.7763/IJCEA.2016.V7.536

Lee, H. Z., Hsu, S. L., Liu, M. C., & Wu, C. H. (2001). Effects and mechanisms of aloe-emodin on cell death in human lung squamous cell carcinoma. *European Journal of Pharmacology*, *431*(3), 287–295. doi:10.1016/S0014-2999(01)01467-4 PMID:11730720

Lee, S. M., Song, K. C., & Lee, B. S. (2010). Antibacterial activity of silver nanoparticles prepared by a chemical reduction method. *Korean Journal of Chemical Engineering*, *27*(2), 688–692. doi:10.100711814-010-0067-0

Leroux, F., Campagne, C., Perwuelz, A., & Gengembre, L. (2009). Atmospheric air plasma treatment of polyester textile materials. Textile structure influence on surface oxidation and silicon resin adhesion. *Surface and Coatings Technology*, *203*(20-21), 3178–3183. doi:10.1016/j.surfcoat.2009.03.045

Moghimifar, V., Raisi, A., & Aroujalian, A. (2014). Surface modification of polyethersulfone ultrafiltration membranes by corona plasma-assisted coating TiO_2 nanoparticles. *Journal of Membrane Science*, *461*, 69–80. doi:10.1016/j.memsci.2014.02.012

Montazer, M., Pakdel, E., & Moghadam, M. B. (2010). Nano titanium dioxide on wool keratin as UV absorber stabilized by butane tetra carboxylic acid (BTCA): A statistical prospect. *Fibers and Polymers*, *11*(7), 967–975. doi:10.100712221-010-0967-y

Montazer, P. H., Pawade, V. B., & Bhanvase, B. A. (2018). Characterization Tools and Techniques for Nanomaterials. Nanomaterials for Green Energy, 16(12), 3331-3334.

Nabi, G., Raza, W., & Tahir, M. M. (2020). Green Synthesis of TiO_2 Nanoparticle Using Cinnamon Powder Extract and the Study of Optical Properties. *Journal of Inorganic and Organometallic Polymers and Materials*, *30*(4), 1425–1429. doi:10.100710904-019-01248-3

Nalini, K. S. P., & Vijayaraghavan, K. (2020). Green Synthesis of Silver and Gold Nanoparticles using Aloe Vera Gel and Determining its Antimicrobial Properties on Nanoparticle Impregnated Cotton Fabric. *Journal of Nanotechnology Research*, *2*(3), 42–50. doi:10.26502/jnr.2688-85210015

Naz, M. Y., Shukrullah, S., Ghaffar, A., Rehman, N. U., & Sagir, M. (2016). A Low-Frequency Dielectric Barrier Discharge System Design for Textile Treatment. *Synthesis and Reactivity in Inorganic, Metal-Organic, and Nano-Metal Chemistry*, *46*(1), 104–109. doi:10.1080/15533174.2014.900789

Pakdel, E., Daoud, W. A., Sun, L., & Wang, X. (2014). Visible and UV functionality of TiO_2 ternary nanocomposites on cotton. *Applied Surface Science*, *321*, 447–456. doi:10.1016/j.apsusc.2014.10.018

Palaskar, S. S., Desai, A. N., & Shukla, S. R. (2016). Development of multifunctional cotton fabricusing atmospheric pressure plasma and nano-finishing. *Journal of the Textile Institute*, *107*(3), 405–412. doi:10.1080/00405000.2015.1034932

Qi, K., Xin, J. H., Daoud, W. A., & Mak, C. L. (2007). Functionalizing polyester fiber with a self-cleaning property using anatase TiO_2 and low-temperature plasma treatment. *International Journal of Applied Ceramic Technology*, *4*(6), 554–563. doi:10.1111/j.1744-7402.2007.02168.x

Selvi, B. T., Rajendren, R., Nithyalakshmi, B., & Gayathirignaneswari, S. (2011). Antimicrobial activity of cotton fabric treated with Aloevera extract. *International Journal of Applied Environmental Sciences*, *6*(2), 127–131.

Shahidi, S., Ahmadi, M., Rashidi, A., & Ghoranneviss, M. (2015). Effect of plasma treatment on self-cleaning of textile fabric using titanium dioxide. *Micro & Nano Letters*, *10*(8), 408–413. doi:10.1049/mnl.2015.0053

Sinica, D. P., Mathur, A., Kushwaha, A., Dalakoti, V., & Dalakoti, G. (2014). Pelagia Research Library Green synthesis of silver nanoparticles using medicinal plant and its characterization. *Der Pharmacia Sinica*, *5*, 118–122.

Vigneshwaran, N., Kumar, S., Kathe, A. A., Varadarajan, P. V., & Prasad, V. (2006). Functional finishing of cotton fabrics using zinc oxide–soluble starch nanocomposites. *Nanotechnology*, *20*(20), 3087–3099. doi:10.1088/0957-4484/17/20/008

Vu, N. K., Zille, A., Oliveira, F. R., Carneiro, N., & Souto, A. P. (2013). Effect of particle size on silver nanoparticle deposition onto dielectric barrier discharge (DBD) plasma functionalized polyamide fabric. *Plasma Processes and Polymers*, *10*(3), 285–296. doi:10.1002/ppap.201200089

Xia, N., Cai, Y., Jiang, T., & Yao, J. (2011). Green synthesis of silver nanoparticles by chemical reduction with hyaluronan. *Carbohydrate Polymers*, *86*(2), 956–961. doi:10.1016/j.carbpol.2011.05.053

Chapter 9
Plasma Formation and Its Parameters Used in Calibration-Free Laser-Induced Breakdown Spectroscopy

Alina Saleem
University of Agriculture, Faisalabad, Pakistan

Yasir Jamil
University of Agriculture, Faisalabad, Pakistan

ABSTRACT

A rapidly developing technique over the last two years is laser induced break-down spectroscopy (LIBS), also known as laser-induced plasma spectroscopy (LIPS), a non-destructive spectroscopy technique that is mostly used for the analytical study of samples. With this technique, multiple elemental composition of elements are analyzed simultaneously without considering the form of sample that may be solid, liquid, or gas. Moreover, economically it is a very beneficial and valuable technique because no sample preparation is required, and sample consumption is very small. This technique is powerful enough that it can bore a microscopic crater usually in the solid samples to target. This technique has great sensitivity to find the resolution down to a single grain. It has a variety of applications in the field of science such as archeology, soils, environmental protection, and so on.

DOI: 10.4018/978-1-7998-8398-2.ch009

INTRODUCTION

Calibration-Free LIBS (CF-LIBS) method was presented by Ciucci et al. in 1999 to analyze composition in LIBS spectra. (Aguilera, Aragón, Cristoforetti, & Tognoni, 2009). The configuration of different elements are find out by using some mathematical models without calibration curves as well as reference samples availability and ignore the matrix effect. The accuracy of the result depends on the intensities of spectral lines and temperature of plasma .There are some important parameters of CF-LIBS method such as plasma temperature which is find out on the basis of Boltzmann plot. As the plasma temperature is considered to be same for the different species in the identical sample there are radiations which are emitted from the plasma when laser light falls on it and these radiations are usually depends in the temperature of the atmosphere. These emitted radiations are not affect the sample properties but after some time these radiations are begin to absorb in the sample then Local Thermal Equilibrium occurs.

Local Thermal Equilibrium

To fulfill his condition, introduce the number density

Electron Number Density

$$N_0 \geq 1.6 \times 10^{12} T^{\frac{1}{2}} (\Delta E)^3$$

Where T = temperature of plasma
(ΔE) = energy difference between energy levels

According to this condition, quantitative analysis of the sample is done when plasma is kept in local thermal equilibrium and some of the radiations are just scattered across the sample.

Optically Thin Plasma

To check the optically thin plasma condition, intensity ratios are required which are given below:

$$\frac{I_2}{I_1} = \frac{\lambda_2 A_2 g_2}{\lambda_1 A_1 g_1} \exp(E_2 - E_1)/k_\beta T$$

Where is the intensity of second line in the spectrum with wavelength and energy of upper level is the intensity of first line in the spectrum with wavelength

Plasma Formation

When the plasma is optically thick the light emitted from the hot region of plasma and goes towards cold region and light is absorbed by cold atoms which causes self-absorption, the spectral produced will be weakened because of absorption of light itself plasma emission and this effects the quantity of elements. While in optically thin plasma there are only few scattering across the plasma. In the LIBS experiment the elemental composition analysis from the line intensities becomes simple when the plasma is in LTE and also optically thin. . By comparing this ratio values which are measured at different energies the plasma is optically thin.

LASER INDUCED BREAKDOWN SPECTROSCOPY(LIBS)

Laser Induced Breakdown Spectroscopy(LIBS) is a versatile and quick chemical analysis technology it creates a micro plasma on the sample surface as it uses a short laser pulse. For the heavy metallic elements which all are lie in the PM range LIBS gives a suitable classical detection limits. LIBS is valid to a wide range of sample matrices that consist of metals, semi-conductors, soil, plants, tissues and electronics materials etc. As including air and water soil is also considered the one of the major source of nature. Provision is one the essential means for plant growth which is include it's one of the chief functions. The uppermost layer of the Earth atmosphere is made of the mixtures of organic matter, gases, liquids and minerals which includes plants and organisms, all of which are entirely necessary for life on earth. Hence, it also offers a habitat of microorganisms (Hu et al., 2013).It is an significant source for life, therefore, it has been the subject of the great concern for scientists in preceding decades, basically for environmental monitoring, as it has been infected by numerous sources. That is, anthropogenic heavyweight metals resources, soil corrosion and depletion etc.

HISTORY OF LIBS

laser induced breakdown spectroscopy (LIBS) is a new emerging technology it is also consider a latest analysis tool for the all multi types elements which are present in the gases, solids, and also either use in the aerosols and liquids. To generate a micro- plasma over the sample surface, the process which is used surely depends upon the atomic emission spectroscopy, this lased pulse have a high energy and it must be focus on the sample. To know about the basic characteristics of all the elemental configuration, all the sample constituents which are excited and vaporized over the plasma surface, this creates the various spectrum such as atomic and ionic. After words, it helps in the qualitative determination of the various samples it is happen due to the emission lines gathered by spectrally of all the available elements. It also divert the results towards the quantitative analysis. For the elemental examination LIBS offers number of benefits as compared to the conservative methods. By acid-digestion of the various silicate which is include in the soils, it helps out in the whole elemental analysis of the soil and also gives information about the organic matrix which have an eminent pressure and also temperature.

Figure 1. Schematic diagram of LIBS experimental system.

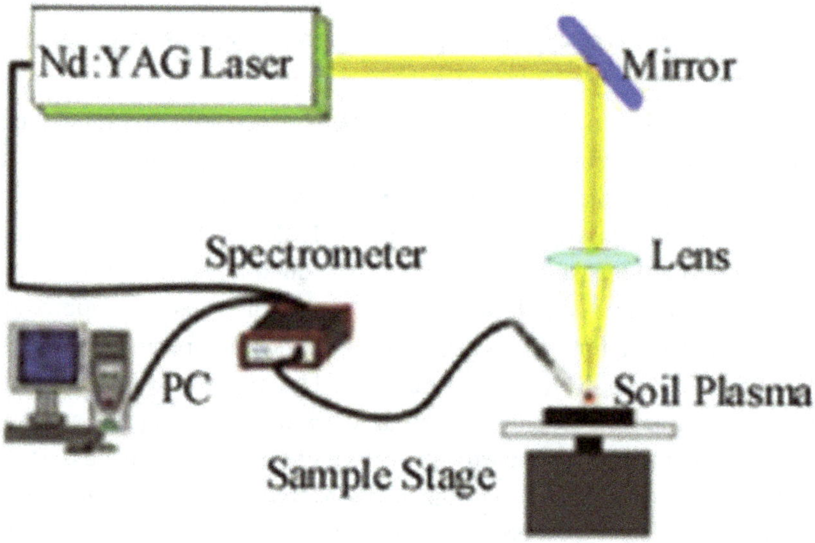

PLASMA FORMATION IN LIBS

Once thawed, the elemental concentration of the various samples can be examined using the traditional and established tools, such as (ICP-MS), hydride generation AAS, (ICP-OES), and including all other atomic absorbance spectroscopy techniques. Hence, these techniques attain a great importance due to its well-done performances related to the characteristics. As, these digestion methods are seemed to be challenging in the case of the determining the errors related to the sample manipulation, all the great number of the dilution factors, these are also considered the time taken and proved overwhelmed due to harmful chemical uses and its disposal. Furthermore, LIBS is consider one of the emerging tool which shows its vital role in non-destructive type of surface analysis which needs a no and minimum amount of the sample preparation. It is observed that to prepare the samples which terms shows significance are its pelletization and its homogenization when a certain pressure apply on them, these are necessary for the dried samples and also about someone else called sediments which are also termed as unconsolidated materials. Afterword, the gathered pellets are battered using the laser pulses directly under the standard atmospheric conditions in the presence of the laboratory air. This process can be accompolished in between a second and number of hours, this is totally depends upon the emitted pulse from the laser according to the per sample taken under observation.

Although, it is observed after the comparsion of LIBS with those of the typical methods regarding the sample analysis and their manipulation, it is seen that normally there is a loses related to its analytical capabilities. These emitted laser pulse interrelates with the little amount of the soil also many factors distress on the plasma formation over the surface of the samples also manipulate the dependant spectral autographs of all the used samples. Hence, it is seen that it takes appropriates conditions which are as, ablated volume, thermal properties, its configuration, surface roughness and all related physical and chemical properties. It is concluded that to diminish these all intangibles a number of mechanism can be determined by all the spectral and the experimental analysis.

LIBS SET-UP COMPONENTS

The major devices used in the LIBS set-up dominated in the above figure, firstly a pulsed laser beam which lie in the range of few nanosecond having very high energy as well always directed towards the sample material (Cremers et al., 2006). This sample vaporized due to falling the respective light energy which become the cause to induced the plasma. Moreover, to get the reasonable spectral signature spectrometer is used to collect the diffracted light then either its have a more and also a less optical system at all. So, by taking in use the different devices such as CCD, PMT, and also PDA the light can be detected easily. After that, for the further examination these detected spectrum usually processed using the computer. Although it is analyzed that LIBS set-up required a reasonable time control to minimize the various plasma life stages which help out in the advancement of these spectral signature used in the experiment. So, it is concluded that the combination of laser with that of the spectrometer-detector as well as with the time control with the satisfactory environmental conditions decides the success also the failure of the respective experiment (Miziolek et al., 2006).

LASER CONFIGURATIONS FOR LIBS

The device which is necessary for the LIBS technique is laser. To examine the different kind of the plasma features it usually produced energy to induced the plasma. There are various factors which are required for the laser action such as wavelength, pulse generation time, no. of pulses during every pulse and also the energy used in every pulse. Although, it is concluded that every type of the application works well due to this type of the combination with various parameters. It is seen that nanosecond pulsed laser is the most appropriate for the LIBS technique. That's the reason the major part of this section is linked this type of the laser (Onge et al., 1998).

Laser Wavelength

The laser wavelength depends upon the two main factors i) Energy absorption ii) plasma-material interaction. Photon ionization happens when the photon emitted and its energy must be larger than that of the bond energy. Plasma influences on the wavelength in the nanosecond range in the LIBS set-up (Abdellatif et al., 2002). As, the short range wavelengths plays a vital behavior during the laser coupling process by keeping the threshold must be higher at that time for the formation of the plasma et al (Cabalin et al., 1998).

The photoionization phenomena is required in the case, when these shorter wavelengths lie in the the range between 266 and 157nm (Russo et al., 2000). Where, at the same time the longer wavelengths follow the inverse Bremesstrahlung process for the various kind of the plasma shielding which become the cause to minimize the ablation rate as well as on the same time it help out in the enhancement of the elemental fractionation (Mao et al.,1996). The Nd: YAG pulsed laser proved to be the most efficient laser for the LIBS analysis . As, this type of the laser proves beneficial as well as the most appropriate and efficient way to generate the plasma during the LIBS experiments. The major mode of this laser is 1064nm range although, its pulse width lie in 6ns to 15ns range. Fundamental of first harmonics considered to be very common and precise wavelength used during the LIBS experiments, As, these can also work easily with the various kind of the wavelengths keeping the same type of the environmental conditions. Hence, it is observed that LIBS during the CO2 and also excimer laser most probably work in the ranges of the UV and in far IR respectively.

Laser Energy

The laser energy factors mostly linked with that of the laser material interaction which depends upon its energy per unit time and area, W/cm2, and the other one is the energy per unit area is represented as j/cm2. Furthermore, there are various kind of the ablation mechanisms such as explosion, melting, the erosion and also the sublimation, as these are process have a various threshold conditions (Mao et al., 1998). So, by altering the laser pulse time rate and also the its wavelength the laser energy would be changed accordingly. Hence, it is seems to be tough to observe this energy effect alone. Where, there are different threshold levels for solids, liquids, gases and also for the aerosols lie around about 1011W/cm2 and 1010W/cm2 (Cremers et al., 2006).

ADVANTAGES AND DISADVANTAGES OF LIBS

Advantages

There are following advantages of emerging technology Laser induced breakdown spectroscopy(LIBS).

- LIBS technique plays a vital role in the material disbursed processes, which is behaved sometimes as a non- destructive or usually destructive in some extent having a average power density which is normally less than that of one-watt radiated of different type of the specimen although, one of them doesn't heated by the ablation sites.
- Naturally this technique gives a satisfactory outcomes in the sample preparation which usually minimized to homogenization, and also mostly redundant where its heterogeneity is need to discovered either a specimen taken to be homogenous, which although minimized the maximum opportunities of adulteration through every chemically preparation steps.
- Furthermore, the one of the main advantage of this LIBS tool having a capability of finding its depth profile related to any specimen in the case of every frequently discharging of laser beam which are lie in the same position that impacted greatly whenever it goes deeper towards the specimen by every shot.
- LIBS technique is used in the elimination of every type of the surface contamination completely.
- Although, it is considered to be a most effective tool that shows the required ouutcomes within a second, which makes it significantly useful in the case of high volume examination also in the field of on-line industrial monitoring.
- LIBS is termed as one of the optical tool which needs mostly optical access for any specimen.

- For the determination of every remote analysis LIBS have major connotation as fibre optics at all, therefore, as Optical tool it is considered to be a non-invasive which employed as stand-off analytical tool which are usually coupled by a telescopic apparatus.
- These all features shows a vitality whenever they are required in the areas from dangerous environment for the exploration of space and time.
- So, in the analytical flexibility case LIBS system normally united with an optical microscope about all types of the micro-sampling.
- LIBS is advantageous in the chemical imaging analysis which is more precious for in the field of science and technology.
- Moreover, the portable LIBS system are seems to be more quick and also the more sensitive which have a ability to detect the elements which are wider inn range.
- LIBS technique doesn't use ionizing radiations which are necessary for the excitation of sample.

Disadvantages

- Like all other types of the analytical tools LIBS is also have a certain limitations.
- It creates variations during the laser spark also the consequential plasma having the certain limits reproducibility.
- Although, it represents 10% better accuracy in the LIBS measurements and same as in the case of precision which usually better than 5% at all.
- Furthermore, the detection of limit of emerging technology LIBS mainly fluctuate from one element to the another which contingent on the specimen different types and also all their experimental apparatus which are being used during the process.
- Lastly, LIBS have a specific detection limit lie in between the 1 to 30 ppm variated by mass which are normally not infrequent, Hence, it has a range limit lie from >100ppm<1ppm at all.

The Additional Advantages and Applications of LIBS

1. Using an optical fibre the optical connection can be performed in between the target and the instruments therefore, the whole interaction between them is seen to be optical at all.
2. Although, in the case of remote applications target put 100m away from that of the mobile instrument.

3. The different production phases are needed in the detection of LIBS excels cases as real-time, on-site and also on-line measurements.
4. LIBS technique can be used in the determination of different elements such as Al, Mg, Mn, V, Ti, Hg, U, Ca, I, Mo, Ni, Be etc.
5. LIBS applications which met the chemical and all nuclear industry bt the quantitative and qualitative remote examination when the physical contacts got preferred.
6. To labelled the target materials the different engineering systems can be used for the aim to designed and assembled every analytical problem, although, the diverse methods used for the case of environmental monitoring.
7. The beam which is used for the material ablation the all LIBS analysis gathered quantitatively even on the very small samples.

Figure 2. Additional LIBS applications.

LASER ABLATION AND PLASMA PHYSICS OF LIBS

To illustrate that how the different number of the emitted photons absorbed or emitted due to the atoms the quantum mechanics plays a vital role to govern about their laser matter interactions. As, when an electron usually absorbs the photon it goes to the higher states, similarly the electron jumps back to the ground level in this decay process electron emits a photon too which called the atomic de-excitation. Although, these various type of the atoms having the various energy in each state or level showing the narrowband emissions representing quantization phenomena so that their uncertainty lies which explain using the Heisenberg uncertainty principle (Cagnac et al., 1975). The emission observed in this case are termed as the spectral emitted lines present in the LIBS spectra whereas it features are usually linked with that of the different energy levels which are well known for each type of the atom (Singh et al., 2007).

Figure 3. Schematic of laser induced breakdown process

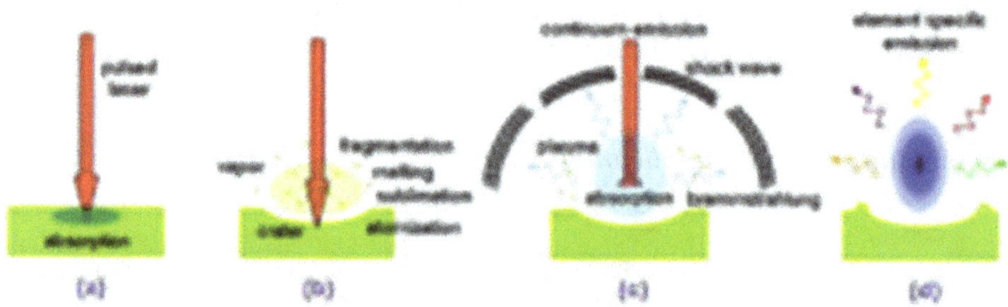

PLASMA EMISSION SPECTRA

The lines emitted from the various kind of the atomic species easily hide by the continuum radiations which usually due to the number of two emerging processes. One of them is known as radiative recombination. In every these type of the recombination process continuum and line photons usually produces which jumps from the upper excited level to the ground state this is called its free transition whereas, the photon of energy which emit have the different type of the wavelengths and energy. Hence, it is seen that this continuum emission normally fleeces the atomically emitted peaks as well, so its beneficial to avoid it (Singh et al.,2007). This emission depends upon the on two factors such as temperature and its density (Angel et al., 2001).

Plasma Temperature

The population of these energy levels follows the specific Boltzmann distribution.

$$\ln(I\,\lambda/g\,A) = -E/k_B T + \ln(4\pi z/hcN_0)$$

Where the term $(4\pi z/hcN_0)$taken constant. And similarly the other terms defined as,

E= Energy level of the upper state for the emission.

Z= it is known as partition function
A= transition probability
I= intensity of the spectral line
g =statistical weight of the energy level
T= temperature
c = speed of the light
h= planks constant
λ=wavelength of the spectral line
$N_{0=}$ total species population
K_B=Boltzmann constant

Figure 4. Laser induced plasma.

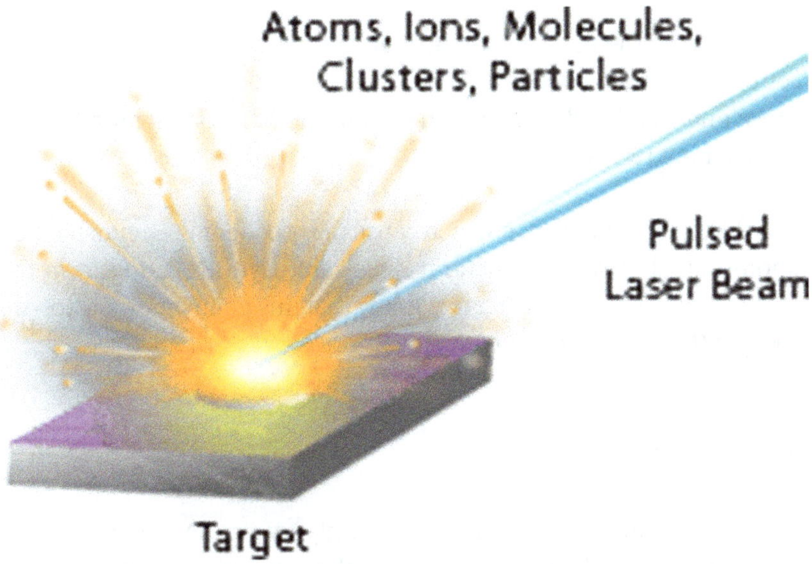

It is analyzed that the temperature was usually realized using Boltzmann plot process. Plasma during the LTE conditions, we draw a plot which normally proportional to intensity vs the energy regarding the upper level emission always gives the straight line and although, the given slope always linked with the temperature.

Hence, **m=-1/KT**

So, it is examined from the obtained graph that it seems to be excellent fit which gives the plasma temperature as well as the better accurate results et al., moreover, to achieve the assumptions related to the Maxwellian electron distribution which always required for the LET conditions too.

Electron Number Density

The electron number density could be determined using a various process such that McWriter criteria, Stark broadening process and also saha-equation. Although the stark Broadening process seems to be most of the accurate and easy one.

Hence, this electron number density summarized as a full width as well as half Max. related to the stark broadening lines given using the following relation as write below.

$$\Lambda\lambda_{1/2}=2w(N_e/10^{16})+ 3.5A(N_e/10^{16})^{1/4}[1-3N_D^{-1/3}/4]w(N_e/10^{16})$$

N_D= The number of the particles in the Debye sphere
Ne= electron number density
W= electron impact related to the width parameters of the stark broadening process
A= ion broadening terms

$2w(N_e/10^{16})$, it is a actual term which cause the broadening occurs by the electron contributions called always the neutral lines same as the second term forwarded towards the ion broadening effect as we know that this contribution is too minimum that the reason this term can be neglected.

PLASMA LIGHT

To collect and focusing the lightthere are mainly two parts which are linked with light acquisition and its focusing. For the enhancement of irradiance during the plasma formation we have to focus the laser spots in it. To escort the spectrometer there are number of devices such as lenses, optical fibers or their combinations which help out in the collection of light emitted from the plasma.

Majority of solid- state laser beam widths lie in the range of 6mm to 8mm, the commonly used lenses which are shows a great act in focusing the laser pulse usually lie in the 25mm to 50mm diameter range et al, as well as their focal lengths always lie in 50 to 150mm. The other LIBS applications such as standoff set-ups need various kind of the lens systems. Although, these lenses provide a large number of transmissions at the time of wavelength measurements so that coating these lenses reduces the maximum number of chances of transmission losses which occur in around about 8 to 10%. By coating the lenses through antireflection their ratio could be reduce up-to 0.5%.

Plasma Light Focusing

To obtain the plasma light a very common set-up used. As, first lens shows vitality in collimating the light in order to increase the focalization effect on the second one which lie on the fiber probe and work well to optimize the ratio of obtained light. So, due to the some of its specific features regarding its multi-lens devices or having one lens as well as fiber probe which lie in that place where spark occurs this LIBS set-up usually adapted to various system by using them. Although, using their various features regarding environmental set-ups the features related to the lenses proved itself useful in this purpose in order to focus plasma light.

CONCLUSION

The configuration of different elements are find out by using some mathematical models without calibration curves as well as reference samples availability and ignore the matrix effect. The accuracy of the result depends on the intensities of spectral lines and temperature of plasma .There are some important parameters of CF-LIBS method such as plasma temperature which is find out on the basis of Boltzmann plot. As the plasma temperature is considered to be same for the different species in the identical sample there are radiations which are emitted from the plasma when laser light falls on it and these radiations are usually depends in the temperature of the atmosphere

REFERENCES

Abdellatif, G., & Imran, H. (2002). A study of the laser plasma parameters at different laser wavelengths. *Spectrochimica Acta. Part B, Atomic Spectroscopy*, *57*(7), 1155–1165. doi:10.1016/S0584-8547(02)00057-5

Aguilera, J. A., Aragón, C., Cristoforetti, G., & Tognoni, E. (2009). Spectrochimica Acta Part B Application of calibration-free laser-induced breakdown spectroscopy to radially resolved spectra from a copper-based alloy laser-induced plasma. *Spectrochimica Acta. Part B, Atomic Spectroscopy*, *64*(7), 685–689. doi:10.1016/j.sab.2009.06.002

Angel, S. M., Stratis, D. N., Eland, K. L., Lai, T., Berg, M. A., & Gold, D. M. (2001). LIBS using dual-and ultra-short laser pulses. *Fresenius' Journal of Analytical Chemistry, 369*(3-4), 320–327. doi:10.1007002160000656 PMID:11293711

Cabalin, L. M., & Laserna, J. J. (1998). Experiment determination of laser induced breakdown thresholds of metals under nanosecond Q-switched laser operation. *Spectrochimica Acta. Part B, Atomic Spectroscopy, 53*(5), 723–730. doi:10.1016/S0584-8547(98)00107-4

Cremers & Radziemski. (2006). *History and fundamental of LIBS. Laser induced break down spectroscopy: Fundamentals and applications*. Academic Press.

Cremers, D. A., Ebinger, M. H., Breshears, D. D., Unkefer, P. J., Kammerdiener, S. A., Ferris, M. J., & Brown, J. R. (2001). Measuring total soil carbon with laser-induced breakdown spectroscopy (LIBS). *Journal of Environmental Quality, 30*(6), 2202–2206. doi:10.2134/jeq2001.2202 PMID:11790033

Mao, X. L., Ciocan, A. C., Borisov, O. V., & Russo, R. E. (1998). Laser ablation processes investigated using inductively coupled plasma-atomic emission spectroscopy(ICP-AES). *Applied Surface Science, 127*, 262–268. doi:10.1016/S0169-4332(97)00787-3

Miziolek, Palleschi, & Schechter. (Eds.). (2006). *Laser induced breakdown spectroscopy*. Cambridge University Press.

Onge, L., Sabsabi, M., & Cielo, P. (1998). Analysis of solids using laser induced breakdown plasma spectroscopy in double-pulse mode. *Spectrochimica Acta. Part B, Atomic Spectroscopy, 53*(3), 407–415. doi:10.1016/S0584-8547(98)00080-9

Russo, R. E., Mao, X. L., Borisov, O. V., & Liu, H. (2000). Influence of wavelength on fractionation in laser ablation ICP-MS. *Journal of Analytical Atomic Spectrometry, 15*(9), 1115–1120. doi:10.1039/b004243i

Singh, J. P., & Thakur, S. N. (Eds.). (2007). *Laser-induced breakdown spectroscopy*. Elsevier.

Compilation of References

Kim, Y. A., Muramatsu, H., Hayashi, T., & Endo, M. (2012). Catalytic metal-free formation of multi-walled carbon nanotubes in atmospheric arc discharge. *Carbon*, *50*(12), 4588–4595. doi:10.1016/j.carbon.2012.05.044

Lu, P., Cullen, P. J., & Ostrikov, K. (2016). Atmospheric Pressure Nonthermal Plasma Sources. In Cold Plasma in Food and Agriculture: Fundamentals and Applications. doi:10.1016/B978-0-12-801365-6.00004-4

Shahzad, A., & He, M.-G. (2012a). Thermal conductivity calculation of complex (dusty) plasmas. *Physics of Plasmas*, *19*(8). doi:10.1063/1.4748526

Cai, X., Cong, H., & Liu, C. (2012). Synthesis of vertically-aligned carbon nanotubes without a catalyst by hydrogen arc discharge. *Carbon*, *50*(8), 2726–2730. doi:10.1016/j.carbon.2012.02.031

Shahzad, A., & He, M.-G. (2014). *Computer Simulation of Complex Plasmas: Molecular Modeling and Elementary Processes in Complex Plasmas* (1st ed.). Scholar's Press.

Kia, K. K., & Bonabi, F. (2012). Electric field induced needle-pulsed arc discharge carbon nanotube production apparatus: Circuitry and mechanical design. *The Review of Scientific Instruments*, *83*(12), 123907. doi:10.1063/1.4772575 PMID:23278005

Shahzad, A., & He, M.-G. (2012b). Thermodynamic Characteristics of Dusty Plasma Studied by Using Molecular Dynamics Simulation. *Plasma Science & Technology*, *14*(9), 771–777. doi:10.1088/1009-0630/14/9/01

Fang, L., Sheng, L., An, K., Yu, L., Ren, W., Ando, Y., & Zhao, X. (2013). Effect of adding W to Fe catalyst on the synthesis of SWCNTs by arc discharge. *Physica E, Low-Dimensional Systems and Nanostructures*, *50*, 116–121. doi:10.1016/j.physe.2013.03.005

Shahzad, A., & He, M.-G. (2012c). Thermal Conductivity of Three-Dimensional Yukawa Liquids (Dusty Plasmas). *Contributions to Plasma Physics*, *52*(8), 667. doi:10.1002/ctpp.201200002

Shahzad, A., & He, M.-G. (2013). Interaction contributions in thermal conductivity of three dimensional complex liquids. AIP Conf. Proc, 173-180. Doi:10.1063/1.4816866

Su, Y., Zhang, Y., Wei, H., Qian, B., Yang, Z., & Zhang, Y. (2012). Length-controlled synthesis of single-walled carbon nanotubes by arc discharge with variable cathode diameters. *Physica E, Low-Dimensional Systems and Nanostructures*, *44*(7-8), 1548–1551. doi:10.1016/j.physe.2012.03.025

Chau, M. Q., Hoang, A. T., Truong, T. T., & Nguyen, X. P. (2020). Endless story about the alarming reality of plastic waste in Vietnam. *Energy Sources. Part A, Recovery, Utilization, and Environmental Effects*, 1–9. doi:10.1080/15567036.2020.1802535

Arora, N., & Sharma, N. N. (2014). Arc discharge synthesis of carbon nanotubes: Comprehensive review. *Diamond and Related Materials*, *50*, 135–150. doi:10.1016/j.diamond.2014.10.001

Berry, T. D., Filley, T. R., & Blanchette, R. A. (2014). Oxidative enzymatic response of white-rot fungi to single-walled carbon nanotubes. *Environmental Pollution*, *193*, 197–204. doi:10.1016/j.envpol.2014.06.013 PMID:25047356

Scalese, S., Scuderi, V., Bagiante, S., Gibilisco, S., Faraci, G., & Privitera, V. (2010). Order and disorder of carbon deposit produced by arc discharge in liquid nitrogen. *Journal of Applied Physics*, *108*(6), 064305. doi:10.1063/1.3475726

He, G., Duan, Y., & Pang, H. (2020). Microwave Absorption of Crystalline Fe/MnO@C Nanocapsules Embedded in Amorphous Carbon. *Nano-Micro Letters*, *12*, 1-16.

Chen, F. F. (2006). *Introduction to plasma physics and controlled fusion* (2nd ed.). New York Springer Verlag.

Matra, K. (2016). Non-thermal Plasma for Germination Enhancement of Radish Seeds. *Procedia Computer Science*, *86*, 132–135. doi:10.1016/j.procs.2016.05.033

Zhao, J., Su, Y., Yang, Z., Wei, L., Wang, Y., & Zhang, Y. (2013). Arc synthesis of doublewalled carbon nanotubes in low pressure air and their superior field emission properties. *Carbon*, *58*, 92–98. doi:10.1016/j.carbon.2013.02.036

Zhao, J., Zhang, J., Su, Y., Yang, Z., Wei, L., & Zhang, Y. (2012). Synthesis of straight multiwalled carbon nanotubes by arc discharge in air and their field emission properties. *Journal of Materials Science*, *47*(18), 6535–6541. doi:10.100710853-012-6583-z

Takekoshi, K., Kizu, T., Aikawa, S., & Nishikawa, E. (2012). One-step synthesis of metalencapsulated carbon nanotubes by pulsed arc discharge in water. *e-Journal of Surface Science and Nanotechnology, 10*, 414-416.

Shukrullah, S., Mohamed, N. M., Naz, M. Y., & Shaharun, M. S. (2015). Synthesis and Characterization of Multiwalled Carbon Nanotubes Using Ferrocene and Aluminum Oxide/Iron Nitrate Catalysts. *Analytical Letters*, *48*(2), 340–348. doi:10.1080/00032719.2014.938342

Nishizaka, H., Namura, M., Motomiya, K., Ogawa, Y., Udagawa, Y., Tohji, K., & Sato, Y. (2011). Influence of carbon structure of the anode on the production of graphite in single-walled carbon nanotube soot synthesized by arc discharge using a Fe–Ni–S catalyst. *Carbon*, *49*(11), 3607–3614. doi:10.1016/j.carbon.2011.04.063

Su, J. S. (2010). Direct route to grow CNTs by micro-electrodischarge machining without catalyst in open air. *Materials and Manufacturing Processes*, *25*(12), 1432–1436. doi:10.1080/10426914.2010.500339

Łabędź, O., Lange, H., Huczko, A., Borysiuk, J., Szybowicz, M., & Bystrzejewski, M. (2009). Influence of carbon structure of the anode on the synthesis of single-walled carbon nanotubes in a carbon arc plasma. *Carbon*, *47*(12), 2847–2854. doi:10.1016/j.carbon.2009.06.031

Qiu, J., Chen, G., Li, Z., & Zhao, Z. (2010). Preparation of double-walled carbon nanotubes from fullerene waste soot by arc-discharge. *Carbon*, *48*(4), 1312–1315. doi:10.1016/j.carbon.2009.01.036

Bahgat, M., Farghali, A. A., El Rouby, W. M. A., & Khedr, M. H. (2011). Synthesis andmodification of multi-walled carbon nano-tubes (MWCNTs) for water treatment applications. *Journal of Analytical and Applied Pyrolysis*, *92*(2), 307–313. doi:10.1016/j.jaap.2011.07.002

Bellan, P. M. (2008). *Fundamentals of plasma physics*. Cambridge University Press.

Lotfy, K. (2017). *Effects of Cold Atmospheric Plasma Jet Treatment on the Seed Germination and Enhancement Growth of Watermelon*. Academic Press.

Ghaleb, Z. A. A., & Jaafar, M. (2018). Carbon Nanotube-Metal Oxide Hybrid Nanocomposites Synthesis and Applications. *1-Dimensional Metal Oxide Nanostructures*, 281-312.

Zhong, B., Wang, C. J., Wen, G. W., Yu, Y. L., & Xia, L. (2018). Facile fabrication of boron and nitrogen co-doped carbon@ Fe_2O_3/Fe_3C/Fe nanoparticle decorated carbon nanotubes three-dimensional structure with excellent microwave absorption properties. *Composites. Part B, Engineering*, *132*, 141–150. doi:10.1016/j.compositesb.2017.09.001

Su, Y., Yang, Z., Wei, H., Kong, E. S. W., & Zhang, Y. (2011). Synthesis of single-walled carbon nanotubes with selective diameter distributions using DC arc discharge under CO mixed atmosphere. *Applied Surface Science*, *257*(7), 3123–3127. doi:10.1016/j.apsusc.2010.10.127

Yousef, S., Khattab, A., Osman, T. A., & Zaki, M. (2013). Effects of increasing electrodes on CNTs yield synthesized by using arc-discharge technique. *Journal of Nanomaterials*. doi:10.1155/2013/392126

Boulos, M. I., Fauchais, P., & Pfender, E. (2013). *Thermal plasmas: fundamentals and applications*. Springer Science & Business Media. doi:10.1007/978-1-1337-1

de Groot, G. J. J. B., Hundt, A., Murphy, A. B., Bange, M. P., & Mai-Prochnow, A. (2018). Cold plasma treatment for cotton seed germination improvement. *Scientific Reports*, *8*(1), 1–10. doi:10.103841598-018-32692-9 PMID:30258075

Yoshinori, A. (2010). Carbon nanotube: Story in ando laboratory. *Journal of Siberian Federal University. Mathematics and Physics*, *3*(1), 3–22.

Shahzad, A., & Bashir, S. (2019a). *Nuclear Medicine Physics*. IntechOpen. doi:10.5772/intechopen.75255

Shukla, P. K., & Mamun, A. A. (2015). *Introduction to dusty plasma physics*. CRC Press.

Tessonnier, J. P., & Su, D. S. (2011). Recent progress on the growth mechanism of carbon nanotubes: A review. *ChemSusChem*, *4*(7), 824–847. doi:10.1002/cssc.201100175 PMID:21732543

Journet, C., Picher, M., & Jourdain, V. (2012). Carbon nanotube synthesis: From large-scale production to atom-by-atom growth. *Nanotechnology*, *23*(14), 142001. doi:10.1088/0957-4484/23/14/142001 PMID:22433510

Shahzad, A., & Bashir, S. (2019b). *Applications of Diagnostic Reference Levels of Standard Doses in Nuclear Medicine. In Nuclear medicine physics*. IntechOpen.

Shahzad, A., Shakoori, M. A., Mao-Gang, H. E., & Feng, Y. (2019). *Numerical Approach to Dynamical Structure Factor of Dusty Plasmas. In Plasma Science and Technology-Basic Fundamentals and Modern Applications*. Intech Open.

Shahzad, A. (2019). *Introductory Chapter: Role of Nuclear Medicine in Medical Science*. IntechOpen.

Shahzad, A., Shakoori, M. A., He, M. G., & Yang, F. (2019). Dynamical structure factor of complex plasmas for varying wave vectors. *Physics of Plasmas*, *26*(2), 023704.

Zhang, Y. (2012). Synthesis of few-walled carbon nanotube–Rh nanoparticles by arc discharge: Effect of selective oxidation. *Materials Characterization*, *68*, 102–109. doi:10.1016/j.matchar.2012.03.013

Moradi, O., Yari, M., Zare, K., Mirza, B., & Najafi, F. (2012). Carbon nanotubes: A review of chemistry principles and reactions. *Fullerenes, Nanotubes, and Carbon Nanostructures*, *20*(2), 138–151. doi:10.1080/1536383X.2010.533312

Sera, B., Stranák, V., Serý, M., Tichý, M., & Spatenka, P. (2008). Germination of Chenopodium album in response to microwave plasma treatment. *Plasma Science & Technology*, *10*(4), 506.

Shahzad, A., Bashir, S., & Anwar, A. (2019). Establishment of age-speciðc reference levels and achievable doses for children and adults undergoing nuclear medicine exams. *Radioprotection*, *54*(3), 187–194. doi:10.1051/radiopro/2019019

Misra, N.N., Schlüter, O., & Cullen, P.J. (2016). *Cold plasma in food and agriculture: Fundamentals and applications*. Academic Press.

Sparavigna, A. (2008). *Plasma treatment advantages for textiles*. arXiv preprint arXiv:0801.3727.

Su, Y., Zhou, P., Zhao, J., Yang, Z., & Zhang, Y. (2013). Large-scale synthesis of few-walled carbon nanotubes by DC arc discharge in low-pressure flowing air. *Materials Research Bulletin*, *48*(9), 3232–3235. doi:10.1016/j.materresbull.2013.04.092

Attri, P., Arora, B., & Choi, E. H. (2013). Utility of Plasma: A New Road from Physics to Chemistry. *Sensors*, *6*(207890), 7.

Zhao, T., Hou, C., Zhang, H., Zhu, R., She, S., Wang, J., & Wei, B. (2014). Electromagnetic wave absorbing properties of amorphous carbon nanotubes. *Scientific Reports*, *4*(1), 1–7. PMID:25007783

Aalipourmohammadi, M., Davodiroknabadi, A., & Nazari, A. (2019). Homogeneous Coatings of Titanium Dioxide Nanoparticles on Corona-Treated Cotton Fabric for Enhanced Self-Cleaning and Antibacterial Properties. *AUTEX Research Journal*, *3*(7), 797–799. doi:10.2478/aut-2019-0058

Abdellatif, G., & Imran, H. (2002). A study of the laser plasma parameters at different laser wavelengths. *Spectrochimica Acta. Part B, Atomic Spectroscopy*, *57*(7), 1155–1165. doi:10.1016/S0584-8547(02)00057-5

Adamovich, I., Baalrud, S. D., Bogaerts, A., Bruggeman, P. J., Cappelli, M., Colombo, V., Czarnetzki, U., Ebert, U., Eden, J. G., Favia, P., Graves, D. B., Hamaguchi, S., Hieftje, G., Hori, M., Kaganovich, I. D., Kortshagen, U., Kushner, M. J., Mason, N. J., Mazouffre, S., ... Vardelle, A. (2017). The 2017 Plasma Roadmap: Low temperature plasma science and technology. *Journal of Physics. D, Applied Physics*, *50*(32), 323001. doi:10.1088/1361-6463/aa76f5

Aguilera, J. A., Aragón, C., Cristoforetti, G., & Tognoni, E. (2009). Spectrochimica Acta Part B Application of calibration-free laser-induced breakdown spectroscopy to radially resolved spectra from a copper-based alloy laser-induced plasma. *Spectrochimica Acta. Part B, Atomic Spectroscopy*, *64*(7), 685–689. doi:10.1016/j.sab.2009.06.002

Agun, L., Ahmad, N., Redzuan, N., Idirs, N. A. S., Taib, S. M., Zakaria, Z., & Ibrahim, R. K. R. (2021). Sterilization of oyster mushroom crop residue substrate by using cold plasma technology. *Materials Today: Proceedings*, *39*, 903–906. doi:10.1016/j.matpr.2020.03.584

Agun, L., Sabirin, S., Ahmad, N., Zakaria, Z., Redzuan, N., & Ibrahim, R. R. (2021, April). The Efficiency of DBD Cold Plasma Pen Treatment on the Oyster Mushroom Bacterial Decontamination. *Journal of Physics: Conference Series*, *1892*(1), 012036. doi:10.1088/1742-6596/1892/1/012036

Alemdar, S., & Agaoglu, S. (2009). Investigation of in vitro antimicrobial activity of aloe vera juice. *Journal of Animal and Veterinary Advances*, *8*(1), 99–102.

Ali, A., Nguyen, N. H. A., Baheti, V., Ashraf, M., Militky, J., Mansoor, T., Noman, M. T., & Ahmad, S. (2018). Electrical conductivity and physiological comfort of silver coated cotton fabrics. *Journal of the Textile Institute*, *109*(5), 620–628. doi:10.1080/00405000.2017.1362148

Ali, S., Moslem, W., Shukla, P., & Kourakis, I. (2007). Fully nonlinear ion-sound waves in a dense Fermi magnetoplasma. *Physics Letters. [Part A]*, *366*(6), 606–610. doi:10.1016/j.physleta.2007.05.073

Ali, S., Shukla, N., Shukla, P. K., & Kourakis, I. (2007). Instability of drift-like waves and cross-field charged particle transport in a nonuniform collisional quantum magnetoplasma. *Europhysics Letters*, *78*(4), 45001–p4. doi:10.1209/0295-5075/78/45001

Ali, S., Terças, H., & Mendonça, J. T. (2011). Nonlocal plasmon excitation in metallic nanostructures. *Physical Review. B*, *83*(15), 153401–153404. doi:10.1103/PhysRevB.83.153401

Ali, S., Ur Rehman, S., & Ding, Z. J. (2018). Stable plasmon excitations in quantum nanowires. *Physics of Plasmas*, *25*(8), 082115–082117. doi:10.1063/1.5041297

Alkawareek, M. Y., Algwari, Q. T., Laverty, G., Gorman, S. P., Graham, W. G., O'Connell, D., & Gilmore, B. F. (2012). Eradication of Pseudomonas aeruginosa Biofilms by Atmospheric Pressure Non-Thermal Plasma. *PLoS One*, *7*(8), e44289. doi:10.1371/journal.pone.0044289 PMID:22952948

Allen, M. A., & Rowlands, G. (1993). Determination of the growth rate for the linearized Zakharov-Kuznetsov equation. *Journal of Plasma Physics*, *50*(3), 413–424. doi:10.1017/S0022377800017244

Anderegg, M., Feuerbacher, B., & Fitton, B. (1971). Optically excited longitudinal plasmons in potassium. *Physical Review Letters*, *27*(23), 1565–1568. doi:10.1103/PhysRevLett.27.1565

Angel, S. M., Stratis, D. N., Eland, K. L., Lai, T., Berg, M. A., & Gold, D. M. (2001). LIBS using dual-and ultra-short laser pulses. *Fresenius' Journal of Analytical Chemistry*, *369*(3-4), 320–327. doi:10.1007002160000656 PMID:11293711

Ang, L. K., Kwan, T. J. T., & Lau, Y. Y. (2003). New scaling of child-Langmuir law in the quantum regime. *Physical Review Letters*, *91*(20), 208303. Advance online publication. doi:10.1103/PhysRevLett.91.208303 PMID:14683407

Anwar, M., Shukrullah, S., Haq, I. U., Saleem, M., AbdEl-Salam, N. M., Ibrahim, K. A., Mohamed, H. F., & Khan, Y. (2021). Ultrasonic Bioconversion of Silver Ions into Nanoparticles with Azadirachta indica Extract and Coating over Plasma-Functionalized Cotton Fabric. *ChemistrySelect*, *6*(8), 1920–1928. doi:10.1002lct.202004623

Attri, P., Arora, B., & Choi, E. H. (2017). Retraction: Utility of plasma: a new road from physics to chemistry. *RSC Advances*, *7*(26), 15735–15735. doi:10.1039/C7RA90034A

Atungulu, G. G., & Pan, Z. (2012). Microbial decontamination of nuts and spices. In Microbial Decontamination in the Food Industry: Novel Methods and Applications. Woodhead Publishing Series in Food Science, Technology and Nutrition. doi:10.1533/9780857095756.1.125

Bafoil, M., Jemmat, A., Martinez, Y., Merbahi, N., Eichwald, O., Dunand, C., & Yousfi, M. (2018). Effects of low temperature plasmas and plasma activated waters on Arabidopsis thaliana germination and growth. *PLoS One*, *13*(4), e0195512. doi:10.1371/journal.pone.0195512 PMID:29630641

Baniya, H. B., Guragain, R. P., Panta, G. P., Safaai, S. S., Dhungana, S., Qin, G., & Subedi, D. P. (2020). Investigation on parameters of atmospheric pressure plasma jet by electrical and optical methods. *Journal of Nepal Physical Society*, *6*(2), 50–56. doi:10.3126/jnphyssoc.v6i2.34857

Barba, F. J., Koubaa, M., Prado-silva, L., Orlien, V., De Souza, A., & Ana, S. (2017). Mild processing applied to the inactivation of the main foodborne bacterial pathogens: A review. *Trends in Food Science & Technology*, *66*, 20–35. doi:10.1016/j.tifs.2017.05.011

Bastien, M., Vaniscotte, A., Combes, B., Umhang, G., Raton, V., Germain, E., Villena, I., Aubert, D., Boué, F., & Poulle, M. L. (2019). Identifying drivers of fox and cat faecal deposits in kitchen gardens in order to evaluate measures for reducing contamination of fresh fruit and vegetables. *Food and Waterborne Parasitology*, *14*, e00034. doi:10.1016/j.fawpar.2018.e00034 PMID:32095604

Bayliss, D. L. (2012). *Understanding the inactivation mechanism of foodborne pathogens using cold atmospheric plasma*. Academic Press.

Becker, S., Zimmermann, J. L., Baumeister, P., Brunner, T. F., Shimizu, T., Li, Y. F., ... Welz, C. (2019). Effects of cold atmospheric plasma (CAP) on bacteria and mucosa of the upper aerodigestive tract. *Auris, Nasus, Larynx*, *46*, 294–301.

Begum, M., & Das, N. (2016). Self-diffusion as a criterion for melting of dust crystal in the presence of magnetic field. *The European Physical Journal Plus*, *131*(2), 1–12. doi:10.1140/epjp/i2016-16046-2

Boddu, S. R., Gutti, V. R., Ghosh, T. K., Tompson, R. V., & Loyalka, S. K. (2011). Gold, silver, and palladium nanoparticle/nano-agglomerate generation, collection, and characterization. *Journal of Nanoparticle Research*, *13*(12), 6591–6601. doi:10.100711051-011-0566-x

Bormashenko, E., Grynyov, R., Bormashenko, Y., & Drori, E. (2012). Cold radiofrequency plasma treatment modifies wettability and germination speed of plant seeds. *Scientific Reports*, *2*(1), 1–8. doi:10.1038rep00741 PMID:23077725

Boufendi, L., & Bouchoule, A. (2002). Industrial developments of scientific insights in dusty plasmas. *Plasma Sources Science & Technology*, *11*(3A), A211–A218. doi:10.1088/0963-0252/11/3A/332

Cabalin, L. M., & Laserna, J. J. (1998). Experiment determination of laser induced breakdown thresholds of metals under nanosecond Q-switched laser operation. *Spectrochimica Acta. Part B, Atomic Spectroscopy*, *53*(5), 723–730. doi:10.1016/S0584-8547(98)00107-4

Cabral, J. A. C., Lapão, L. M., & Mendonça, J. T. (1993). Nonlinear Trivelpiece–Gould waves: Recurrence, harmonic cascade, and sidebands. *Physics of Fluids. B, Plasma Physics*, *5*(3), 787–795. doi:10.1063/1.860975

Castro-Ibáñez, I., Gil, M. I., & Allende, A. (2017). Ready-to-eat vegetables: Current problems and potential solutions to reduce microbial risk in the production chain. *Lebensmittel-Wissenschaft + Technologie*, *85*, 284–292. doi:10.1016/j.lwt.2016.11.073

Chabrier, G., Douchin, F., & Potekhin, A. Y. (2002). Dense astrophysical plasmas. *Journal of Physics Condensed Matter*, *14*(40), 9133–9139. doi:10.1088/0953-8984/14/40/307

Charoux, C. M. G., Free, L., Hinds, L. M., Vijayaraghavan, R. K., Daniels, S., O'Donnell, C. P., & Tiwari, B. K. (2020). Effect of non-thermal plasma technology on microbial inactivation and total phenolic content of a model liquid food system and black pepper grains. *Lebensmittel-Wissenschaft + Technologie*, *118*, 108716. doi:10.1016/j.lwt.2019.108716

Chen, F. F. (1984). *Introduction to Plasma Physics and Controlled Fusion.* Springer. doi:10.1007/978-1-4757-5595-4

Cheng-yang, L., Zhang, N., Guan, B., Zhou, Z., & Mei, F. (2019). Reactive oxygen species are involved in cell death in wheat roots against powdery mildew. *Journal of Integrative Agriculture, 18*(9), 1961–1970. doi:10.1016/S2095-3119(18)62092-1

Chinnammal, S. K., & Arunkumar, K. (2014). Effect of plasma treatment on plain woven cotton fabric. *International Journal of Scientific Research, 20*, 83–88.

Chizoba Ekezie, F.-G., Sun, D.-W., & Cheng, J.-H. (2017). A review on recent advances in cold plasma technology for the food industry: Current applications and future trends. *Trends in Food Science & Technology*, *69*, 46–58. doi:10.1016/j.tifs.2017.08.007

Chu, J. H., & Lin, I. (1994). Direct observation of Coulomb crystals and liquids in strongly coupled rf dusty plasmas. *Physical Review Letters*, *72*(25), 4009–4012. doi:10.1103/PhysRevLett.72.4009 PMID:10056356

Clark, R., Von Domaros, M., Mcintosh, A. J. S., Luzar, A., Kirchner, B., & Welton, T. (2019). Effect of an external electric field on the dynamics and intramolecular structures of ions in an ionic liquid. *The Journal of Chemical Physics, 151*(16), 164503. doi:10.1063/1.5129367 PMID:31675893

Cremers & Radziemski. (2006). *History and fundamental of LIBS. Laser induced break down spectroscopy: Fundamentals and applications*. Academic Press.

Cremers, D. A., Ebinger, M. H., Breshears, D. D., Unkefer, P. J., Kammerdiener, S. A., Ferris, M. J., & Brown, J. R. (2001). Measuring total soil carbon with laser-induced breakdown spectroscopy (LIBS). *Journal of Environmental Quality*, *30*(6), 2202–2206. doi:10.2134/jeq2001.2202 PMID:11790033

Crouseilles, N., Hervieux, P. A., & Manfredi, G. (2008). Quantum hydrodynamic model for the nonlinear electron dynamics in thin metal films. *Physical Review. B*, *78*(15), 155412. Advance online publication. doi:10.1103/PhysRevB.78.155412

Daligault, J. (2012a). Diffusion in ionic mixtures across coupling regimes. *Physical Review Letters*, *108*(22), 1–5. doi:10.1103/PhysRevLett.108.225004 PMID:23003608

Daligault, J. (2012b). Practical model for the self-diffusion coefficient in Yukawa one-component plasmas. *Physical Review E: Statistical, Nonlinear, and Soft Matter Physics*, *86*(4), 1–2. doi:10.1103/PhysRevE.86.047401 PMID:23214710

Dirac, P. A. M. (1930). Note on exchange phenomena in the Thomas atom. *Mathematical Proceedings of the Cambridge Philosophical Society*, *26*(3), 376–385. doi:10.1017/S0305004100016108

Dong, P., Nie, X., Jin, Z., Huang, Z., Wang, X., & Zhang, X. (2019). Dual dielectric barrier discharge plasma treatments for synthesis of Ag–TiO_2 functionalized polypropylene fabrics. *Industrial & Engineering Chemistry Research*, *58*(19), 7734–7741. doi:10.1021/acs.iecr.9b00047

Duan, X., Huang, Y., Cui, Y., Wang, J., & Lieber, C. M. (2001). Indium phosphide nanowires as building blocks for nanoscale electronic and optoelectronic devices. *Nature*, *409*(6816), 66–69. doi:10.1038/35051047 PMID:11343112

Dubey, S. P., Lahtinen, M., & Sillanpää, M. (2010). Tansy fruit mediated greener synthesis of silver and gold nanoparticles. *Process Biochemistry*, *45*(7), 1065–1071. doi:10.1016/j.procbio.2010.03.024

Dzhumagulova, K. N., Ramazanov, T. S., & Masheeva, R. U. (2012). Velocity Autocorrelation Functions and Diffusion Coefficient of Dusty Component in Complex Plasmas. *Contributions to Plasma Physics*, *52*(3), 182–185. doi:10.1002/ctpp.201100070

Ekman, R., Zamanian, J., & Brodin, G. (2015). Exchange corrections in a low-temperature plasma. *Physical Review. E*, *92*(1), 013104. Advance online publication. doi:10.1103/PhysRevE.92.013104 PMID:26274291

Fadhlalmawla, S. A., Mohamed, A. A. H., Almarashi, J. Q., & Boutraa, T. (2019). The impact of cold atmospheric pressure plasma jet on seed germination and seedlings growth of fenugreek (Trigonella foenum-graecum). *Plasma Science & Technology*, *21*(10), 105503. doi:10.1088/2058-6272/ab2a3e

Feizollahi, E., Misra, N. N., & Roopesh, M. S. (2021). Factors influencing the antimicrobial efficacy of Dielectric Barrier Discharge (DBD) Atmospheric Cold Plasma (ACP) in food processing applications. *Critical Reviews in Food Science and Nutrition*, *61*(4), 666–689. doi:10.1080/10408398.2020.1743967 PMID:32208859

Fermi, E. (1927). Un metodo statistico per la determinazione di alcune priorietà dell'atome. *Rend. Accad. Naz. Lincei*, *6*, 602–607.

Fernández, A., & Thompson, A. (2012). The inactivation of Salmonella by cold atmospheric plasma treatment. *Food Research International*, *45*(2), 678–684. doi:10.1016/j.foodres.2011.04.009

Feynman, R. P. (1972). *Statistical Mechanics, A Set of Lectures*. Benjamin.

Frunza, L., Diamandescu, L., Zgura, I., Frunza, S., Ganea, C. P., Negrila, C. C., & Birzu, M. (2018). Photocatalytic activity of wool fabrics deposited at low temperature with ZnO or TiO_2 nanoparticles: Methylene blue degradation as a test reaction. *Catalysis Today*, *306*, 251–259. doi:10.1016/j.cattod.2017.02.044

Gabriel, A. A., Ugay, M. C. C. F., Siringan, M. A. T., Rosario, L. M. D., Tumlos, R. B., & Ramos, H. J. (2016). Atmospheric pressure plasma jet inactivation of Pseudomonas aeruginosa biofilms on stainless steel surfaces. *Innovative Food Science & Emerging Technologies*, *36*, 311–319. doi:10.1016/j.ifset.2016.07.015

Gaminian, H., & Montazer, M. (2015). Enhanced self-cleaning properties on polyester fabric under visible light through single-step synthesis of cuprous oxide doped nano-TiO_2. *Photochemistry and Photobiology*, *91*(5), 1078–1087. doi:10.1111/php.12478 PMID:26073930

Gardner, C. L. (1994). The quantum hydrodynamic model for semiconductor devices. *SIAM Journal on Applied Mathematics*, *54*(2), 409–427. doi:10.1137/S0036139992240425

Giannoglou, M., Xanthou, Z. M., Chanioti, S., Stergiou, P., Christopoulos, M., Dimitrakellis, P., Efthimiadou, A., Gogolides, E., & Katsaros, G. (2021). Effect of cold atmospheric plasma and pulsed electromagnetic fields on strawberry quality and shelf-life. *Innovative Food Science & Emerging Technologies*, *68*, 102631. doi:10.1016/j.ifset.2021.102631

Giaouris, E. E., & Simões, M. V. (2018). *Pathogenic Biofilm Formation in the Food Industry and Alternative Control Strategies.* In A. M. Holban & A. M. Grumezescu (Eds.), *Foodborne Diseases: Handbook of Food Bioengineering* (pp. 309–377). Academic Press.

Gilmore, B. F., Flynn, P. B., O'Brien, S., Hickok, N., Freeman, T., & Bourke, P. (2018). Cold Plasmas for Biofilm Control: Opportunities and Challenges. *Trends in Biotechnology*, *36*(6), 627–638. doi:10.1016/j.tibtech.2018.03.007 PMID:29729997

Gök, V., Aktop, S., Özkan, M., & Tomar, O. (2019). The effects of atmospheric cold plasma on inactivation of Listeria monocytogenes and Staphylococcus aureus and some quality characteristics of pastırma—A dry-cured beef product. *Innovative Food Science and Emerging Technologies, 56*(November), 102188.

Guo, C. A., Tang, F., Chen, J., Wang, X., Zhang, S., & Zhang, X. (2015). Development of dielectric-barrier-discharge ionization. *Analytical and Bioanalytical Chemistry*, *407*(9), 2345–2364. doi:10.100700216-014-8281-y PMID:25510973

Haas, F., Garcia, L. G., Goedert, J., & Manfredi, G. (2003). Quantum ion-acoustic waves. *Physics of Plasmas*, *10*(10), 3858–3866. doi:10.1063/1.1609446

Haas, F., Manfredi, G., & Feix, M. (2000). Multistream model for quantum plasmas. *Physical Review. E*, *62*(2), 2763–2772. doi:10.1103/PhysRevE.62.2763 PMID:11088757

Ha, K. L., Kim, H., Sik, K. W., Jo, C., Kim, J., Hun, S. K., ... Han, W. K. (2016). Evaluation of cold plasma treatments for improved microbial and physicochemical qualities of brown rice. *Lebensmittel-Wissenschaft + Technologie*, *73*, 442–447. doi:10.1016/j.lwt.2016.06.055

Han, L., Patil, S., Boehm, D., Milosavljević, V., Cullen, P. J., & Bourke, P. (2016). Mechanisms of inactivation by high-voltage atmospheric cold plasma differ for Escherichia coli and Staphylococcus aureus. *Applied and Environmental Microbiology*, *82*(2), 450–458. doi:10.1128/AEM.02660-15 PMID:26519396

Han, Y. X., Cheng, J. H., & Sun, D. W. (2019). Changes in activity, structure and morphology of horseradish peroxidase induced by cold plasma. *Food Chemistry*, *301*, 125240. doi:10.1016/j.foodchem.2019.125240 PMID:31387040

Hati, S., Patel, M., & Yadav, D. (2018). Food bioprocessing by non-thermal plasma technology. *Current Opinion in Food Science*, *19*, 85–91. doi:10.1016/j.cofs.2018.03.011

Hebeish, A., El-Naggar, M. E., Fouda, M. M. G., Ramadan, M. A., Al-Deyab, S. S., & El-Rafie, M. H. (2011). Highly effective antibacterial textiles containing green synthesized silver nanoparticles. *Carbohydrate Polymers*, *86*(2), 936–940. doi:10.1016/j.carbpol.2011.05.048

Hertwig, C., Leslie, A., Meneses, N., Reineke, K., Rauh, C., & Schlüter, O. (2017). Inactivation of Salmonella Enteritidis PT30 on the surface of unpeeled almonds by cold plasma. *Innovative Food Science & Emerging Technologies*, *44*, 242–248. doi:10.1016/j.ifset.2017.02.007

Hertwig, C., Meneses, N., & Mathys, A. (2018). Cold atmospheric pressure plasma and low energy electron beam as alternative nonthermal decontamination technologies for dry food surfaces: A review. *Trends in Food Science & Technology*, *77*, 131–142. doi:10.1016/j.tifs.2018.05.011

Hohenberg, P., & Kohn, W. (1964). Inhomogeneous electron gas. *Physical Review*, *136*(3B), B864–B871. doi:10.1103/PhysRev.136.B864

Hollingsworth, S. A., & Dror, R. O. (2018). Molecular Dynamics Simulation for All. *Neuron*, *99*(6), 1129–1143. doi:10.1016/j.neuron.2018.08.011 PMID:30236283

Huang, M., Zhuang, H., Zhao, J., Wang, J., Yan, W., & Zhang, J. (2020). Differences in cellular damages induced by dielectric barrier discharge plasma between Salmonella typhimurium and Staphylococcus aureus. *Bioelectrochemistry (Amsterdam, Netherlands)*, *132*, 107445. doi:10.1016/j.bioelechem.2019.107445 PMID:31918057

Huang, Y., Duan, X., Cui, Y., Lauhon, L. J., Kim, K.-H., & Lieber, C. M. (2001). Logic gates and computation from assembled nanowire building blocks. *Science*, *294*(5545), 1313–1317. doi:10.1126cience.1066192 PMID:11701922

Infeld, E., & Rowlands, G. (2000). *Nonlinear Waves, Solitons and Chaos*. Cambridge University Press. doi:10.1017/CBO9781139171281

Isihara, A. (1968). *Electron Liquids*. Springer Verlag.

Ivlev, A. V., Morfill, G. E., Thomas, H. M., Räth, C., Joyce, G., Huber, P., Kompaneets, R., Fortov, V. E., Lipaev, A. M., Molotkov, V. I., Reiter, T., Turin, M., & Vinogradov, P. (2008). First observation of electrorheological plasmas. *Physical Review Letters*, *100*(9), 1–4. doi:10.1103/PhysRevLett.100.095003 PMID:18352717

Ivlev, A. V., Thoma, M. H., Räth, C., Joyce, G., & Morfill, G. E. (2011). Complex plasmas in external fields: The role of non-hamiltonian interactions. *Physical Review Letters*, *106*(15), 1–4. doi:10.1103/PhysRevLett.106.155001 PMID:21568565

Jiang, J., Lu, Y., Li, J., Li, L., He, X., Shao, H., & Dong, Y. (2014). Effect of seed treatment by cold plasma on the resistance of tomato to Ralstonia solanacearum (bacterial wilt). *PLoS One*, *9*(5), e97753. doi:10.1371/journal.pone.0097753 PMID:24840508

Jiang, Y., Sokorai, K., Pyrgiotakis, G., Demokritou, P., Li, X., Mukhopadhyay, S., Jin, T., & Fan, X. (2017). Cold plasma-activated hydrogen peroxide aerosol inactivates Escherichia coli O157: H7, Salmonella Typhimurium, and Listeria innocua and maintains quality of grape tomato, spinach and cantaloupe ☆. *International Journal of Food Microbiology*, *249*, 53–60. doi:10.1016/j.ijfoodmicro.2017.03.004 PMID:28319798

Ji, S. H., Ki, S. H., Ahn, J. H., Shin, J. H., Hong, E. J., Kim, Y. J., & Choi, E. H. (2018). Inactivation of Escherichia coli and Staphylococcus aureus on contaminated perilla leaves by Dielectric Barrier Discharge (DBD) plasma treatment. *Archives of Biochemistry and Biophysics*, *643*, 32–41. doi:10.1016/j.abb.2018.02.010 PMID:29454864

Johnson, J. C., Choi, H. J., Knutsen, K. P., Schaller, R. D., Yang, P., & Saykally, R. J. (2002). Single gallium nitride nanowire lasers. *Nature Materials*, *1*(2), 106–110. doi:10.1038/nmat728 PMID:12618824

Jones, W. E., Kliewer, K. L., & Fuchs, R. (1969). Nonlocal theory of the optical properties of thin metallic films. *Physical Review*, *178*(3), 1201–1203. doi:10.1103/PhysRev.178.1201

Joshi, H., Kumar, R., & Meena, D. (2018). *Explosion of plasma technology in agriculture*. Academic Press.

Jung, Y. D. (2001). Quantum-mechanical effects on electron–electron scattering in dense high-temperature plasmas. *Physics of Plasmas*, *8*(8), 3842–3844. doi:10.1063/1.1386430

Kandregula, W., Sarwar, Z., Abid, S., Munir, U., & Azeem, A. (2017). Aloe vera leaf gel extract for antibacterial and softness properties of cotton. *Journal of Textile Science & Engineering*, *7*(3), 1–6.

Kanu, R. C., & Shaw, M. T. (1995). Effects of DC and AC Electric Fields on the Response of Electrorheological Fluids Comprising Cylindrical PBZT Particles. *Progress in Electrorheology*, 303–323.

Katewaraphorn, J., & Aldred, A. K. (2016). A study of microcapsules containing Psidium guajava leaf extract for antibacterial agent on cotton fabric. *International Journal of Chemical Engineering and Applications*, *71*(1), 27–41. doi:10.7763/IJCEA.2016.V7.536

Khamsen, N., Onwimol, D., Teerakawanich, N., Dechanupaprittha, S., Kanokbannakorn, W., Hongesombut, K., & Srisonphan, S. (2016). Rice (Oryza sativa L.) Seed Sterilization and Germination Enhancement via Atmospheric Hybrid Nonthermal Discharge Plasma. *ACS Applied Materials & Interfaces*, *8*(30), 19268–19275. doi:10.1021/acsami.6b04555 PMID:27404121

Khan, S. A., & Hassan, S. (2014). Effects of electron exchange-correlation potential on electrostatic oscillations in single-walled carbon nanotubes. *Journal of Applied Physics*, *115*(20), 204304. doi:10.1063/1.4878936

Khan, T. M., Khan, S. U. D., Khan, S. U. D., Ahmad, A., & Abbasi, S. A. (2020). A new strategy of using dielectric barrier discharge plasma in tubular geometry for surface coating and extension to biomedical application. *The Review of Scientific Instruments*, *91*(7), 073902. doi:10.1063/5.0011890 PMID:32752834

Khelissa, S. O., Abdallah, M., Jama, C., Faille, C., & Chihib, N. E. (2017). Bacterial contamination and biofilm formation on abiotic surfaces and strategies to overcome their persistence. *Journal of Materials and Environmental Science*, *8*(9), 3326–3346.

Kljujev, I., Raicevic, V., Jovicic-Petrovic, J., Vujovic, B., Mirkovic, M., & Rothballer, M. (2018). Listeria monocytogenes – Danger for health safety vegetable production. *Microbial Pathogenesis*, *120*, 23–31. doi:10.1016/j.micpath.2018.04.034 PMID:29684542

Kohn, W., & Sham, L. J. (1965). Self-consistent equations including exchange and correlation effects. *Physical Review*, *140*(4A), A1133–A1138. doi:10.1103/PhysRev.140.A1133

Kremp, D., Bornath, T., Bonitz, M., & Schlanges, M. (1999). Quantum kinetic theory of plasmas in strong laser fields. *Physical Review. E*, *60*(4), 4725–4732. doi:10.1103/PhysRevE.60.4725 PMID:11970338

Kremp, D., Schlanges, M., & Kraft, W. D. (2005). *Quantum Statistics of Nonideal Plasmas*. Springer.

Kumar, R., Singh, P., & Kumar, S. (2017). Plasma Treatment–A Tool to Improve Seed Quality–A Review. *Advances in Research*, *12*(4), 1–6. doi:10.9734/AIR/2017/36703

Lafarga, T., Colás-Medà, P., Abadías, M., Aguiló-Aguayo, I., Bobo, G., & Viñas, I. (2019). Strategies to reduce microbial risk and improve quality of fresh and processed strawberries: A review. *Innovative Food Science and Emerging Technologies*, *52*(May), 197–212.

Laroussi, M. (2020). Cold Plasma in Medicine and Healthcare: The New Frontier in Low Temperature Plasma Applications. *Frontiers in Physics*, *8*, 1–7. doi:10.3389/fphy.2020.00074

Laroussi, M., Mendis, D. A., & Rosenberg, M. (2003). Plasma interaction with microbes. *New Journal of Physics*, *5*, 41. doi:10.1088/1367-2630/5/1/341

Lee, H. Z., Hsu, S. L., Liu, M. C., & Wu, C. H. (2001). Effects and mechanisms of aloe-emodin on cell death in human lung squamous cell carcinoma. *European Journal of Pharmacology*, *431*(3), 287–295. doi:10.1016/S0014-2999(01)01467-4 PMID:11730720

Lee, M. R., Eckert, R. D., Forberich, K., Dennler, G., Brabec, C. J., & Gaudiana, R. A. (2009). Solar power wires based on organic photovoltaic materials. *Science*, *324*(5924), 232–235. doi:10.1126cience.1168539 PMID:19286521

Leemans, W. P., Nagler, B., Gonsalves, A. J., Tóth, C., Nakamura, K., Geddes, C. G. R., Esarey, E., Schroeder, C. B., & Hooker, S. M. (2006). GeV electron beams from a centimeter-scale accelerator. *Nature Physics*, *2*(10), 696–699. doi:10.1038/nphys418

Lee, S. M., Song, K. C., & Lee, B. S. (2010). Antibacterial activity of silver nanoparticles prepared by a chemical reduction method. *Korean Journal of Chemical Engineering*, *27*(2), 688–692. doi:10.100711814-010-0067-0

Leroux, F., Campagne, C., Perwuelz, A., & Gengembre, L. (2009). Atmospheric air plasma treatment of polyester textile materials. Textile structure influence on surface oxidation and silicon resin adhesion. *Surface and Coatings Technology*, *203*(20-21), 3178–3183. doi:10.1016/j.surfcoat.2009.03.045

Liao, X., Liu, D., Xiang, Q., Ahn, J., Chen, S., Ye, X., & Ding, T. (2017). Inactivation mechanisms of non-thermal plasma on microbes: A review. *Food Control*, *75*, 83–91. doi:10.1016/j.foodcont.2016.12.021

Lindau, I., & Nilsson, P. (1970). Experimental evidence for excitation of longitudinal plasmons by photons. *Physics Letters. [Part A]*, *31*(7), 352–353. doi:10.1016/0375-9601(70)90978-3

Ling, L., Jiafeng, J., Jiangang, L., Minchong, S., Xin, H., Hanliang, S., & Yuanhua, D. (2014). Effects of cold plasma treatment on seed germination and seedling growth of soybean. *Scientific Reports*, *4*(1), 1–7. PMID:25080862

Liu, M., Guyot-Sionnest, P., Lee, T. W., & Gray, S. K. (2007). Optical properties of rodlike and bipyramidal gold nanoparticles from three-dimensional computations. *Physical Review. B*, *76*(23), 235428. Advance online publication. doi:10.1103/PhysRevB.76.235428

Loffredo, M. I., & Morato, L. M. (1993). On the creation of quantized vortex lines in rotating He II. *Il Nuovo Cimento B*, *108*(2), 205–215. doi:10.1007/BF02874411

López, M., Calvo, T., Prieto, M., Múgica-Vidal, R., Muro-Fraguas, I., Alba-Elías, F., & Alvarez-Ordóñez, A. (2019). A review on non-thermal atmospheric plasma for food preservation: Mode of action, determinants of effectiveness, and applications. *Frontiers in Microbiology*, *10*, 622. doi:10.3389/fmicb.2019.00622 PMID:31001215

Los, A., Ziuzina, D., Boehm, D., Cullen, P. J., & Bourke, P. (2017). The potential of atmospheric air cold plasma for control of bacterial contaminants relevant to cereal grain production. *Innovative Food Science and Emerging Technologies, 44*(2016), 36–45.

Lotfy, K. (2017). Effects of cold atmospheric plasma jet treatment on the seed germination and enhancement growth of watermelon. *Ozean Journal of Applied Sciences*, *7*(12), 705–719. doi:10.4236/ojapps.2017.712050

Lu, H., Patil, S., Keener, K. M., Cullen, P. J., & Bourke, P. (2014). Bacterial inactivation by high-voltage atmospheric cold plasma: Influence of process parameters and effects on cell leakage and DNA. *Journal of Applied Microbiology*, *116*(4), 784–794. doi:10.1111/jam.12426 PMID:24372804

Mahdavian Mehr, H., & Koocheki, A. (2020). Effect of atmospheric cold plasma on structure, interfacial and emulsifying properties of Grass pea (Lathyrus sativus L.) protein isolate. *Food Hydrocolloids*, *106*, 105899. doi:10.1016/j.foodhyd.2020.105899

Mahnot, N. K., Mahanta, C. L., Keener, K. M., & Misra, N. N. (2019). Strategy to achieve a 5-log Salmonella inactivation in tender coconut water using high voltage atmospheric cold plasma (HVACP). *Food Chemistry, 284*, 303–311. doi:10.1016/j.foodchem.2019.01.084 PMID:30744862

Mandal, R., Singh, A., & Pratap Singh, A. (2018). Recent developments in cold plasma decontamination technology in the food industry. *Trends in Food Science & Technology, 80*, 93–103. doi:10.1016/j.tifs.2018.07.014

Manfredi, G. (2005). How to model quantum plasmas. *Fields Inst. Commun., 46*, 263–287.

Manfredi, G., & Haas, F. (2001). Self-consistent fluid model for a quantum electron gas. *Physical Review. B, 64*(7), 075316. Advance online publication. doi:10.1103/PhysRevB.64.075316

Mao, X. L., Ciocan, A. C., Borisov, O. V., & Russo, R. E. (1998). Laser ablation processes investigated using inductively coupled plasma-atomic emission spectroscopy(ICP-AES). *Applied Surface Science, 127*, 262–268. doi:10.1016/S0169-4332(97)00787-3

Markowich, P. A., Ringhofer, C. A., & Schmeiser, C. (1990). *Semiconductor Equations*. Springer-Verlag. doi:10.1007/978-3-7091-6961-2

McMahon, J. M., Gray, S. K., & Schatz, G. C. (2009). Nonlocal optical response of metal nanostructures with arbitrary shape. *Physical Review Letters, 103*(9), 097403. Advance online publication. doi:10.1103/PhysRevLett.103.097403 PMID:19792829

Mehmood, F., Kamal, T., & Ashraf, U. (2018). *Generation and applications of plasma (an academic review)*. Academic Press.

Meng, Y., Qu, G., Wang, T., Sun, Q., Liang, D., & Hu, S. (2017). Enhancement of germination and seedling growth of wheat seed using dielectric barrier discharge plasma with various gas sources. *Plasma Chemistry and Plasma Processing, 37*(4), 1105–1119. doi:10.100711090-017-9799-5

Min, S. C., Hyeon, S., Niemira, B. A., Boyd, G., Sites, J. E., Fan, X., ... Jin, T. Z. (2018). In-package atmospheric cold plasma treatment of bulk grape tomatoes for microbiological safety and preservation ☆. *Food Research International, 108*, 378–386. doi:10.1016/j.foodres.2018.03.033 PMID:29735070

Min, S. C., Hyeon, S., Niemira, B. A., Sites, J. E., Boyd, G., & Lacombe, A. (2016). Dielectric barrier discharge atmospheric cold plasma inhibits Escherichia coli O157: H7, Salmonella, Listeria monocytogenes, and Tulane virus in Romaine lettuce ☆. *International Journal of Food Microbiology, 237*, 114–120. doi:10.1016/j.ijfoodmicro.2016.08.025 PMID:27562348

Mir, S. A., Siddiqui, M. W., Dar, B. N., Shah, M. A., Wani, M. H., Roohinejad, S., ... Ali, A. (2019). Promising applications of cold plasma for microbial safety, chemical decontamination and quality enhancement in fruits. *Journal of Applied Microbiology*, 1–13. doi:10.1111/jam.14541 PMID:31800143

Misra, N. N. (2016). Quality of Cold Plasma Treated Plant Foods. In Cold Plasma in Food and Agriculture: Fundamentals and Applications (pp. 253–271). Elsevier Inc. doi:10.1016/B978-0-12-801365-6.00010-X

Misra, N. N., & Jo, C. (2017). Applications of Cold Plasma Technology for Microbiological Safety in Meat Industry. *Trends in Food Science & Technology*, *64*, 74–86. doi:10.1016/j.tifs.2017.04.005

Misra, N. N., Schlüter, O., & Cullen, P. J. (Eds.). (2016). *Cold plasma in food and agriculture: fundamentals and applications*. Academic Press.

Miziolek, Palleschi, & Schechter. (Eds.). (2006). *Laser induced breakdown spectroscopy*. Cambridge University Press.

Moghimifar, V., Raisi, A., & Aroujalian, A. (2014). Surface modification of polyethersulfone ultrafiltration membranes by corona plasma-assisted coating TiO_2 nanoparticles. *Journal of Membrane Science*, *461*, 69–80. doi:10.1016/j.memsci.2014.02.012

Moisan, M., Barbeau, J., Crevier, M., Pelletier, J., Philip, N., & Saoudi, B. (2002). Plasma sterilization. Methods and mechanisms. *Pure and Applied Chemistry*, *74*(3), 349–358. doi:10.1351/pac200274030349

Montazer, P. H., Pawade, V. B., & Bhanvase, B. A. (2018). Characterization Tools and Techniques for Nanomaterials. Nanomaterials for Green Energy, 16(12), 3331-3334.

Montazer, M., Pakdel, E., & Moghadam, M. B. (2010). Nano titanium dioxide on wool keratin as UV absorber stabilized by butane tetra carboxylic acid (BTCA): A statistical prospect. *Fibers and Polymers*, *11*(7), 967–975. doi:10.100712221-010-0967-y

Moradi, A. (2015). Plasmon modes of metallic nanowires including quantum nonlocal effects. *Physics of Plasmas*, *22*(3), 032112–032115. doi:10.1063/1.4916059

Moslem, W. M., Ali, S., Shukla, P. K., & Eliasson, B. (2008). Three-dimensional electrostatic waves in a nonuniform quantum electron-positron magnetoplasma. *Physics Letters. [Part A]*, *372*(19), 3471–3475. doi:10.1016/j.physleta.2007.09.080

Moslem, W. M., Ali, S., Shukla, P. K., Tang, X. Y., & Rowlands, G. (2007). Solitary, explosive, and periodic solutions of the quantum Zakharov-Kuznetsov equation and its transverse instability. *Physics of Plasmas*, *14*(8), 082308. doi:10.1063/1.2757612

Nabi, G., Raza, W., & Tahir, M. M. (2020). Green Synthesis of TiO_2 Nanoparticle Using Cinnamon Powder Extract and the Study of Optical Properties. *Journal of Inorganic and Organometallic Polymers and Materials*, *30*(4), 1425–1429. doi:10.100710904-019-01248-3

Nalini, K. S. P., & Vijayaraghavan, K. (2020). Green Synthesis of Silver and Gold Nanoparticles using Aloe Vera Gel and Determining its Antimicrobial Properties on Nanoparticle Impregnated Cotton Fabric. *Journal of Nanotechnology Research*, *2*(3), 42–50. doi:10.26502/jnr.2688-85210015

Nantapan, J., Sarapirom, S., & Janpong, K. (2019, November). The effects of atmospheric plasma jet treatment to the germination and enhancement growth of sunflower seeds. *Journal of Physics: Conference Series*, *1380*(1), 012131. doi:10.1088/1742-6596/1380/1/012131

Naz, M. Y., Shukrullah, S., Ghaffar, A., Rehman, N. U., & Sagir, M. (2016). A Low-Frequency Dielectric Barrier Discharge System Design for Textile Treatment. *Synthesis and Reactivity in Inorganic, Metal-Organic, and Nano-Metal Chemistry*, *46*(1), 104–109. doi:10.1080/15533174.2014.900789

Neetoo, H., & Chen, H. (2014). *Alternative Food Processing Technologies.* In S. Clark, S. Jung, & B. Lamsal (Eds.), *Food Processing: Principles and Applications* (pp. 137–169). John Wiley & Sons, Ltd. doi:10.1002/9781118846315.ch7

Ng, S. W., Lu, P., Rulikowska, A., Boehm, D., O'Neill, G., & Bourke, P. (2021). The effect of atmospheric cold plasma treatment on the antigenic properties of bovine milk casein and whey proteins. *Food Chemistry*, *342*, 128283. doi:10.1016/j.foodchem.2020.128283 PMID:33067041

Niemira, B. A. (2012). Cold Plasma Reduction of Salmonella and Escherichia coli O157: H7 on Almonds Using Ambient Pressure Gases. *Journal of Food Science*, *77*(3), 171–175. doi:10.1111/j.1750-3841.2011.02594.x PMID:22384964

Nishime, T. M. C., Borges, A. C., Koga-ito, C. Y., Machida, M., Hein, L. R. O., & Kostov, K. G. (2017). Non-thermal atmospheric pressure plasma jet applied to inactivation of different microorganisms. *Surface and Coatings Technology*, *312*, 19–24. doi:10.1016/j.surfcoat.2016.07.076

Norhayati, M. N., Lee, B. K., Yap, S. S., Thong, K. L., & Yap, S. L. (2016). Cold Plasma Inactivation of Chronic Wound Bacteria. *Archives of Biochemistry and Biophysics*. PMID:27046340

Nunomura, S., Samsonov, D., Zhdanov, S., & Morfill, G. (2006). Self-diffusion in a liquid complex plasma. *Physical Review Letters*, *96*(1), 13–16. doi:10.1103/PhysRevLett.96.015003 PMID:16486468

O'Shea, H., Blacklaws, B. A., Collins, P. J., McKillen, J., & Fitzgerald, R. (2019). Viruses Associated With Foodborne Infections. *Reference Module in Life Sciences*, *2*, 1–17.

Ohta, H., & Hamaguchi, S. (2000). Molecular dynamics evaluation of self-diffusion in Yukawa systems. *Physics of Plasmas*, *7*(11), 4506–4514. doi:10.1063/1.1316084

Ohta, T. (2016). Plasma in Agriculture. In N. Misra, O. Schlüter, & P. Cullen (Eds.), *Cold Plasma in Food and Agriculture: Fundamentals and Applications* (pp. 205–222). Academic Press. doi:10.1016/B978-0-12-801365-6.00008-1

Olatunde, O. O., Benjakul, S., & Vongkamjan, K. (2019). High voltage cold atmospheric plasma: Antibacterial properties and its effect on quality of Asian sea bass slices. *Innovative Food Science & Emerging Technologies*, *52*, 305–312. doi:10.1016/j.ifset.2019.01.011

Onge, L., Sabsabi, M., & Cielo, P. (1998). Analysis of solids using laser induced breakdown plasma spectroscopy in double-pulse mode. *Spectrochimica Acta. Part B, Atomic Spectroscopy*, *53*(3), 407–415. doi:10.1016/S0584-8547(98)00080-9

Ourabah, K., & Tribeche, M. (2013). Quantum ion-acoustic solitary waves: The effect of exchange correlation. *Physical Review. E*, *88*(4), 045101. doi:10.1103/PhysRevE.88.045101 PMID:24229311

Pakdel, E., Daoud, W. A., Sun, L., & Wang, X. (2014). Visible and UV functionality of TiO_2 ternary nanocomposites on cotton. *Applied Surface Science*, *321*, 447–456. doi:10.1016/j.apsusc.2014.10.018

Palaskar, S. S., Desai, A. N., & Shukla, S. R. (2016). Development of multifunctional cotton fabricusing atmospheric pressure plasma and nano-finishing. *Journal of the Textile Institute*, *107*(3), 405–412. doi:10.1080/00405000.2015.1034932

Pamyatnykh, E. A. (2020). Electron quantum waves in metal nanowires. *The Physics of Metals and Metallography*, *121*(5), 405–407. doi:10.1134/S0031918X20050099

Pankaj, S. K., Bueno-ferrer, C., Misra, N. N., Milosavljevic, C., Donnell, C. P. O., Bourke, P., ... Cullen, P. J. (2013). Applications of cold plasma technology in food packaging. *Trends in Food Science & Technology*, 1–13.

Pankaj, S. K., & Kevin, M. (2017). Cold Plasma Applications in Food Packaging. *Reference Module in Food Science*.

Parkes, E. J., & Munro, S. (2005). The stability of obliquely-propagating solitary-wave solutions to Zakharov–Kuznetsov-type equations. *Journal of Plasma Physics*, *71*(05), 695. doi:10.1017/S0022377805003727

Park, Y., Oh, K. S., Oh, J., Seok, D. C., Kim, S. B., Yoo, S. J., & Lee, M. J. (2018). The biological effects of surface dielectric barrier discharge on seed germination and plant growth with barley. *Plasma Processes and Polymers*, *15*(2), 1600056. doi:10.1002/ppap.201600056

Patange, A., Boehm, D., Giltrap, M., Lu, P., Cullen, P. J., & Bourke, P. (2018). Assessment of the disinfection capacity and eco-toxicological impact of atmospheric cold plasma for treatment of food industry effluents. *The Science of the Total Environment*, *631–632*, 298–307. doi:10.1016/j.scitotenv.2018.02.269 PMID:29525709

Pawłat, J., Starek, A., Sujak, A., Terebun, P., Kwiatkowski, M., Budzeń, M., & Andrejko, D. (2018). Effects of atmospheric pressure plasma jet operating with DBD on Lavatera thuringiaca L. seeds' germination. *PLoS One*, *13*(4), e0194349. doi:10.1371/journal.pone.0194349 PMID:29630623

Pawlat, J., Terebun, P., Kwiatkowski, M., Starek, A., Kiczorowski, P., Andrejko, D., & Kopacki, M. (2017, December). Effects of Helium-Air RF plasma jet on onion seeds' germination. In *2017 International Conference on Electromagnetic Devices and Processes in Environment Protection with Seminar Applications of Superconductors (ELMECO & AoS)* (pp. 1-4). IEEE. 10.1109/ELMECO.2017.8267743

Petersilka, M., Gossmann, U. J., & Gross, E. K. U. (1996). Excitation Energies from Time-Dependent Density-Functional Theory. *Physical Review Letters*, *76*(8), 1212–1215. doi:10.1103/PhysRevLett.76.1212 PMID:10061664

Phan, C. Q., Gangal, U., Granick, J. L., Kondeti, V. S. S. K., Hunter, R. C., Jablonowski, H., ... Bruggeman, P. J. (2018). Long-lived and short-lived reactive species produced by a cold atmospheric pressure plasma jet for the inactivation of Pseudomonas aeruginosa and Staphylococcus aureus. *Free Radical Biology & Medicine*, *124*, 275–287. doi:10.1016/j.freeradbiomed.2018.05.083 PMID:29864482

Pignata, C., D'Angelo, D., Fea, E., & Gilli, G. (2017). A review on microbiological decontamination of fresh produce with nonthermal plasma. *Journal of Applied Microbiology*, *122*(6), 1438–1455. doi:10.1111/jam.13412 PMID:28160353

Prasad, P., Mehta, D., Bansal, V., & Sangwan, R. S. (2017). Effect of atmospheric cold plasma (ACP) with its extended storage on the inactivation of Escherichia coli inoculated on tomato. *Food Research International*, *102*, 402–408. doi:10.1016/j.foodres.2017.09.030 PMID:29195965

Qi, K., Xin, J. H., Daoud, W. A., & Mak, C. L. (2007). Functionalizing polyester fiber with a self-cleaning property using anatase TiO_2 and low-temperature plasma treatment. *International Journal of Applied Ceramic Technology*, *4*(6), 554–563. doi:10.1111/j.1744-7402.2007.02168.x

Ramos, B., Brandão, T. R. S., Teixeira, P., & Silva, C. L. M. (2020). Biopreservation approaches to reduce Listeria monocytogenes in fresh vegetables. *Journal of Food Microbiology*, *85*, 103282. doi:10.1016/j.fm.2019.103282 PMID:31500713

Redzuan, N., Rabbi, M. R., Agun, L., Ahmad, N., Zakaria, K., Sirat, R. M., ... Yasak, R. (2020). Development of simultaneous substrate mixer and cold atmospheric plasma treatment for mushroom cultivation. In *AIP Conference Proceedings* (Vol. 2217, pp. 030125–1, 7). AIP Publishing.

Reineke, K., Langer, K., Hertwig, C., Ehlbeck, J., & Schlüter, O. (2015). The impact of different process gas compositions on the inactivation effect of an atmospheric pressure plasma jet on Bacillus spores. *Innovative Food Science & Emerging Technologies*, *30*, 112–118. doi:10.1016/j.ifset.2015.03.019

Rodríguez, Ó., Gomes, W. F., Rodrigues, S., & Fernandes, F. A. N. (2017). Effect of indirect cold plasma treatment on cashew apple juice (Anacardium occidentale L.). *Lebensmittel-Wissenschaft + Technologie*, *84*, 457–463. doi:10.1016/j.lwt.2017.06.010

Rosenfeld, Y. (1977). Relation between the transport coefficients and the internal entropy of simple systems. *Physical Review A*, *15*(6), 2545–2549. doi:10.1103/PhysRevA.15.2545

Rossi, F., Kylian, O., & Hasiwa, M. (2008). Mechanisms of Sterilization and Decontamination of Surfaces by Low-Pressure Plasma. In R. D'Agostino, P. Favia, Y. Kawai, H. Ikegami, N. Sato, & F. Arefi-Khonsari (Eds.), Advanced Plasma Technology (pp. 319–340). Wiley-VCH Verlag GmbH & Co. KGaA.

Russo, R. E., Mao, X. L., Borisov, O. V., & Liu, H. (2000). Influence of wavelength on fractionation in laser ablation ICP-MS. *Journal of Analytical Atomic Spectrometry*, *15*(9), 1115–1120. doi:10.1039/b004243i

Sabillón, L., Stratton, J., Rose, D., & Bianchini, A. (2020). Reduction in pathogenic load of wheat by tempering with saline organic acid solutions at different seasonal temperatures. *International Journal of Food Microbiology*, *313*, 108381. doi:10.1016/j.ijfoodmicro.2019.108381 PMID:31670167

Sabry, R., Moslem, W., & Shukla, P. (2008). Explosive and solitary excitations in a very dense magnetoplasma. *Physics Letters. [Part A]*, *372*(35), 5691–5694. doi:10.1016/j.physleta.2008.06.090

Sahari, N. (2013). *Generation of Homogeneous Glow Discharge Using a Combination of Fine Wire Mesh Perforated Aluminium Alectrode* (Doctoral dissertation). Universiti Teknologi Malaysia.

Sakudo, A. (2017). Recent advances in gas plasma technology for decontamination of food surfaces. In A. M. Grumezescu (Ed.), *Food Preservation: Nanotechnology in the Agri-Food Industry* (pp. 197–228). Acedemic Press. doi:10.1016/B978-0-12-804303-5.00006-7

Sakudo, A., Yagyu, Y., & Onodera, T. (2019). Disinfection and sterilization using plasma technology: Fundamentals and future perspectives for biological applications. *International Journal of Molecular Sciences*, *20*(20), 1–17. doi:10.3390/ijms20205216 PMID:31640211

Sanlisoy, A., & Carpinlioglu, M. O. (2017). A review on plasma gasification for solid waste disposal. *International Journal of Hydrogen Energy*, *42*(2), 1361–1365. doi:10.1016/j.ijhydene.2016.06.008

Schnabel, U., Niquet, R., Krohmann, U., Winter, J., Schlüter, O., Weltmann, K. D., & Ehlbeck, J. (2012). Decontamination of microbiologically contaminated specimen by direct and indirect plasma treatment. *Plasma Processes and Polymers*, *9*(6), 569–575. doi:10.1002/ppap.201100088

Schwabe, M., Khrapak, S. A., Zhdanov, S. K., Pustylnik, M. Y., Räth, C., Fink, M., Kretschmer, M., Lipaev, A. M., Molotkov, V. I., Schmitz, A. S., Thoma, M. H., Usachev, A. D., Zobnin, A. V., Padalka, G. I., Fortov, V. E., Petrov, O. F., & Thomas, H. M. (2020). Slowing of acoustic waves in electrorheological and string-fluid complex plasmas. *New Journal of Physics*, *22*(8), 083079. doi:10.1088/1367-2630/aba91b

Selcuk, M., Oksuz, L., & Basaran, P. (2008). Decontamination of grains and legumes infected with Aspergillus spp. and Penicillum spp. by cold plasma treatment. *Bioresource Technology*, *99*(11), 5104–5109. doi:10.1016/j.biortech.2007.09.076 PMID:17993274

Selvi, B. T., Rajendren, R., Nithyalakshmi, B., & Gayathirignaneswari, S. (2011). Antimicrobial activity of cotton fabric treated with Aloevera extract. *International Journal of Applied Environmental Sciences*, *6*(2), 127–131.

Sen, Y., Onal-Ulusoy, B., & Mutlu, M. (2019). Aspergillus decontamination in hazelnuts: Evaluation of atmospheric and low-pressure plasma technology. *Innovative Food Science & Emerging Technologies*, *54*, 235–242. doi:10.1016/j.ifset.2019.04.014

Shahidi, S., Ahmadi, M., Rashidi, A., & Ghoranneviss, M. (2015). Effect of plasma treatment on self-cleaning of textile fabric using titanium dioxide. *Micro & Nano Letters*, *10*(8), 408–413. doi:10.1049/mnl.2015.0053

Shahriari, S., Monajjemi, M., & Zare, K. (2018). Increasing the efficiency of some antibiotics on penetrating bacteria cell membrane. *Ukrainian Journal of Ecology*, *8*(1), 671–679. doi:10.15421/2018_265

Shah, U., Ranieri, P., Zhou, Y., Schauer, C. L., Miller, V., Fridman, G., & Sekhon, J. K. (2019). Effects of cold plasma treatments on spot-inoculated Escherichia coli O157:H7 and quality of baby kale (Brassica oleracea) leaves. *Innovative Food Science & Emerging Technologies*, *57*, 102104. doi:10.1016/j.ifset.2018.12.010

Shahzad, A., He, M.-G., & Shakoori, M. A. (2017). Thermal Transport and Non-Newtonian behaviors of3D Complex Liquids Using Molecular Simulations. *2017 14th International Bhurban Conference on Applied Sciences and Technology (IBCAST)*, 472–474.

Shahzad, A., Khan, M. Q., Shakoori, M. A., He, M.-G., & Feng, Y. (2020). Thermal Conductivity of Dusty Plasmas through Molecular Dynamics Simulations. *Thermophysical Properties of Complex Materials*, 13.

Shahzad, A., Shakoori, M. A., & He, M.-G. (2020). Wave Spectra in Dusty Plasmas of Nuclear Fusion Devices. *Fusion Energy*, 13. https://www.intechopen.com/books/advanced-biometric-technologies/liveness-detection-in-biometrics

Shahzad, A., Aslam, A., & He, M.-G. (2014). Equilibrium molecular dynamics simulation of shear viscosity of two-dimensional complex (dusty) plasmas. *Radiation Effects and Defects in Solids*, *169*(11), 931–941. doi:10.1080/10420150.2014.968852

Shahzad, A., & He, M.-G. (2012c). Shear viscosity and diffusion motion of two-dimensional dusty plasma liquids. *Physica Scripta*, *86*(1), 015502. doi:10.1088/0031-8949/86/01/015502

Shahzad, A., & He, M.-G. (2015a). Nonlinear Effects of Dusty Plasmas using Homogenous Nonequilibrium Molecular Dynamics Simulations. *Journal of Applied Nonlinear Dynamics*, *4*(3), 259–265. doi:10.5890/JAND.2015.09.006

Shahzad, A., & He, M.-G. (2015b). Calculations of thermal conductivity of complex (dusty) plasmas using homogenous nonequilibrium molecular simulations. *Radiation Effects and Defects in Solids*, *170*(9), 758–770. doi:10.1080/10420150.2015.1108316

Shahzad, A., & He, M.-G. (2015c). Numerical experiment of thermal conductivity in two-dimensional Yukawa liquids. *Physics of Plasmas*, *22*(12), 123707. doi:10.1063/1.4938275

Shahzad, A., He, M.-G., Haider, S. I., & Feng, Y. (2017). Studies of force field effects on thermal conductivity of complex plasmas. *Physics of Plasmas*, *24*(9), 093701. doi:10.1063/1.4993992

Shahzad, A., He, M.-G., & He, K. (2013). Diffusion motion of two-dimensional weakly coupled complex (dusty) plasmas. *Physica Scripta*, *87*(3), 035501. doi:10.1088/0031-8949/87/03/035501

Shahzad, A., Shakoori, M. A., He, M.-G., & Bashir, S. (2018). *Sound Waves in Complex (Dusty) Plasmas*. In *Computational and Experimental Studies of Acoustic Waves*. Intech Open. doi:10.5772/intechopen.71203

Shahzad, A., Shakoori, M. A., He, M.-G., & Yan, F. (2020). *Numerical Approach to Dynamical Structure Factor of Dusty Plasmas*. Intechopen. doi:10.5772/intechopen.78334

Shen, J., Zhang, H., Xu, Z., Zhang, Z., Cheng, C., Ni, G., Lan, Y., Meng, Y., Xia, W., & Chu, P. K. (2019). Preferential production of reactive species and bactericidal efficacy of gas-liquid plasma discharge. *Chemical Engineering Journal*, *362*, 402–412. doi:10.1016/j.cej.2019.01.018

Shintani, H., Sakudo, A., Burke, P., & Mcdonnell, G. (2010). Gas plasma sterilization of microorganisms and mechanisms of action [Review]. *Experimental and Therapeutic Medicine*, *1*(5), 731–738. doi:10.3892/etm.2010.136 PMID:22993596

Shpatakovskaya, G. V. (2006). Semiclassical model of a one-dimensional quantum dot. *Journal of Experimental and Theoretical Physics*, *102*(3), 466–474. doi:10.1134/S1063776106030095

Shukla, P. K., & Akbari-Moghanjoughi, M. (2013). Hydrodynamic theory for ion structure and stopping power in quantum plasmas. *Physical Review. E*, *87*(4), 043106. Advance online publication. doi:10.1103/PhysRevE.87.043106 PMID:23679529

Shukla, P. K., Ali, S., Stenflo, L., & Marklund, M. (2006). Nonlinear wave interactions in quantum magnetoplasmas. *Physics of Plasmas*, *13*(11), 112111–112116. doi:10.1063/1.2390688

Shukla, P. K., & Eliasson, B. (2006). Formation and Dynamics of Dark Solitons and Vortices in Quantum Electron Plasmas. *Physical Review Letters*, *96*(24), 245001. Advance online publication. doi:10.1103/PhysRevLett.96.245001 PMID:16907249

Shukla, P., Brodin, G., Marklund, M., & Stenflo, L. (2009). Excitation of multiple wakefields by short laser pulses in quantum plasmas. *Physics Letters. [Part A]*, *373*(35), 3165–3168. doi:10.1016/j.physleta.2009.06.053

Singh, J. P., & Thakur, S. N. (Eds.). (2007). *Laser-induced breakdown spectroscopy*. Elsevier.

Sinica, D. P., Mathur, A., Kushwaha, A., Dalakoti, V., & Dalakoti, G. (2014). Pelagia Research Library Green synthesis of silver nanoparticles using medicinal plant and its characterization. *Der Pharmacia Sinica*, *5*, 118–122.

Spreiter, Q., & Walter, M. (1999). Classical molecular dynamics simulation with the velocity verlet algorithm at strong external magnetic fields. *Journal of Computational Physics*, *152*(1), 102–119. doi:10.1006/jcph.1999.6237

Sukhinin, G. I., Fedoseev, A. V., Khokhlov, R. O., & Suslov, S. Y. U. (2012). Plasma Polarization Around Dust Particle in an External Electric Field. *Contributions to Plasma Physics*, *52*(1), 62–65. doi:10.1002/ctpp.201100063

Sukhinin, G. I., Fedoseev, A. V., Salnikov, M. V., Rostom, A., Vasiliev, M. M., & Petrov, O. F. (2017). Plasma anisotropy around a dust particle placed in an external electric field. *Physical Review. E*, *95*(6), 1–8. doi:10.1103/PhysRevE.95.063207 PMID:28709289

Tanino, T., Arisaka, T., Iguchi, Y., Matsui, M., & Ohshima, T. (2019). Inactivation of Aspergillus sp. spores on whole black peppers by nonthermal plasma and quality evaluation of the treated peppers. *Food Control*, *97*, 94–99. doi:10.1016/j.foodcont.2018.10.023

Tappi, S., Gozzi, G., Vannini, L., Berardinelli, A., Romani, S., Ragni, L., & Rocculi, P. (2016). Cold plasma treatment for fresh-cut melon stabilization. *Innovative Food Science & Emerging Technologies*, *33*, 225–233. doi:10.1016/j.ifset.2015.12.022

Terças, H., Mendonça, J. T., & Shukla, P. K. (2008). Quantum Trivelpiece–Gould waves in a magnetized dense plasma. *Physics of Plasmas*, *15*(7), 072109. doi:10.1063/1.2947235

Thomas, L. H. (1927). The calculation of atomic fields. *Mathematical Proceedings of the Cambridge Philosophical Society*, *23*(5), 542–548. doi:10.1017/S0305004100011683

Ukuku, D. O., Niemira, B. A., & Ukanalis, J. (2019). LWT - Food Science and Technology Nisin-based antimircobial combination with cold plasma treatment inactivate Listeria monocytogenes on Granny Smith apples ☆. *Lebensmittel-Wissenschaft + Technologie*, *104*, 120–127. doi:10.1016/j.lwt.2018.12.049

Usta, Y. H., Çukur, E., Yıldırım, Ç., & Ercan, U. K. (2019). Design of a portable, battery-powered non-thermal atmospheric plasma device and characterization of its antibacterial efficacies. *Journal of Electrostatics*, *99*, 1–8. doi:10.1016/j.elstat.2019.03.002

Vasickova, P., Dvorska, L., Lorencova, A., & Pavlik, I. (2005). Viruses as a cause of foodborne diseases: A review of the literature. *Veterinarni Medicina*, *50*(3), 89–104. doi:10.17221/5601-VETMED

Vaulina, O. S., Nefedov, A. P., Petrov, O. F., & Fortov, V. E. (2001). Transport properties of macroparticles in dust plasma induced by solar radiation. *Journal of Experimental and Theoretical Physics*, *92*(6), 979–985. doi:10.1134/1.1385637

Vigneshwaran, N., Kumar, S., Kathe, A. A., Varadarajan, P. V., & Prasad, V. (2006). Functional finishing of cotton fabrics using zinc oxide–soluble starch nanocomposites. *Nanotechnology*, *20*(20), 3087–3099. doi:10.1088/0957-4484/17/20/008

von Roos, O., & Zmuidzinas, J. S. (1961). Effect of electron exchange on the dispersion relation of plasmons. *Physical Review*, *121*(4), 941–942. doi:10.1103/PhysRev.121.941

Vu, N. K., Zille, A., Oliveira, F. R., Carneiro, N., & Souto, A. P. (2013). Effect of particle size on silver nanoparticle deposition onto dielectric barrier discharge (DBD) plasma functionalized polyamide fabric. *Plasma Processes and Polymers*, *10*(3), 285–296. doi:10.1002/ppap.201200089

Wang, Y., Wang, T., Yuan, Y., Fan, Y., Guo, K., & Yue, T. (2018). Inactivation of yeast in apple juice using gas-phase surface discharge plasma treatment with a spray reactor. *Lebensmittel-Wissenschaft + Technologie*, *97*, 530–536. doi:10.1016/j.lwt.2018.07.049

Wei, L., & Wang, Y.-N. (2007). Quantum Ion-Acoustic Waves in Single-Walled Carbon Nanotubes Studied with a Quantum Hydrodynamic Model. *Physical Review. B*, *75*(19), 193407. doi:10.1103/PhysRevB.75.193407

Weizsäcker, C. F. V. (1935). Zur Theorie der Kernmassen. *Zeitschrift für Physik*, *96*(7–8), 431–458. doi:10.1007/BF01337700

Willets, K. A., & Van Duyne, R. P. (2007). Localized surface plasmon resonance spectroscopy and sensing. *Annual Review of Physical Chemistry*, *58*(1), 267–297. doi:10.1146/annurev.physchem.58.032806.104607 PMID:17067281

Xia, N., Cai, Y., Jiang, T., & Yao, J. (2011). Green synthesis of silver nanoparticles by chemical reduction with hyaluronan. *Carbohydrate Polymers*, *86*(2), 956–961. doi:10.1016/j.carbpol.2011.05.053

Xiang, J., Lu, W., Hu, Y., Wu, Y., Yan, H., & Lieber, C. M. (2006). Ge/Si nanowire heterostructures as high-performance field-effect transistors. *Nature*, *441*(7092), 489–493. doi:10.1038/nature04796 PMID:16724062

Xu, Y., Tian, Y., Ma, R., Liu, Q., & Zhang, J. (2016). Effect of plasma activated water on the postharvest quality of button mushrooms, Agaricus bisporus. *Food Chemistry*, *197*, 436–444. doi:10.1016/j.foodchem.2015.10.144 PMID:26616972

Yeon, M., Jo, S., & Min, S. C. (2017). Mandarin preservation by microwave-powered cold plasma treatment. *Innovative Food Science & Emerging Technologies*, *39*, 25–32. doi:10.1016/j.ifset.2016.10.021

Zainal, M. N. F., Redzuan, N., & Misnal, M. F. I. (2015). Brief Review: Cold Plasma. *Jurnal Teknologi*, *10*, 57–61.

Zakharov, V. E., & Kuznetsov, E. A. (1974). Three-dimensional solitons. *Soviet Physics, JETP*, *39*, 285–286.

Zhang, H., Ma, D., Qiu, R., Tang, Y., & Du, C. (2017). Non-thermal plasma technology for organic contaminated soil remediation: A review. *Chemical Engineering Journal*, *313*, 157–170. doi:10.1016/j.cej.2016.12.067

Zhang, X., Liu, D., Zhou, R., Song, Y., Sun, Y., Zhang, Q., Niu, J., Fan, H., & Yang, S. Z. (2014). Atmospheric cold plasma jet for plant disease treatment. *Applied Physics Letters*, *104*(4), 043702. doi:10.1063/1.4863204

Zhao, N., Ge, L., Huang, Y., Wang, Y., Wang, Y., Lai, H., Wang, Y., Zhu, Y., & Zhang, J. (2020). Impact of cold plasma processing on quality parameters of packaged fermented vegetable (radish paocai) in comparison with pasteurization processing: Insight into safety and storage stability of products. *Innovative Food Science & Emerging Technologies*, *60*, 102300. doi:10.1016/j.ifset.2020.102300

Zhu, X. G., Long, S. P., & Ort, D. R. (2010). Improving photosynthetic efficiency for greater yield. *Annual Review of Plant Biology*, *61*(1), 235–261. doi:10.1146/annurev-arplant-042809-112206 PMID:20192734

Ziuzina, D., Misra, N. N., Han, L., Cullen, P. J., Moiseev, T., Mosnier, J. P., Keener, K., Gaston, E., Vilaró, I., & Bourke, P. (2020). Investigation of a large gap cold plasma reactor for continuous in-package decontamination of fresh strawberries and spinach. *Innovative Food Science & Emerging Technologies*, *59*, 102229. doi:10.1016/j.ifset.2019.102229

Ziuzina, D., Patil, S., Cullen, P. J., Keener, K. M., & Bourke, P. (2014). Atmospheric cold plasma inactivation of Escherichia coli, Salmonella enterica serovar Typhimurium and Listeria monocytogenes inoculated on fresh produce. *Food Microbiology*, *42*, 109–116. doi:10.1016/j.fm.2014.02.007 PMID:24929725

Related References

To continue our tradition of advancing academic research, we have compiled a list of recommended IGI Global readings. These references will provide additional information and guidance to further enrich your knowledge and assist you with your own research and future publications.

Abed, S., Khir, T., & Ben Brahim, A. (2016). Thermodynamic and Energy Study of a Regenerator in Gas Turbine Cycle and Optimization of Performances. *International Journal of Energy Optimization and Engineering*, *5*(2), 25–44. doi:10.4018/IJEOE.2016040102

Abu Bakar, W. A., Abdullah, W. N., Ali, R., & Mokhtar, W. N. (2016). Polymolybdate Supported Nano Catalyst for Desulfurization of Diesel. In T. Saleh (Ed.), *Applying Nanotechnology to the Desulfurization Process in Petroleum Engineering* (pp. 263–280). Hershey, PA: IGI Global. doi:10.4018/978-1-4666-9545-0.ch009

Addo-Tenkorang, R., Helo, P., & Kantola, J. (2016). Engineer-To-Order Product Development: A Communication Network Analysis for Supply-Chain's Sustainable Competitive Advantage. In R. Addo-Tenkorang, J. Kantola, P. Helo, & A. Shamsuzzoha (Eds.), *Supply Chain Strategies and the Engineer-to-Order Approach* (pp. 43–59). Hershey, PA: IGI Global. doi:10.4018/978-1-5225-0021-6.ch003

Adebiyi, I. D., Popoola, P. A., & Pityana, S. (2016). Mitigation of Wear Damage by Laser Surface Alloying Technique. In E. Akinlabi, R. Mahamood, & S. Akinlabi (Eds.), *Advanced Manufacturing Techniques Using Laser Material Processing* (pp. 172–196). Hershey, PA: IGI Global. doi:10.4018/978-1-5225-0329-3.ch007

Ahmad, W. (2016). Sulfur in Petroleum: Petroleum Desulfurization Techniques. In T. Saleh (Ed.), *Applying Nanotechnology to the Desulfurization Process in Petroleum Engineering* (pp. 1–52). Hershey, PA: IGI Global. doi:10.4018/978-1-4666-9545-0.ch001

Ahmed, I., Ahmad, N., Mehmood, I., Haq, I. U., Hassan, M., & Khan, M. U. (2016). Applications of Nanotechnology in Transportation Engineering. In A. Khitab & W. Anwar (Eds.), *Advanced Research on Nanotechnology for Civil Engineering Applications* (pp. 180–207). Hershey, PA: IGI Global. doi:10.4018/978-1-5225-0344-6.ch006

Aikhuele, D. (2018). A Study of Product Development Engineering and Design Reliability Concerns. *International Journal of Applied Industrial Engineering*, *5*(1), 79–89. doi:10.4018/IJAIE.2018010105

Al-Najar, B. T., & Bououdina, M. (2016). Bioinspired Nanoparticles for Efficient Drug Delivery System. In M. Bououdina (Ed.), *Emerging Research on Bioinspired Materials Engineering* (pp. 69–103). Hershey, PA: IGI Global. doi:10.4018/978-1-4666-9811-6.ch003

Al-Shebeeb, O. A., Rangaswamy, S., Gopalakrishan, B., & Devaru, D. G. (2017). Evaluation and Indexing of Process Plans Based on Electrical Demand and Energy Consumption. *International Journal of Manufacturing, Materials, and Mechanical Engineering*, *7*(3), 1–19. doi:10.4018/IJMMME.2017070101

Alexakis, H., & Makris, N. (2016). Validation of the Discrete Element Method for the Limit Stability Analysis of Masonry Arches. In V. Sarhosis, K. Bagi, J. Lemos, & G. Milani (Eds.), *Computational Modeling of Masonry Structures Using the Discrete Element Method* (pp. 292–325). Hershey, PA: IGI Global. doi:10.4018/978-1-5225-0231-9.ch012

AlMegren, H. A., Gonzalez-Cortes, S., Huang, Y., Chen, H., Qian, Y., Alkinany, M., ... Xiao, T. (2016). Preparation of Deep Hydrodesulfurzation Catalysts for Diesel Fuel using Organic Matrix Decomposition Method. In H. Al-Megren & T. Xiao (Eds.), *Petrochemical Catalyst Materials, Processes, and Emerging Technologies* (pp. 216–253). Hershey, PA: IGI Global. doi:10.4018/978-1-4666-9975-5.ch009

Alshammari, A., Kalevaru, V. N., Bagabas, A., & Martin, A. (2016). Production of Ethylene and its Commercial Importance in the Global Market. In H. Al-Megren & T. Xiao (Eds.), *Petrochemical Catalyst Materials, Processes, and Emerging Technologies* (pp. 82–115). Hershey, PA: IGI Global. doi:10.4018/978-1-4666-9975-5.ch004

Amel, M. (2016). Synthesis, Characterizations, and Biological Effects Study of Some Quinoline Family. In M. Bououdina (Ed.), *Emerging Research on Bioinspired Materials Engineering* (pp. 160–196). Hershey, PA: IGI Global. doi:10.4018/978-1-4666-9811-6.ch006

Amna, T., Haasan, M. S., Khil, M., & Hwang, I. (2016). Impact of Electrospun Biomimetic Extracellular Environment on Proliferation and Intercellular Communication of Muscle Precursor Cells: An Overview – Intercellular Communication of Muscle Precursor Cells with Extracellular Environment. In M. Bououdina (Ed.), *Emerging Research on Bioinspired Materials Engineering* (pp. 247–265). Hershey, PA: IGI Global. doi:10.4018/978-1-4666-9811-6.ch009

Amuda, M. O., Lawal, T. F., & Akinlabi, E. T. (2017). Research Progress on Rheological Behavior of AA7075 Aluminum Alloy During Hot Deformation. *International Journal of Materials Forming and Machining Processes*, *4*(1), 53–96. doi:10.4018/IJMFMP.2017010104

An, M., & Qin, Y. (2016). Challenges of Railway Safety Risk Assessment and Maintenance Decision Making. In B. Rai (Ed.), *Handbook of Research on Emerging Innovations in Rail Transportation Engineering* (pp. 173–211). Hershey, PA: IGI Global. doi:10.4018/978-1-5225-0084-1.ch009

Anil, M., Ayyildiz-Tamis, D., Tasdemir, S., Sendemir-Urkmez, A., & Gulce-Iz, S. (2016). Bioinspired Materials and Biocompatibility. In M. Bououdina (Ed.), *Emerging Research on Bioinspired Materials Engineering* (pp. 294–322). Hershey, PA: IGI Global. doi:10.4018/978-1-4666-9811-6.ch011

Armutlu, H. (2018). Intelligent Biomedical Engineering Operations by Cloud Computing Technologies. In U. Kose, G. Guraksin, & O. Deperlioglu (Eds.), *Nature-Inspired Intelligent Techniques for Solving Biomedical Engineering Problems* (pp. 297–317). Hershey, PA: IGI Global. doi:10.4018/978-1-5225-4769-3.ch015

Arokiyaraj, S., Saravanan, M., Bharanidharan, R., Islam, V. I., Bououdina, M., & Vincent, S. (2016). Green Synthesis of Metallic Nanoparticles Using Plant Compounds and Their Applications: Metallic Nanoparticles Synthesis Using Plants. In M. Bououdina (Ed.), *Emerging Research on Bioinspired Materials Engineering* (pp. 1–34). Hershey, PA: IGI Global. doi:10.4018/978-1-4666-9811-6.ch001

Atik, M., Sadek, M., & Shahrour, I. (2017). Single-Run Adaptive Pushover Procedure for Shear Wall Structures. In V. Plevris, G. Kremmyda, & Y. Fahjan (Eds.), *Performance-Based Seismic Design of Concrete Structures and Infrastructures* (pp. 59–83). Hershey, PA: IGI Global. doi:10.4018/978-1-5225-2089-4.ch003

Aydin, A., Akyol, E., Gungor, M., Kaya, A., & Tasdelen, S. (2018). Geophysical Surveys in Engineering Geology Investigations With Field Examples. In N. Ceryan (Ed.), *Handbook of Research on Trends and Digital Advances in Engineering Geology* (pp. 257–280). Hershey, PA: IGI Global. doi:10.4018/978-1-5225-2709-1.ch007

Azevedo, N. M., Lemos, J. V., & Rocha de Almeida, J. (2016). Discrete Element Particle Modelling of Stone Masonry. In V. Sarhosis, K. Bagi, J. Lemos, & G. Milani (Eds.), *Computational Modeling of Masonry Structures Using the Discrete Element Method* (pp. 146–170). Hershey, PA: IGI Global. doi:10.4018/978-1-5225-0231-9.ch007

Bamufleh, H. S., Noureldin, M. M., & El-Halwagi, M. M. (2016). Sustainable Process Integration in the Petrochemical Industries. In H. Al-Megren & T. Xiao (Eds.), *Petrochemical Catalyst Materials, Processes, and Emerging Technologies* (pp. 150–163). Hershey, PA: IGI Global. doi:10.4018/978-1-4666-9975-5.ch006

Banerjee, S., Gautam, R. K., Gautam, P. K., Jaiswal, A., & Chattopadhyaya, M. C. (2016). Recent Trends and Advancement in Nanotechnology for Water and Wastewater Treatment: Nanotechnological Approach for Water Purification. In A. Khitab & W. Anwar (Eds.), *Advanced Research on Nanotechnology for Civil Engineering Applications* (pp. 208–252). Hershey, PA: IGI Global. doi:10.4018/978-1-5225-0344-6.ch007

Bas, T. G. (2017). Nutraceutical Industry with the Collaboration of Biotechnology and Nutrigenomics Engineering: The Significance of Intellectual Property in the Entrepreneurship and Scientific Research Ecosystems. In T. Bas & J. Zhao (Eds.), *Comparative Approaches to Biotechnology Development and Use in Developed and Emerging Nations* (pp. 1–17). Hershey, PA: IGI Global. doi:10.4018/978-1-5225-1040-6.ch001

Beale, R., & André, J. (2017). *Design Solutions and Innovations in Temporary Structures*. Hershey, PA: IGI Global. doi:10.4018/978-1-5225-2199-0

Behnam, B. (2017). Simulating Post-Earthquake Fire Loading in Conventional RC Structures. In P. Samui, S. Chakraborty, & D. Kim (Eds.), *Modeling and Simulation Techniques in Structural Engineering* (pp. 425–444). Hershey, PA: IGI Global. doi:10.4018/978-1-5225-0588-4.ch015

Ben Hamida, I., Salah, S. B., Msahli, F., & Mimouni, M. F. (2018). Distribution Network Reconfiguration Using SPEA2 for Power Loss Minimization and Reliability Improvement. *International Journal of Energy Optimization and Engineering*, *7*(1), 50–65. doi:10.4018/IJEOE.2018010103

Benjamin, S. R., de Lima, F., & Rathoure, A. K. (2016). Genetically Engineered Microorganisms for Bioremediation Processes: GEMs for Bioremediaton. In A. Rathoure & V. Dhatwalia (Eds.), *Toxicity and Waste Management Using Bioremediation* (pp. 113–140). Hershey, PA: IGI Global. doi:10.4018/978-1-4666-9734-8.ch006

Bhaskar, S. V., & Kudal, H. N. (2017). Effect of TiCN and AlCrN Coating on Tribological Behaviour of Plasma-nitrided AISI 4140 Steel. *International Journal of Surface Engineering and Interdisciplinary Materials Science*, *5*(2), 1–17. doi:10.4018/IJSEIMS.2017070101

Bhowmik, S., Sahoo, P., Acharyya, S. K., Dhar, S., & Chattopadhyay, J. (2016). Effect of Microstructure Degradation on Fracture Toughness of 20MnMoNi55 Steel in DBT Region. *International Journal of Manufacturing, Materials, and Mechanical Engineering*, *6*(3), 11–27. doi:10.4018/IJMMME.2016070102

Bhutto, A. W., Abro, R., Abbas, T., Yu, G., & Chen, X. (2016). Desulphurization of Fuel Oils Using Ionic Liquids. In H. Al-Megren & T. Xiao (Eds.), *Petrochemical Catalyst Materials, Processes, and Emerging Technologies* (pp. 254–284). Hershey, PA: IGI Global. doi:10.4018/978-1-4666-9975-5.ch010

Bhuyan, D. (2018). Designing of a Twin Tube Shock Absorber: A Study in Reverse Engineering. In K. Kumar & J. Davim (Eds.), *Design and Optimization of Mechanical Engineering Products* (pp. 83–104). Hershey, PA: IGI Global. doi:10.4018/978-1-5225-3401-3.ch005

Bouloudenine, M., & Bououdina, M. (2016). Toxic Effects of Engineered Nanoparticles on Living Cells. In M. Bououdina (Ed.), *Emerging Research on Bioinspired Materials Engineering* (pp. 35–68). Hershey, PA: IGI Global. doi:10.4018/978-1-4666-9811-6.ch002

Brunetti, A., Sellaro, M., Drioli, E., & Barbieri, G. (2016). Membrane Engineering and its Role in Oil Refining and Petrochemical Industry. In H. Al-Megren & T. Xiao (Eds.), *Petrochemical Catalyst Materials, Processes, and Emerging Technologies* (pp. 116–149). Hershey, PA: IGI Global. doi:10.4018/978-1-4666-9975-5.ch005

Bügler, M., & Borrmann, A. (2016). Simulation Based Construction Project Schedule Optimization: An Overview on the State-of-the-Art. In F. Miranda & C. Abreu (Eds.), *Handbook of Research on Computational Simulation and Modeling in Engineering* (pp. 482–507). Hershey, PA: IGI Global. doi:10.4018/978-1-4666-8823-0.ch016

Calderon, F. A., Giolo, E. G., Frau, C. D., Rengel, M. G., Rodriguez, H., Tornello, M., ... Gallucci, R. (2018). Seismic Microzonation and Site Effects Detection Through Microtremors Measures: A Review. In N. Ceryan (Ed.), *Handbook of Research on Trends and Digital Advances in Engineering Geology* (pp. 326–349). Hershey, PA: IGI Global. doi:10.4018/978-1-5225-2709-1.ch009

Carmona-Murillo, J., & Valenzuela-Valdés, J. F. (2016). Motivation on Problem Based Learning. In D. Fonseca & E. Redondo (Eds.), *Handbook of Research on Applied E-Learning in Engineering and Architecture Education* (pp. 179–203). Hershey, PA: IGI Global. doi:10.4018/978-1-4666-8803-2.ch009

Ceryan, N. (2016). A Review of Soft Computing Methods Application in Rock Mechanic Engineering. In P. Samui (Ed.), *Handbook of Research on Advanced Computational Techniques for Simulation-Based Engineering* (pp. 1–70). Hershey, PA: IGI Global. doi:10.4018/978-1-4666-9479-8.ch001

Ceryan, N., & Can, N. K. (2018). Prediction of The Uniaxial Compressive Strength of Rocks Materials. In N. Ceryan (Ed.), *Handbook of Research on Trends and Digital Advances in Engineering Geology* (pp. 31–96). Hershey, PA: IGI Global. doi:10.4018/978-1-5225-2709-1.ch002

Ceryan, S. (2018). Weathering Indices Used in Evaluation of the Weathering State of Rock Material. In N. Ceryan (Ed.), *Handbook of Research on Trends and Digital Advances in Engineering Geology* (pp. 132–186). Hershey, PA: IGI Global. doi:10.4018/978-1-5225-2709-1.ch004

Chandrasekaran, S., Silva, B., Patil, A., Oo, A. M., & Campbell, M. (2016). Evaluating Engineering Students' Perceptions: The Impact of Team-Based Learning Practices in Engineering Education. *International Journal of Quality Assurance in Engineering and Technology Education*, *5*(4), 42–59. doi:10.4018/IJQAETE.2016100103

Chen, H., Padilla, R. V., & Besarati, S. (2017). Supercritical Fluids and Their Applications in Power Generation. In L. Chen & Y. Iwamoto (Eds.), *Advanced Applications of Supercritical Fluids in Energy Systems* (pp. 369–402). Hershey, PA: IGI Global. doi:10.4018/978-1-5225-2047-4.ch012

Chen, L. (2017). Principles, Experiments, and Numerical Studies of Supercritical Fluid Natural Circulation System. In L. Chen & Y. Iwamoto (Eds.), *Advanced Applications of Supercritical Fluids in Energy Systems* (pp. 136–187). Hershey, PA: IGI Global. doi:10.4018/978-1-5225-2047-4.ch005

Clementi, F., Di Sciascio, G., Di Sciascio, S., & Lenci, S. (2017). Influence of the Shear-Bending Interaction on the Global Capacity of Reinforced Concrete Frames: A Brief Overview of the New Perspectives. In V. Plevris, G. Kremmyda, & Y. Fahjan (Eds.), *Performance-Based Seismic Design of Concrete Structures and Infrastructures* (pp. 84–111). Hershey, PA: IGI Global. doi:10.4018/978-1-5225-2089-4.ch004

Cortés-Polo, D., Calle-Cancho, J., Carmona-Murillo, J., & González-Sánchez, J. (2017). Future Trends in Mobile-Fixed Integration for Next Generation Networks: Classification and Analysis. *International Journal of Vehicular Telematics and Infotainment Systems*, *1*(1), 33–53. doi:10.4018/IJVTIS.2017010103

Cui, X., Zeng, S., Li, Z., Zheng, Q., Yu, X., & Han, B. (2018). Advanced Composites for Civil Engineering Infrastructures. In K. Kumar & J. Davim (Eds.), *Composites and Advanced Materials for Industrial Applications* (pp. 212–248). Hershey, PA: IGI Global. doi:10.4018/978-1-5225-5216-1.ch010

Dalgıç, S., & Kuşku, İ. (2018). Geological and Geotechnical Investigations in Tunneling. In N. Ceryan (Ed.), *Handbook of Research on Trends and Digital Advances in Engineering Geology* (pp. 482–529). Hershey, PA: IGI Global. doi:10.4018/978-1-5225-2709-1.ch014

de la Varga, D., Soto, M., Arias, C. A., van Oirschot, D., Kilian, R., Pascual, A., & Álvarez, J. A. (2017). Constructed Wetlands for Industrial Wastewater Treatment and Removal of Nutrients. In Á. Val del Río, J. Campos Gómez, & A. Mosquera Corral (Eds.), *Technologies for the Treatment and Recovery of Nutrients from Industrial Wastewater* (pp. 202–230). Hershey, PA: IGI Global. doi:10.4018/978-1-5225-1037-6.ch008

del Valle-Zermeño, R., Chimenos, J. M., & Formosa, J. (2016). Flue Gas Desulfurization: Processes and Technologies. In T. Saleh (Ed.), *Applying Nanotechnology to the Desulfurization Process in Petroleum Engineering* (pp. 337–377). Hershey, PA: IGI Global. doi:10.4018/978-1-4666-9545-0.ch011

Delgado, J. M., Henriques, A. A., & Delgado, R. M. (2016). Structural Non-Linear Models and Simulation Techniques: An Efficient Combination for Safety Evaluation of RC Structures. In F. Miranda & C. Abreu (Eds.), *Handbook of Research on Computational Simulation and Modeling in Engineering* (pp. 540–584). Hershey, PA: IGI Global. doi:10.4018/978-1-4666-8823-0.ch018

Delgado, P. S., Arêde, A., Pouca, N. V., & Costa, A. (2016). Numerical Modeling of RC Bridges for Seismic Risk Analysis. In F. Miranda & C. Abreu (Eds.), *Handbook of Research on Computational Simulation and Modeling in Engineering* (pp. 457–481). Hershey, PA: IGI Global. doi:10.4018/978-1-4666-8823-0.ch015

Deng, Y., & Liu, S. (2016). Catalysis with Room Temperature Ionic Liquids Mediated Metal Nanoparticles. In H. Al-Megren & T. Xiao (Eds.), *Petrochemical Catalyst Materials, Processes, and Emerging Technologies* (pp. 285–329). Hershey, PA: IGI Global. doi:10.4018/978-1-4666-9975-5.ch011

Deperlioglu, O. (2018). Intelligent Techniques Inspired by Nature and Used in Biomedical Engineering. In U. Kose, G. Guraksin, & O. Deperlioglu (Eds.), *Nature-Inspired Intelligent Techniques for Solving Biomedical Engineering Problems* (pp. 51–77). Hershey, PA: IGI Global. doi:10.4018/978-1-5225-4769-3.ch003

Dias, G. L., Magalhães, R. R., Ferreira, D. D., & Vitoriano, F. A. (2016). The Use of a Robotic Arm for Displacement Measurements in a Cantilever beam. *International Journal of Manufacturing, Materials, and Mechanical Engineering*, *6*(3), 45–57. doi:10.4018/IJMMME.2016070104

Dimitratos, N., Villa, A., Chan-Thaw, C. E., Hammond, C., & Prati, L. (2016). Valorisation of Glycerol to Fine Chemicals and Fuels. In H. Al-Megren & T. Xiao (Eds.), *Petrochemical Catalyst Materials, Processes, and Emerging Technologies* (pp. 352–384). Hershey, PA: IGI Global. doi:10.4018/978-1-4666-9975-5.ch013

Dixit, A. (2018). Application of Silica-Gel-Reinforced Aluminium Composite on the Piston of Internal Combustion Engine: Comparative Study of Silica-Gel-Reinforced Aluminium Composite Piston With Aluminium Alloy Piston. In K. Kumar & J. Davim (Eds.), *Composites and Advanced Materials for Industrial Applications* (pp. 63–98). Hershey, PA: IGI Global. doi:10.4018/978-1-5225-5216-1.ch004

Drei, A., Milani, G., & Sincraian, G. (2016). Application of DEM to Historic Masonries, Two Case-Studies in Portugal and Italy: Aguas Livres Aqueduct and Arch-Tympana of a Church. In V. Sarhosis, K. Bagi, J. Lemos, & G. Milani (Eds.), *Computational Modeling of Masonry Structures Using the Discrete Element Method* (pp. 326–366). Hershey, PA: IGI Global. doi:10.4018/978-1-5225-0231-9.ch013

Dutta, S., Roy, P. K., & Nandi, D. (2016). Optimal Allocation of Static Synchronous Series Compensator Controllers using Chemical Reaction Optimization for Reactive Power Dispatch. *International Journal of Energy Optimization and Engineering*, *5*(3), 43–62. doi:10.4018/IJEOE.2016070103

Dutta, S., Roy, P. K., & Nandi, D. (2016). Quasi Oppositional Teaching-Learning based Optimization for Optimal Power Flow Incorporating FACTS. *International Journal of Energy Optimization and Engineering*, *5*(2), 64–84. doi:10.4018/IJEOE.2016040104

Eloy, S., Dias, M. S., Lopes, P. F., & Vilar, E. (2016). Digital Technologies in Architecture and Engineering: Exploring an Engaged Interaction within Curricula. In D. Fonseca & E. Redondo (Eds.), *Handbook of Research on Applied E-Learning in Engineering and Architecture Education* (pp. 368–402). Hershey, PA: IGI Global. doi:10.4018/978-1-4666-8803-2.ch017

Elsayed, A. M., Dakkama, H. J., Mahmoud, S., Al-Dadah, R., & Kaialy, W. (2017). Sustainable Cooling Research Using Activated Carbon Adsorbents and Their Environmental Impact. In T. Kobayashi (Ed.), *Applied Environmental Materials Science for Sustainability* (pp. 186–221). Hershey, PA: IGI Global. doi:10.4018/978-1-5225-1971-3.ch009

Ercanoglu, M., & Sonmez, H. (2018). General Trends and New Perspectives on Landslide Mapping and Assessment Methods. In N. Ceryan (Ed.), *Handbook of Research on Trends and Digital Advances in Engineering Geology* (pp. 350–379). Hershey, PA: IGI Global. doi:10.4018/978-1-5225-2709-1.ch010

Erinosho, M. F., Akinlabi, E. T., & Pityana, S. (2016). Enhancement of Surface Integrity of Titanium Alloy with Copper by Means of Laser Metal Deposition Process. In E. Akinlabi, R. Mahamood, & S. Akinlabi (Eds.), *Advanced Manufacturing Techniques Using Laser Material Processing* (pp. 60–91). Hershey, PA: IGI Global. doi:10.4018/978-1-5225-0329-3.ch004

Farag, H., & Kishida, M. (2016). Kinetic Models for Complex Parallel–Consecutive Reactions Assessment of Reaction Network and Product Selectivity. In H. Al-Megren & T. Xiao (Eds.), *Petrochemical Catalyst Materials, Processes, and Emerging Technologies* (pp. 330–351). Hershey, PA: IGI Global. doi:10.4018/978-1-4666-9975-5.ch012

Faroz, S. A., Pujari, N. N., Rastogi, R., & Ghosh, S. (2017). Risk Analysis of Structural Engineering Systems Using Bayesian Inference. In P. Samui, S. Chakraborty, & D. Kim (Eds.), *Modeling and Simulation Techniques in Structural Engineering* (pp. 390–424). Hershey, PA: IGI Global. doi:10.4018/978-1-5225-0588-4.ch014

Fernando, P. R., Hamigah, T., Disne, S., Wickramasingha, G. G., & Sutharshan, A. (2018). The Evaluation of Engineering Properties of Low Cost Concrete Blocks by Partial Doping of Sand with Sawdust: Low Cost Sawdust Concrete Block. *International Journal of Strategic Engineering*, *1*(2), 26–42. doi:10.4018/IJoSE.2018070103

Fragiadakis, M., Stefanou, I., & Psycharis, I. N. (2016). Vulnerability Assessment of Damaged Classical Multidrum Columns. In V. Sarhosis, K. Bagi, J. Lemos, & G. Milani (Eds.), *Computational Modeling of Masonry Structures Using the Discrete Element Method* (pp. 235–253). Hershey, PA: IGI Global. doi:10.4018/978-1-5225-0231-9.ch010

Gaines, T. W., Williams, K. R., & Wagener, K. B. (2016). ADMET: Functionalized Polyolefins. In H. Al-Megren & T. Xiao (Eds.), *Petrochemical Catalyst Materials, Processes, and Emerging Technologies* (pp. 1–21). Hershey, PA: IGI Global. doi:10.4018/978-1-4666-9975-5.ch001

Garg, H. (2016). Bi-Criteria Optimization for Finding the Optimal Replacement Interval for Maintaining the Performance of the Process Industries. In P. Vasant, G. Weber, & V. Dieu (Eds.), *Handbook of Research on Modern Optimization Algorithms and Applications in Engineering and Economics* (pp. 643–675). Hershey, PA: IGI Global. doi:10.4018/978-1-4666-9644-0.ch025

Gaspar, P. D., Dinho da Silva, P., Gonçalves, J. P., & Carneiro, R. (2016). Computational Modelling and Simulation to Assist the Improvement of Thermal Performance and Energy Efficiency in Industrial Engineering Systems: Application to Cold Stores. In F. Miranda & C. Abreu (Eds.), *Handbook of Research on Computational Simulation and Modeling in Engineering* (pp. 1–68). Hershey, PA: IGI Global. doi:10.4018/978-1-4666-8823-0.ch001

Ge, H., Tang, M., & Wen, X. (2016). Ni/ZnO Nano Sorbent for Reactive Adsorption Desulfurization of Refinery Oil Streams. In T. Saleh (Ed.), *Applying Nanotechnology to the Desulfurization Process in Petroleum Engineering* (pp. 216–239). Hershey, PA: IGI Global. doi:10.4018/978-1-4666-9545-0.ch007

Ghosh, S., Mitra, S., Ghosh, S., & Chakraborty, S. (2017). Seismic Reliability Analysis in the Framework of Metamodelling Based Monte Carlo Simulation. In P. Samui, S. Chakraborty, & D. Kim (Eds.), *Modeling and Simulation Techniques in Structural Engineering* (pp. 192–208). Hershey, PA: IGI Global. doi:10.4018/978-1-5225-0588-4.ch006

Gil, M., & Otero, B. (2017). Learning Engineering Skills through Creativity and Collaboration: A Game-Based Proposal. In R. Alexandre Peixoto de Queirós & M. Pinto (Eds.), *Gamification-Based E-Learning Strategies for Computer Programming Education* (pp. 14–29). Hershey, PA: IGI Global. doi:10.4018/978-1-5225-1034-5.ch002

Gill, J., Ayre, M., & Mills, J. (2017). Revisioning the Engineering Profession: How to Make It Happen! In M. Gray & K. Thomas (Eds.), *Strategies for Increasing Diversity in Engineering Majors and Careers* (pp. 156–175). Hershey, PA: IGI Global. doi:10.4018/978-1-5225-2212-6.ch008

Gopal, S., & Al-Hazmi, M. H. (2016). Advances in Catalytic Technologies for Selective Oxidation of Lower Alkanes. In H. Al-Megren & T. Xiao (Eds.), *Petrochemical Catalyst Materials, Processes, and Emerging Technologies* (pp. 22–52). Hershey, PA: IGI Global. doi:10.4018/978-1-4666-9975-5.ch002

Goyal, N., Ram, M., Bhardwaj, A., & Kumar, A. (2016). Thermal Power Plant Modelling with Fault Coverage Stochastically. *International Journal of Manufacturing, Materials, and Mechanical Engineering*, *6*(3), 28–44. doi:10.4018/IJMMME.2016070103

Goyal, N., Ram, M., & Kumar, P. (2017). Welding Process under Fault Coverage Approach for Reliability and MTTF. In M. Ram & J. Davim (Eds.), *Mathematical Concepts and Applications in Mechanical Engineering and Mechatronics* (pp. 222–245). Hershey, PA: IGI Global. doi:10.4018/978-1-5225-1639-2.ch011

Gray, M., & Lundy, C. (2017). Engineering Study Abroad: High Impact Strategy for Increasing Access. In M. Gray & K. Thomas (Eds.), *Strategies for Increasing Diversity in Engineering Majors and Careers* (pp. 42–59). Hershey, PA: IGI Global. doi:10.4018/978-1-5225-2212-6.ch003

Guha, D., Roy, P. K., & Banerjee, S. (2016). Application of Modified Biogeography Based Optimization in AGC of an Interconnected Multi-Unit Multi-Source AC-DC Linked Power System. *International Journal of Energy Optimization and Engineering*, *5*(3), 1–18. doi:10.4018/IJEOE.2016070101

Guha, D., Roy, P. K., & Banerjee, S. (2016). Grey Wolf Optimization to Solve Load Frequency Control of an Interconnected Power System: GWO Used to Solve LFC Problem. *International Journal of Energy Optimization and Engineering*, *5*(4), 62–83. doi:10.4018/IJEOE.2016100104

Gupta, A. K., Dey, A., & Mukhopadhyay, A. K. (2016). Micromechanical and Finite Element Modeling for Composites. In S. Datta & J. Davim (Eds.), *Computational Approaches to Materials Design: Theoretical and Practical Aspects* (pp. 101–162). Hershey, PA: IGI Global. doi:10.4018/978-1-5225-0290-6.ch005

Guraksin, G. E. (2018). Internet of Things and Nature-Inspired Intelligent Techniques for the Future of Biomedical Engineering. In U. Kose, G. Guraksin, & O. Deperlioglu (Eds.), *Nature-Inspired Intelligent Techniques for Solving Biomedical Engineering Problems* (pp. 263–282). Hershey, PA: IGI Global. doi:10.4018/978-1-5225-4769-3.ch013

Hansman, C. A. (2016). Developing Mentoring Programs in Engineering and Technology Education. *International Journal of Quality Assurance in Engineering and Technology Education*, *5*(2), 1–15. doi:10.4018/IJQAETE.2016040101

Hasan, U., Chegenizadeh, A., & Nikraz, H. (2016). Nanotechnology Future and Present in Construction Industry: Applications in Geotechnical Engineering. In A. Khitab & W. Anwar (Eds.), *Advanced Research on Nanotechnology for Civil Engineering Applications* (pp. 141–179). Hershey, PA: IGI Global. doi:10.4018/978-1-5225-0344-6.ch005

Hejazi, T., & Akbari, L. (2017). A Multiresponse Optimization Model for Statistical Design of Processes with Discrete Variables. In M. Ram & J. Davim (Eds.), *Mathematical Concepts and Applications in Mechanical Engineering and Mechatronics* (pp. 17–37). Hershey, PA: IGI Global. doi:10.4018/978-1-5225-1639-2.ch002

Hejazi, T., & Hejazi, A. (2017). Monte Carlo Simulation for Reliability-Based Design of Automotive Complex Subsystems. In M. Ram & J. Davim (Eds.), *Mathematical Concepts and Applications in Mechanical Engineering and Mechatronics* (pp. 177–200). Hershey, PA: IGI Global. doi:10.4018/978-1-5225-1639-2.ch009

Hejazi, T., & Poursabbagh, H. (2017). Reliability Analysis of Engineering Systems: An Accelerated Life Testing for Boiler Tubes. In M. Ram & J. Davim (Eds.), *Mathematical Concepts and Applications in Mechanical Engineering and Mechatronics* (pp. 154–176). Hershey, PA: IGI Global. doi:10.4018/978-1-5225-1639-2.ch008

Henao, J., & Sotelo, O. (2018). Surface Engineering at High Temperature: Thermal Cycling and Corrosion Resistance. In A. Pakseresht (Ed.), *Production, Properties, and Applications of High Temperature Coatings* (pp. 131–159). Hershey, PA: IGI Global. doi:10.4018/978-1-5225-4194-3.ch006

Huirache-Acuña, R., Alonso-Nuñez, G., Rivera-Muñoz, E. M., Gutierrez, O., & Pawelec, B. (2016). Trimetallic Sulfide Catalysts for Hydrodesulfurization. In T. Saleh (Ed.), *Applying Nanotechnology to the Desulfurization Process in Petroleum Engineering* (pp. 240–262). Hershey, PA: IGI Global. doi:10.4018/978-1-4666-9545-0.ch008

Ilori, O. O., Adetan, D. A., & Umoru, L. E. (2017). Effect of Cutting Parameters on the Surface Residual Stress of Face-Milled Pearlitic Ductile Iron. *International Journal of Materials Forming and Machining Processes*, *4*(1), 38–52. doi:10.4018/IJMFMP.2017010103

Imam, M. H., Tasadduq, I. A., Ahmad, A., Aldosari, F., & Khan, H. (2017). Automated Generation of Course Improvement Plans Using Expert System. *International Journal of Quality Assurance in Engineering and Technology Education*, *6*(1), 1–12. doi:10.4018/IJQAETE.2017010101

Injeti, S. K., & Kumar, T. V. (2018). A WDO Framework for Optimal Deployment of DGs and DSCs in a Radial Distribution System Under Daily Load Pattern to Improve Techno-Economic Benefits. *International Journal of Energy Optimization and Engineering*, *7*(2), 1–38. doi:10.4018/IJEOE.2018040101

Ishii, N., Anami, K., & Knisely, C. W. (2018). *Dynamic Stability of Hydraulic Gates and Engineering for Flood Prevention*. Hershey, PA: IGI Global. doi:10.4018/978-1-5225-3079-4

J., J., Chowdhury, S., Goyal, P., Samui, P., & Dalkiliç, Y. (2016). Determination of Bearing Capacity of Shallow Foundation Using Soft Computing. In P. Saxena, D. Singh, & M. Pant (Eds.), *Problem Solving and Uncertainty Modeling through Optimization and Soft Computing Applications* (pp. 292-328). Hershey, PA: IGI Global. doi:10.4018/978-1-4666-9885-7.ch014

Jagan, J., Gundlapalli, P., & Samui, P. (2016). Utilization of Classification Techniques for the Determination of Liquefaction Susceptibility of Soils. In S. Bhattacharyya, P. Banerjee, D. Majumdar, & P. Dutta (Eds.), *Handbook of Research on Advanced Hybrid Intelligent Techniques and Applications* (pp. 124–160). Hershey, PA: IGI Global. doi:10.4018/978-1-4666-9474-3.ch005

Jayapalan, S. (2018). A Review of Chemical Treatments on Natural Fibers-Based Hybrid Composites for Engineering Applications. In K. Kumar & J. Davim (Eds.), *Composites and Advanced Materials for Industrial Applications* (pp. 16–37). Hershey, PA: IGI Global. doi:10.4018/978-1-5225-5216-1.ch002

Jeet, K., & Dhir, R. (2016). Software Module Clustering Using Bio-Inspired Algorithms. In P. Vasant, G. Weber, & V. Dieu (Eds.), *Handbook of Research on Modern Optimization Algorithms and Applications in Engineering and Economics* (pp. 445–470). Hershey, PA: IGI Global. doi:10.4018/978-1-4666-9644-0.ch017

Joshi, S. D., & Talange, D. B. (2016). Fault Tolerant Control for a Fractional Order AUV System. *International Journal of Energy Optimization and Engineering*, *5*(2), 1–24. doi:10.4018/IJEOE.2016040101

Julião, D., Ribeiro, S., de Castro, B., Cunha-Silva, L., & Balula, S. S. (2016). Polyoxometalates-Based Nanocatalysts for Production of Sulfur-Free Diesel. In T. Saleh (Ed.), *Applying Nanotechnology to the Desulfurization Process in Petroleum Engineering* (pp. 426–458). Hershey, PA: IGI Global. doi:10.4018/978-1-4666-9545-0.ch014

Kamthan, P. (2016). On the Nature of Collaborations in Agile Software Engineering Course Projects. *International Journal of Quality Assurance in Engineering and Technology Education*, *5*(2), 42–59. doi:10.4018/IJQAETE.2016040104

Karaman, O., Celik, C., & Urkmez, A. S. (2016). Self-Assembled Biomimetic Scaffolds for Bone Tissue Engineering. In M. Bououdina (Ed.), *Emerging Research on Bioinspired Materials Engineering* (pp. 104–132). Hershey, PA: IGI Global. doi:10.4018/978-1-4666-9811-6.ch004

Karkalos, N. E., Markopoulos, A. P., & Dossis, M. F. (2017). Optimal Model Parameters of Inverse Kinematics Solution of a 3R Robotic Manipulator Using ANN Models. *International Journal of Manufacturing, Materials, and Mechanical Engineering*, *7*(3), 20–40. doi:10.4018/IJMMME.2017070102

Kesimal, A., Karaman, K., Cihangir, F., & Ercikdi, B. (2018). Excavatability Assessment of Rock Masses for Geotechnical Studies. In N. Ceryan (Ed.), *Handbook of Research on Trends and Digital Advances in Engineering Geology* (pp. 231–256). Hershey, PA: IGI Global. doi:10.4018/978-1-5225-2709-1.ch006

Khanh, D. V., Vasant, P. M., Elamvazuthi, I., & Dieu, V. N. (2016). Multi-Objective Optimization of Two-Stage Thermo-Electric Cooler Using Differential Evolution: MO Optimization of TEC Using DE. In F. Miranda & C. Abreu (Eds.), *Handbook of Research on Computational Simulation and Modeling in Engineering* (pp. 139–170). Hershey, PA: IGI Global. doi:10.4018/978-1-4666-8823-0.ch004

Kim, D., Hassan, M. K., Chang, S., & Bigdeli, Y. (2016). Nonlinear Vibration Control of 3D Irregular Structures Subjected to Seismic Loads. In P. Samui (Ed.), *Handbook of Research on Advanced Computational Techniques for Simulation-Based Engineering* (pp. 103–119). Hershey, PA: IGI Global. doi:10.4018/978-1-4666-9479-8.ch003

Knoflacher, H. (2017). The Role of Engineers and Their Tools in the Transport Sector after Paradigm Change: From Assumptions and Extrapolations to Science. In H. Knoflacher & E. Ocalir-Akunal (Eds.), *Engineering Tools and Solutions for Sustainable Transportation Planning* (pp. 1–29). Hershey, PA: IGI Global. doi:10.4018/978-1-5225-2116-7.ch001

Kose, U. (2018). Towards an Intelligent Biomedical Engineering With Nature-Inspired Artificial Intelligence Techniques. In U. Kose, G. Guraksin, & O. Deperlioglu (Eds.), *Nature-Inspired Intelligent Techniques for Solving Biomedical Engineering Problems* (pp. 1–26). Hershey, PA: IGI Global. doi:10.4018/978-1-5225-4769-3.ch001

Kostić, S. (2018). A Review on Enhanced Stability Analyses of Soil Slopes Using Statistical Design. In N. Ceryan (Ed.), *Handbook of Research on Trends and Digital Advances in Engineering Geology* (pp. 446–481). Hershey, PA: IGI Global. doi:10.4018/978-1-5225-2709-1.ch013

Kumar, A., Patil, P. P., & Prajapati, Y. K. (2018). *Advanced Numerical Simulations in Mechanical Engineering*. Hershey, PA: IGI Global. doi:10.4018/978-1-5225-3722-9

Kumar, G. R., Rajyalakshmi, G., & Manupati, V. K. (2017). Surface Micro Patterning of Aluminium Reinforced Composite through Laser Peening. *International Journal of Manufacturing, Materials, and Mechanical Engineering*, *7*(4), 15–27. doi:10.4018/IJMMME.2017100102

Kumari, N., & Kumar, K. (2018). Fabrication of Orthotic Calipers With Epoxy-Based Green Composite. In K. Kumar & J. Davim (Eds.), *Composites and Advanced Materials for Industrial Applications* (pp. 157–176). Hershey, PA: IGI Global. doi:10.4018/978-1-5225-5216-1.ch008

Kuppusamy, R. R. (2018). Development of Aerospace Composite Structures Through Vacuum-Enhanced Resin Transfer Moulding Technology (VERTMTy): Vacuum-Enhanced Resin Transfer Moulding. In K. Kumar & J. Davim (Eds.), *Composites and Advanced Materials for Industrial Applications* (pp. 99–111). Hershey, PA: IGI Global. doi:10.4018/978-1-5225-5216-1.ch005

Lemos, J. V. (2016). The Basis for Masonry Analysis with UDEC and 3DEC. In V. Sarhosis, K. Bagi, J. Lemos, & G. Milani (Eds.), *Computational Modeling of Masonry Structures Using the Discrete Element Method* (pp. 61–89). Hershey, PA: IGI Global. doi:10.4018/978-1-5225-0231-9.ch003

Loy, J., Howell, S., & Cooper, R. (2017). Engineering Teams: Supporting Diversity in Engineering Education. In M. Gray & K. Thomas (Eds.), *Strategies for Increasing Diversity in Engineering Majors and Careers* (pp. 106–129). Hershey, PA: IGI Global. doi:10.4018/978-1-5225-2212-6.ch006

Macher, G., Armengaud, E., Kreiner, C., Brenner, E., Schmittner, C., Ma, Z., ... Krammer, M. (2018). Integration of Security in the Development Lifecycle of Dependable Automotive CPS. In N. Druml, A. Genser, A. Krieg, M. Menghin, & A. Hoeller (Eds.), *Solutions for Cyber-Physical Systems Ubiquity* (pp. 383–423). Hershey, PA: IGI Global. doi:10.4018/978-1-5225-2845-6.ch015

Maghsoodlou, S., & Poreskandar, S. (2016). Controlling Electrospinning Jet Using Microscopic Model for Ideal Tissue Engineering Scaffolds. *International Journal of Chemoinformatics and Chemical Engineering*, *5*(2), 1–16. doi:10.4018/IJCCE.2016070101

Mahendramani, G., & Lakshmana Swamy, N. (2018). Effect of Weld Groove Area on Distortion of Butt Welded Joints in Submerged Arc Welding. *International Journal of Manufacturing, Materials, and Mechanical Engineering*, *8*(2), 33–44. doi:10.4018/IJMMME.2018040103

Maiti, S. (2016). Engineered Gellan Polysaccharides in the Design of Controlled Drug Delivery Systems. In M. Bououdina (Ed.), *Emerging Research on Bioinspired Materials Engineering* (pp. 266–293). Hershey, PA: IGI Global. doi:10.4018/978-1-4666-9811-6.ch010

Majumdar, J. D., Weisheit, A., & Manna, I. (2016). Laser Surface Processing for Tailoring of Properties by Optimization of Microstructure. In E. Akinlabi, R. Mahamood, & S. Akinlabi (Eds.), *Advanced Manufacturing Techniques Using Laser Material Processing* (pp. 121–171). Hershey, PA: IGI Global. doi:10.4018/978-1-5225-0329-3.ch006

Maldonado-Macías, A. A., García-Alcaraz, J. L., Hernández-Arellano, J. L., & Cortes-Robles, G. (2016). An Ergonomic Compatibility Perspective on the Selection of Advanced Manufacturing Technology: A Case Study for CNC Vertical Machining Centers. In G. Alor-Hernández, C. Sánchez-Ramírez, & J. García-Alcaraz (Eds.), *Handbook of Research on Managerial Strategies for Achieving Optimal Performance in Industrial Processes* (pp. 137–165). Hershey, PA: IGI Global. doi:10.4018/978-1-5225-0130-5.ch008

Mamaghani, I. H. (2016). Application of Discrete Finite Element Method for Analysis of Unreinforced Masonry Structures. In V. Sarhosis, K. Bagi, J. Lemos, & G. Milani (Eds.), *Computational Modeling of Masonry Structures Using the Discrete Element Method* (pp. 440–458). Hershey, PA: IGI Global. doi:10.4018/978-1-5225-0231-9.ch017

Mansor, M. R., Sapuan, S. M., Salim, M. A., Akop, M. Z., Musthafah, M. T., & Shaharuzaman, M. A. (2016). Concurrent Design of Green Composites. In D. Verma, S. Jain, X. Zhang, & P. Gope (Eds.), *Green Approaches to Biocomposite Materials Science and Engineering* (pp. 48–75). Hershey, PA: IGI Global. doi:10.4018/978-1-5225-0424-5.ch003

Mansouri, I., & Esmaeili, E. (2016). Nanotechnology Applications in the Construction Industry. In A. Khitab & W. Anwar (Eds.), *Advanced Research on Nanotechnology for Civil Engineering Applications* (pp. 111–140). Hershey, PA: IGI Global. doi:10.4018/978-1-5225-0344-6.ch004

Manzoor, A. (2016). MOOCs for Enhancing Engineering Education. In D. Fonseca & E. Redondo (Eds.), *Handbook of Research on Applied E-Learning in Engineering and Architecture Education* (pp. 204–223). Hershey, PA: IGI Global. doi:10.4018/978-1-4666-8803-2.ch010

Martin, A., Kalevaru, V. N., & Radnik, J. (2016). Palladium in Heterogeneous Oxidation Catalysis. In H. Al-Megren & T. Xiao (Eds.), *Petrochemical Catalyst Materials, Processes, and Emerging Technologies* (pp. 53–81). Hershey, PA: IGI Global. doi:10.4018/978-1-4666-9975-5.ch003

Melnyczuk, J. M., & Palchoudhury, S. (2016). Introduction to Bio-Inspired Hydrogel and Their Application: Hydrogels. In M. Bououdina (Ed.), *Emerging Research on Bioinspired Materials Engineering* (pp. 133–159). Hershey, PA: IGI Global. doi:10.4018/978-1-4666-9811-6.ch005

Mitra-Kirtley, S., Mullins, O. C., & Pomerantz, A. E. (2016). Sulfur and Nitrogen Chemical Speciation in Crude Oils and Related Carbonaceous Materials. In T. Saleh (Ed.), *Applying Nanotechnology to the Desulfurization Process in Petroleum Engineering* (pp. 53–83). Hershey, PA: IGI Global. doi:10.4018/978-1-4666-9545-0.ch002

Moalosi, R., Uziak, J., & Oladiran, M. T. (2016). Using Blended Learning Approach to Deliver Courses in An Engineering Programme. *International Journal of Quality Assurance in Engineering and Technology Education, 5*(1), 23–39. doi:10.4018/IJQAETE.2016010103

Mohammadzadeh, S., & Kim, Y. (2017). Nonlinear System Identification of Smart Buildings. In P. Samui, S. Chakraborty, & D. Kim (Eds.), *Modeling and Simulation Techniques in Structural Engineering* (pp. 328–347). Hershey, PA: IGI Global. doi:10.4018/978-1-5225-0588-4.ch011

Mohanty, I., & Bhattacherjee, D. (2016). Artificial Neural Network and Its Application in Steel Industry. In S. Datta & J. Davim (Eds.), *Computational Approaches to Materials Design: Theoretical and Practical Aspects* (pp. 267–300). Hershey, PA: IGI Global. doi:10.4018/978-1-5225-0290-6.ch010

Mohebkhah, A., & Sarhosis, V. (2016). Discrete Element Modeling of Masonry-Infilled Frames. In V. Sarhosis, K. Bagi, J. Lemos, & G. Milani (Eds.), *Computational Modeling of Masonry Structures Using the Discrete Element Method* (pp. 200–234). Hershey, PA: IGI Global. doi:10.4018/978-1-5225-0231-9.ch009

Molina, G. J., Aktaruzzaman, F., Soloiu, V., & Rahman, M. (2017). Design and Testing of a Jet-Impingement Instrument to Study Surface-Modification Effects by Nanofluids. *International Journal of Surface Engineering and Interdisciplinary Materials Science*, *5*(2), 43–61. doi:10.4018/IJSEIMS.2017070104

Montalvan-Sorrosa, D., de los Cobos-Vasconcelos, D., & Gonzalez-Sanchez, A. (2016). Nanotechnology Applied to the Biodesulfurization of Fossil Fuels and Spent Caustic Streams. In T. Saleh (Ed.), *Applying Nanotechnology to the Desulfurization Process in Petroleum Engineering* (pp. 378–389). Hershey, PA: IGI Global. doi:10.4018/978-1-4666-9545-0.ch012

Montillet, J., Yu, K., Bonenberg, L. K., & Roberts, G. W. (2016). Optimization Algorithms in Local and Global Positioning. In P. Vasant, G. Weber, & V. Dieu (Eds.), *Handbook of Research on Modern Optimization Algorithms and Applications in Engineering and Economics* (pp. 1–53). Hershey, PA: IGI Global. doi:10.4018/978-1-4666-9644-0.ch001

Moreira, F., & Ferreira, M. J. (2016). Teaching and Learning Requirements Engineering Based on Mobile Devices and Cloud: A Case Study. In D. Fonseca & E. Redondo (Eds.), *Handbook of Research on Applied E-Learning in Engineering and Architecture Education* (pp. 237–262). Hershey, PA: IGI Global. doi:10.4018/978-1-4666-8803-2.ch012

Mukherjee, A., Saeed, R. A., Dutta, S., & Naskar, M. K. (2017). Fault Tracking Framework for Software-Defined Networking (SDN). In C. Singhal & S. De (Eds.), *Resource Allocation in Next-Generation Broadband Wireless Access Networks* (pp. 247–272). Hershey, PA: IGI Global. doi:10.4018/978-1-5225-2023-8.ch011

Mukhopadhyay, A., Barman, T. K., & Sahoo, P. (2018). Electroless Nickel Coatings for High Temperature Applications. In K. Kumar & J. Davim (Eds.), *Composites and Advanced Materials for Industrial Applications* (pp. 297–331). Hershey, PA: IGI Global. doi:10.4018/978-1-5225-5216-1.ch013

Náprstek, J., & Fischer, C. (2017). Dynamic Stability and Post-Critical Processes of Slender Auto-Parametric Systems. In V. Plevris, G. Kremmyda, & Y. Fahjan (Eds.), *Performance-Based Seismic Design of Concrete Structures and Infrastructures* (pp. 128–171). Hershey, PA: IGI Global. doi:10.4018/978-1-5225-2089-4.ch006

Nautiyal, L., Shivach, P., & Ram, M. (2018). Optimal Designs by Means of Genetic Algorithms. In M. Ram & J. Davim (Eds.), *Soft Computing Techniques and Applications in Mechanical Engineering* (pp. 151–161). Hershey, PA: IGI Global. doi:10.4018/978-1-5225-3035-0.ch007

Nazir, R. (2017). Advanced Nanomaterials for Water Engineering and Treatment: Nano-Metal Oxides and Their Nanocomposites. In T. Saleh (Ed.), *Advanced Nanomaterials for Water Engineering, Treatment, and Hydraulics* (pp. 84–126). Hershey, PA: IGI Global. doi:10.4018/978-1-5225-2136-5.ch005

Nogueira, A. F., Ribeiro, J. C., Fernández de Vega, F., & Zenha-Rela, M. A. (2018). Evolutionary Approaches to Test Data Generation for Object-Oriented Software: Overview of Techniques and Tools. In M. Khosrow-Pour, D.B.A. (Ed.), Incorporating Nature-Inspired Paradigms in Computational Applications (pp. 162-194). Hershey, PA: IGI Global. doi:10.4018/978-1-5225-5020-4.ch006

Nunes, J. F., Moreira, P. M., & Tavares, J. M. (2016). Human Motion Analysis and Simulation Tools: A Survey. In F. Miranda & C. Abreu (Eds.), *Handbook of Research on Computational Simulation and Modeling in Engineering* (pp. 359–388). Hershey, PA: IGI Global. doi:10.4018/978-1-4666-8823-0.ch012

Ogunlaja, A. S., & Tshentu, Z. R. (2016). Molecularly Imprinted Polymer Nanofibers for Adsorptive Desulfurization. In T. Saleh (Ed.), *Applying Nanotechnology to the Desulfurization Process in Petroleum Engineering* (pp. 281–336). Hershey, PA: IGI Global. doi:10.4018/978-1-4666-9545-0.ch010

Ong, P., & Kohshelan, S. (2016). Performances of Adaptive Cuckoo Search Algorithm in Engineering Optimization. In P. Vasant, G. Weber, & V. Dieu (Eds.), *Handbook of Research on Modern Optimization Algorithms and Applications in Engineering and Economics* (pp. 676–699). Hershey, PA: IGI Global. doi:10.4018/978-1-4666-9644-0.ch026

Osho, M. B. (2018). Industrial Enzyme Technology: Potential Applications. In S. Bharati & P. Chaurasia (Eds.), *Research Advancements in Pharmaceutical, Nutritional, and Industrial Enzymology* (pp. 375–394). Hershey, PA: IGI Global. doi:10.4018/978-1-5225-5237-6.ch017

Padmaja, P., & Marutheswar, G. (2017). Certain Investigation on Secured Data Transmission in Wireless Sensor Networks. *International Journal of Mobile Computing and Multimedia Communications*, *8*(1), 48–61. doi:10.4018/IJMCMC.2017010104

Paixão, S. M., Silva, T. P., Arez, B. F., & Alves, L. (2016). Advances in the Reduction of the Costs Inherent to Fossil Fuels' Biodesulfurization towards Its Potential Industrial Application. In T. Saleh (Ed.), *Applying Nanotechnology to the Desulfurization Process in Petroleum Engineering* (pp. 390–425). Hershey, PA: IGI Global. doi:10.4018/978-1-4666-9545-0.ch013

Palmer, S., & Hall, W. (2017). An Evaluation of Group Work in First-Year Engineering Design Education. In R. Tucker (Ed.), *Collaboration and Student Engagement in Design Education* (pp. 145–168). Hershey, PA: IGI Global. doi:10.4018/978-1-5225-0726-0.ch007

Panneer, R. (2017). Effect of Composition of Fibers on Properties of Hybrid Composites. *International Journal of Manufacturing, Materials, and Mechanical Engineering*, *7*(4), 28–43. doi:10.4018/IJMMME.2017100103

Parker, J. (2016). Hubble's Expanding Universe: A Model for Quality in Technology Infused engineering and Technology Education. *International Journal of Quality Assurance in Engineering and Technology Education*, *5*(2), 16–29. doi:10.4018/IJQAETE.2016040102

Paul, S., & Roy, P. (2018). Optimal Design of Power System Stabilizer Using a Novel Evolutionary Algorithm. *International Journal of Energy Optimization and Engineering*, *7*(3), 24–46. doi:10.4018/IJEOE.2018070102

Pavaloiu, A. (2018). Artificial Intelligence Ethics in Biomedical-Engineering-Oriented Problems. In U. Kose, G. Guraksin, & O. Deperlioglu (Eds.), *Nature-Inspired Intelligent Techniques for Solving Biomedical Engineering Problems* (pp. 219–231). Hershey, PA: IGI Global. doi:10.4018/978-1-5225-4769-3.ch010

Peña, F. (2016). A Semi-Discrete Approach for the Numerical Simulation of Freestanding Blocks. In V. Sarhosis, K. Bagi, J. Lemos, & G. Milani (Eds.), *Computational Modeling of Masonry Structures Using the Discrete Element Method* (pp. 416–439). Hershey, PA: IGI Global. doi:10.4018/978-1-5225-0231-9.ch016

Penchovsky, R., & Traykovska, M. (2016). Synthetic Approaches to Biology: Engineering Gene Control Circuits, Synthesizing, and Editing Genomes. In M. Bououdina (Ed.), *Emerging Research on Bioinspired Materials Engineering* (pp. 323–351). Hershey, PA: IGI Global. doi:10.4018/978-1-4666-9811-6.ch012

Pieroni, A., & Iazeolla, G. (2016). Engineering QoS and Energy Saving in the Delivery of ICT Services. In P. Vasant & N. Voropai (Eds.), *Sustaining Power Resources through Energy Optimization and Engineering* (pp. 208–226). Hershey, PA: IGI Global. doi:10.4018/978-1-4666-9755-3.ch009

Pioro, I., Mahdi, M., & Popov, R. (2017). Application of Supercritical Pressures in Power Engineering. In L. Chen & Y. Iwamoto (Eds.), *Advanced Applications of Supercritical Fluids in Energy Systems* (pp. 404–457). Hershey, PA: IGI Global. doi:10.4018/978-1-5225-2047-4.ch013

Plaksina, T., & Gildin, E. (2017). Rigorous Integrated Evolutionary Workflow for Optimal Exploitation of Unconventional Gas Assets. *International Journal of Energy Optimization and Engineering*, *6*(1), 101–122. doi:10.4018/IJEOE.2017010106

Puppala, A. J., Bheemasetti, T. V., Zou, H., Yu, X., Pedarla, A., & Cai, G. (2016). Spatial Variability Analysis of Soil Properties using Geostatistics. In P. Samui (Ed.), *Handbook of Research on Advanced Computational Techniques for Simulation-Based Engineering* (pp. 195–226). Hershey, PA: IGI Global. doi:10.4018/978-1-4666-9479-8.ch008

Ramdani, N., & Azibi, M. (2018). Polymer Composite Materials for Microelectronics Packaging Applications: Composites for Microelectronics Packaging. In K. Kumar & J. Davim (Eds.), *Composites and Advanced Materials for Industrial Applications* (pp. 177–211). Hershey, PA: IGI Global. doi:10.4018/978-1-5225-5216-1.ch009

Ramesh, M., Garg, R., & Subrahmanyam, G. V. (2017). Investigation of Influence of Quenching and Annealing on the Plane Fracture Toughness and Brittle to Ductile Transition Temperature of the Zinc Coated Structural Steel Materials. *International Journal of Surface Engineering and Interdisciplinary Materials Science*, *5*(2), 33–42. doi:10.4018/IJSEIMS.2017070103

Razavi, A. M., & Ahmad, R. (2016). Agile Software Development Challenges in Implementation and Adoption: Focusing on Large and Distributed Settings – Past Experiences, Emergent Topics. In I. Ghani, D. Jawawi, S. Dorairaj, & A. Sidky (Eds.), *Emerging Innovations in Agile Software Development* (pp. 175–207). Hershey, PA: IGI Global. doi:10.4018/978-1-4666-9858-1.ch010

Reccia, E., Cecchi, A., & Milani, G. (2016). FEM/DEM Approach for the Analysis of Masonry Arch Bridges. In V. Sarhosis, K. Bagi, J. Lemos, & G. Milani (Eds.), *Computational Modeling of Masonry Structures Using the Discrete Element Method* (pp. 367–392). Hershey, PA: IGI Global. doi:10.4018/978-1-5225-0231-9.ch014

Ro, H. K., & McIntosh, K. (2016). Constructing Conducive Environment for Women of Color in Engineering Undergraduate Education. In U. Thomas & J. Drake (Eds.), *Critical Research on Sexism and Racism in STEM Fields* (pp. 23–48). Hershey, PA: IGI Global. doi:10.4018/978-1-5225-0174-9.ch002

Rodulfo-Baechler, S. M. (2016). Dual Role of Perovskite Hollow Fiber Membrane in the Methane Oxidation Reactions. In H. Al-Megren & T. Xiao (Eds.), *Petrochemical Catalyst Materials, Processes, and Emerging Technologies* (pp. 385–430). Hershey, PA: IGI Global. doi:10.4018/978-1-4666-9975-5.ch014

Rudolf, S., Biryuk, V. V., & Volov, V. (2018). Vortex Effect, Vortex Power: Technology of Vortex Power Engineering. In V. Kharchenko & P. Vasant (Eds.), *Handbook of Research on Renewable Energy and Electric Resources for Sustainable Rural Development* (pp. 500–533). Hershey, PA: IGI Global. doi:10.4018/978-1-5225-3867-7.ch021

Sah, A., Bhadula, S. J., Dumka, A., & Rawat, S. (2018). A Software Engineering Perspective for Development of Enterprise Applications. In A. Elçi (Ed.), *Handbook of Research on Contemporary Perspectives on Web-Based Systems* (pp. 1–23). Hershey, PA: IGI Global. doi:10.4018/978-1-5225-5384-7.ch001

Sahoo, P., & Roy, S. (2017). Tribological Behavior of Electroless Ni-P, Ni-P-W and Ni-P-Cu Coatings: A Comparison. *International Journal of Surface Engineering and Interdisciplinary Materials Science*, *5*(1), 1–15. doi:10.4018/IJSEIMS.2017010101

Sahoo, S. (2018). Laminated Composite Hypar Shells as Roofing Units: Static and Dynamic Behavior. In K. Kumar & J. Davim (Eds.), *Composites and Advanced Materials for Industrial Applications* (pp. 249–269). Hershey, PA: IGI Global. doi:10.4018/978-1-5225-5216-1.ch011

Sahu, H., & Hungyo, M. (2018). Introduction to SDN and NFV. In A. Dumka (Ed.), *Innovations in Software-Defined Networking and Network Functions Virtualization* (pp. 1–25). Hershey, PA: IGI Global. doi:10.4018/978-1-5225-3640-6.ch001

Saikia, P., Bharadwaj, S. K., & Miah, A. T. (2016). Peroxovanadates and Its Bio-Mimicking Relation with Vanadium Haloperoxidases. In M. Bououdina (Ed.), *Emerging Research on Bioinspired Materials Engineering* (pp. 197–219). Hershey, PA: IGI Global. doi:10.4018/978-1-4666-9811-6.ch007

Saladino, R., Botta, G., & Crucianelli, M. (2016). Advances in Nanotechnology Transition Metal Catalysts in Oxidative Desulfurization (ODS) Processes: Nanotechnology Applied to ODS Processing. In T. Saleh (Ed.), *Applying Nanotechnology to the Desulfurization Process in Petroleum Engineering* (pp. 180–215). Hershey, PA: IGI Global. doi:10.4018/978-1-4666-9545-0.ch006

Saleh, T. A., Danmaliki, G. I., & Shuaib, T. D. (2016). Nanocomposites and Hybrid Materials for Adsorptive Desulfurization. In T. Saleh (Ed.), *Applying Nanotechnology to the Desulfurization Process in Petroleum Engineering* (pp. 129–153). Hershey, PA: IGI Global. doi:10.4018/978-1-4666-9545-0.ch004

Saleh, T. A., Shuaib, T. D., Danmaliki, G. I., & Al-Daous, M. A. (2016). Carbon-Based Nanomaterials for Desulfurization: Classification, Preparation, and Evaluation. In T. Saleh (Ed.), *Applying Nanotechnology to the Desulfurization Process in Petroleum Engineering* (pp. 154–179). Hershey, PA: IGI Global. doi:10.4018/978-1-4666-9545-0.ch005

Salem, A. M., & Shmelova, T. (2018). Intelligent Expert Decision Support Systems: Methodologies, Applications, and Challenges. In T. Shmelova, Y. Sikirda, N. Rizun, A. Salem, & Y. Kovalyov (Eds.), *Socio-Technical Decision Support in Air Navigation Systems: Emerging Research and Opportunities* (pp. 215–242). Hershey, PA: IGI Global. doi:10.4018/978-1-5225-3108-1.ch007

Samal, M. (2017). FE Analysis and Experimental Investigation of Cracked and Un-Cracked Thin-Walled Tubular Components to Evaluate Mechanical and Fracture Properties. In P. Samui, S. Chakraborty, & D. Kim (Eds.), *Modeling and Simulation Techniques in Structural Engineering* (pp. 266–293). Hershey, PA: IGI Global. doi:10.4018/978-1-5225-0588-4.ch009

Samal, M., & Balakrishnan, K. (2017). Experiments on a Ring Tension Setup and FE Analysis to Evaluate Transverse Mechanical Properties of Tubular Components. In P. Samui, S. Chakraborty, & D. Kim (Eds.), *Modeling and Simulation Techniques in Structural Engineering* (pp. 91–115). Hershey, PA: IGI Global. doi:10.4018/978-1-5225-0588-4.ch004

Santhanakumar, M., Adalarasan, R., & Rajmohan, M. (2016). An Investigation in Abrasive Waterjet Cutting of Al6061/SiC/Al2O3 Composite Using Principal Component Based Response Surface Methodology. *International Journal of Manufacturing, Materials, and Mechanical Engineering*, *6*(4), 30–47. doi:10.4018/IJMMME.2016100103

Sareen, N., & Bhattacharya, S. (2016). Cleaner Energy Fuels: Hydrodesulfurization and Beyond. In T. Saleh (Ed.), *Applying Nanotechnology to the Desulfurization Process in Petroleum Engineering* (pp. 84–128). Hershey, PA: IGI Global. doi:10.4018/978-1-4666-9545-0.ch003

Sarhosis, V. (2016). Micro-Modeling Options for Masonry. In V. Sarhosis, K. Bagi, J. Lemos, & G. Milani (Eds.), *Computational Modeling of Masonry Structures Using the Discrete Element Method* (pp. 28–60). Hershey, PA: IGI Global. doi:10.4018/978-1-5225-0231-9.ch002

Sarhosis, V., Oliveira, D. V., & Lourenco, P. B. (2016). On the Mechanical Behavior of Masonry. In V. Sarhosis, K. Bagi, J. Lemos, & G. Milani (Eds.), *Computational Modeling of Masonry Structures Using the Discrete Element Method* (pp. 1–27). Hershey, PA: IGI Global. doi:10.4018/978-1-5225-0231-9.ch001

Satyam, N. (2016). Liquefaction Modelling of Granular Soils using Discrete Element Method. In P. Samui (Ed.), *Handbook of Research on Advanced Computational Techniques for Simulation-Based Engineering* (pp. 381–441). Hershey, PA: IGI Global. doi:10.4018/978-1-4666-9479-8.ch015

Sawant, S. (2018). Deep Learning and Biomedical Engineering. In U. Kose, G. Guraksin, & O. Deperlioglu (Eds.), *Nature-Inspired Intelligent Techniques for Solving Biomedical Engineering Problems* (pp. 283–296). Hershey, PA: IGI Global. doi:10.4018/978-1-5225-4769-3.ch014

Sezgin, H., & Berkalp, O. B. (2018). Textile-Reinforced Composites for the Automotive Industry. In K. Kumar & J. Davim (Eds.), *Composites and Advanced Materials for Industrial Applications* (pp. 129–156). Hershey, PA: IGI Global. doi:10.4018/978-1-5225-5216-1.ch007

Shah, M. Z., Gazder, U., Bhatti, M. S., & Hussain, M. (2018). Comparative Performance Evaluation of Effects of Modifier in Asphaltic Concrete Mix. *International Journal of Strategic Engineering*, *1*(2), 13–25. doi:10.4018/IJoSE.2018070102

Shah, V. S., Shah, H. R., & Samui, P. (2016). Application of Meta-Models (MPMR and ELM) for Determining OMC, MDD and Soaked CBR Value of Soil. In S. Bhattacharyya, P. Banerjee, D. Majumdar, & P. Dutta (Eds.), *Handbook of Research on Advanced Hybrid Intelligent Techniques and Applications* (pp. 454–482). Hershey, PA: IGI Global. doi:10.4018/978-1-4666-9474-3.ch015

Sharma, N., & Kumar, K. (2018). Fabrication of Porous NiTi Alloy Using Organic Binders. In K. Kumar & J. Davim (Eds.), *Composites and Advanced Materials for Industrial Applications* (pp. 38–62). Hershey, PA: IGI Global. doi:10.4018/978-1-5225-5216-1.ch003

Sharma, T. K. (2016). Application of Shuffled Frog Leaping Algorithm in Software Project Scheduling. In P. Saxena, D. Singh, & M. Pant (Eds.), *Problem Solving and Uncertainty Modeling through Optimization and Soft Computing Applications* (pp. 225–238). Hershey, PA: IGI Global. doi:10.4018/978-1-4666-9885-7.ch011

Shivach, P., Nautiyal, L., & Ram, M. (2018). Applying Multi-Objective Optimization Algorithms to Mechanical Engineering. In M. Ram & J. Davim (Eds.), *Soft Computing Techniques and Applications in Mechanical Engineering* (pp. 287–301). Hershey, PA: IGI Global. doi:10.4018/978-1-5225-3035-0.ch014

Shmelova, T. (2018). Stochastic Methods for Estimation and Problem Solving in Engineering: Stochastic Methods of Decision Making in Aviation. In S. Kadry (Ed.), *Stochastic Methods for Estimation and Problem Solving in Engineering* (pp. 139–160). Hershey, PA: IGI Global. doi:10.4018/978-1-5225-5045-7.ch006

Shukla, R., Anapagaddi, R., Singh, A. K., Allen, J. K., Panchal, J. H., & Mistree, F. (2016). Integrated Computational Materials Engineering for Determining the Set Points of Unit Operations for Production of a Steel Product Mix. In S. Datta & J. Davim (Eds.), *Computational Approaches to Materials Design: Theoretical and Practical Aspects* (pp. 163–191). Hershey, PA: IGI Global. doi:10.4018/978-1-5225-0290-6.ch006

Siero González, L. R., & Romo Vázquez, A. (2017). Didactic Sequences Teaching Mathematics for Engineers With Focus on Differential Equations. In M. Ramírez-Montoya (Ed.), *Handbook of Research on Driving STEM Learning With Educational Technologies* (pp. 129–151). Hershey, PA: IGI Global. doi:10.4018/978-1-5225-2026-9.ch007

Singh, R., & Dutta, S. (2018). Visible Light Active Nanocomposites for Photocatalytic Applications. In K. Kumar & J. Davim (Eds.), *Composites and Advanced Materials for Industrial Applications* (pp. 270–296). Hershey, PA: IGI Global. doi:10.4018/978-1-5225-5216-1.ch012

Singh, R., & Lou, H. H. (2016). Safety and Efficiency Enhancement in LNG Terminals. In H. Al-Megren & T. Xiao (Eds.), *Petrochemical Catalyst Materials, Processes, and Emerging Technologies* (pp. 164–176). Hershey, PA: IGI Global. doi:10.4018/978-1-4666-9975-5.ch007

Sözbilir, H., Özkaymak, Ç., Uzel, B., & Sümer, Ö. (2018). Criteria for Surface Rupture Microzonation of Active Faults for Earthquake Hazards in Urban Areas. In N. Ceryan (Ed.), *Handbook of Research on Trends and Digital Advances in Engineering Geology* (pp. 187–230). Hershey, PA: IGI Global. doi:10.4018/978-1-5225-2709-1.ch005

Stanciu, I. (2018). Stochastic Methods in Microsystems Engineering. In S. Kadry (Ed.), *Stochastic Methods for Estimation and Problem Solving in Engineering* (pp. 161–176). Hershey, PA: IGI Global. doi:10.4018/978-1-5225-5045-7.ch007

Strebkov, D., Nekrasov, A., Trubnikov, V., & Nekrasov, A. (2018). Single-Wire Resonant Electric Power Systems for Renewable-Based Electric Grid. In V. Kharchenko & P. Vasant (Eds.), *Handbook of Research on Renewable Energy and Electric Resources for Sustainable Rural Development* (pp. 449–474). Hershey, PA: IGI Global. doi:10.4018/978-1-5225-3867-7.ch019

Subburaman, D., Jagan, J., Dalkiliç, Y., & Samui, P. (2016). Reliability Analysis of Slope Using MPMR, GRNN and GPR. In F. Miranda & C. Abreu (Eds.), *Handbook of Research on Computational Simulation and Modeling in Engineering* (pp. 208–224). Hershey, PA: IGI Global. doi:10.4018/978-1-4666-8823-0.ch007

Sun, J., Wan, S., Lin, J., & Wang, Y. (2016). Advances in Catalytic Conversion of Syngas to Ethanol and Higher Alcohols. In H. Al-Megren & T. Xiao (Eds.), *Petrochemical Catalyst Materials, Processes, and Emerging Technologies* (pp. 177–215). Hershey, PA: IGI Global. doi:10.4018/978-1-4666-9975-5.ch008

Tüdeş, Ş., Kumlu, K. B., & Ceryan, S. (2018). Integration Between Urban Planning and Natural Hazards For Resilient City. In N. Ceryan (Ed.), *Handbook of Research on Trends and Digital Advances in Engineering Geology* (pp. 591–630). Hershey, PA: IGI Global. doi:10.4018/978-1-5225-2709-1.ch017

Tyukhov, I., Rezk, H., & Vasant, P. (2016). Modern Optimization Algorithms and Applications in Solar Photovoltaic Engineering. In P. Vasant & N. Voropai (Eds.), *Sustaining Power Resources through Energy Optimization and Engineering* (pp. 390–445). Hershey, PA: IGI Global. doi:10.4018/978-1-4666-9755-3.ch016

Ulamis, K. (2018). Soil Liquefaction Assessment by Anisotropic Cyclic Triaxial Test. In N. Ceryan (Ed.), *Handbook of Research on Trends and Digital Advances in Engineering Geology* (pp. 631–664). Hershey, PA: IGI Global. doi:10.4018/978-1-5225-2709-1.ch018

Umar, M. A., Tenuche, S. S., Yusuf, S. A., Abdulsalami, A. O., & Kufena, A. M. (2016). Usability Engineering in Agile Software Development Processes. In I. Ghani, D. Jawawi, S. Dorairaj, & A. Sidky (Eds.), *Emerging Innovations in Agile Software Development* (pp. 208–221). Hershey, PA: IGI Global. doi:10.4018/978-1-4666-9858-1.ch011

Üzüm, O., & Çakır, Ö. A. (2016). A Bio-Inspired Phenomena in Cementitious Materials: Self-Healing. In M. Bououdina (Ed.), *Emerging Research on Bioinspired Materials Engineering* (pp. 220–246). Hershey, PA: IGI Global. doi:10.4018/978-1-4666-9811-6.ch008

Valente, M., & Milani, G. (2017). Seismic Assessment and Retrofitting of an Under-Designed RC Frame Through a Displacement-Based Approach. In V. Plevris, G. Kremmyda, & Y. Fahjan (Eds.), *Performance-Based Seismic Design of Concrete Structures and Infrastructures* (pp. 36–58). Hershey, PA: IGI Global. doi:10.4018/978-1-5225-2089-4.ch002

Vasant, P. (2018). A General Medical Diagnosis System Formed by Artificial Neural Networks and Swarm Intelligence Techniques. In U. Kose, G. Guraksin, & O. Deperlioglu (Eds.), *Nature-Inspired Intelligent Techniques for Solving Biomedical Engineering Problems* (pp. 130–145). Hershey, PA: IGI Global. doi:10.4018/978-1-5225-4769-3.ch006

Vergara, D., Lorenzo, M., & Rubio, M. (2016). On the Use of Virtual Environments in Engineering Education. *International Journal of Quality Assurance in Engineering and Technology Education*, *5*(2), 30–41. doi:10.4018/IJQAETE.2016040103

Verrollot, J., Tolonen, A., Harkonen, J., & Haapasalo, H. J. (2018). Challenges and Enablers for Rapid Product Development. *International Journal of Applied Industrial Engineering*, *5*(1), 25–49. doi:10.4018/IJAIE.2018010102

Wagner, C., & Ryan, C. (2016). Physical and Digital Integration Strategies of Electronic Device Supply Chains and Their Applicability to ETO Supply Chains. In R. Addo-Tenkorang, J. Kantola, P. Helo, & A. Shamsuzzoha (Eds.), *Supply Chain Strategies and the Engineer-to-Order Approach* (pp. 224–245). Hershey, PA: IGI Global. doi:10.4018/978-1-5225-0021-6.ch011

Wang, Z., Wu, P., Lan, L., & Ji, S. (2016). Preparation, Characterization and Desulfurization of the Supported Nickel Phosphide Catalysts. In H. Al-Megren & T. Xiao (Eds.), *Petrochemical Catalyst Materials, Processes, and Emerging Technologies* (pp. 431–458). Hershey, PA: IGI Global. doi:10.4018/978-1-4666-9975-5.ch015

Yardimci, A. G., & Karpuz, C. (2018). Fuzzy Rock Mass Rating: Soft-Computing-Aided Preliminary Stability Analysis of Weak Rock Slopes. In N. Ceryan (Ed.), *Handbook of Research on Trends and Digital Advances in Engineering Geology* (pp. 97–131). Hershey, PA: IGI Global. doi:10.4018/978-1-5225-2709-1.ch003

Zhang, L., Ding, S., Sun, S., Han, B., Yu, X., & Ou, J. (2016). Nano-Scale Behavior and Nano-Modification of Cement and Concrete Materials. In A. Khitab & W. Anwar (Eds.), *Advanced Research on Nanotechnology for Civil Engineering Applications* (pp. 28–79). Hershey, PA: IGI Global. doi:10.4018/978-1-5225-0344-6.ch002

Zindani, D., & Kumar, K. (2018). Industrial Applications of Polymer Composite Materials. In K. Kumar & J. Davim (Eds.), *Composites and Advanced Materials for Industrial Applications* (pp. 1–15). Hershey, PA: IGI Global. doi:10.4018/978-1-5225-5216-1.ch001

Zindani, D., Maity, S. R., & Bhowmik, S. (2018). A Decision-Making Approach for Material Selection of Polymeric Composite Bumper Beam. In K. Kumar & J. Davim (Eds.), *Composites and Advanced Materials for Industrial Applications* (pp. 112–128). Hershey, PA: IGI Global. doi:10.4018/978-1-5225-5216-1.ch006

About the Contributors

Aamir Shahzad has over seventeen years of experience of University research and teaching at home and abroad and it is due to his continuous efforts the Thermo-Physical properties of plasmas/ materials started at the graduate and postgraduate levels. Dr. Shahzad has received the Post Doctorate and Doctoral degrees from Xi'an Jiaotong University (XJTU) P.R. China, in 2015 and 2012. He has been proposed novel methods to explore outcomes of complex materials which show his aptitude to comprehend and grip Computational Physics and Molecular Modeling and Simulations together with Experimental understanding. Moreover, Dr. Shahzad is exploring in the fields of computational Physics, complex fluids/ plasmas, and in addition, currently working on plasma oncology, bio- and energy materials. Currently, Dr. Shahzad is working as a Tenured Professor (Associate) in the department of Physics, GC University Faisalabad (GCUF). Dr. Shahzad is a member of the ThermoPhysical Society of Xian Jiaotong University (Xjtu), China, Member of Physics society GCUF and UAF, Pakistan. (Maogang He - Editor) Maogang He is missing a biography.

Maogang He is a professor in the school of energy and power engineering, Xi'an Jiaotong University, China. He has published more than 370 research articles in different international and Chinese journals.

* * *

Linda Agun is currently a postgraduate Mechanical Engineering student from Universiti Teknologi Malaysia, Johor Bahru, Malaysia. She study is focusing on development of dielectric barrier discharge cold plasma system for oyster mushroom spawn growth enhancement and also plant seeds. She obtained her Master of Mechanical Engineering in Preparation and Durability Study of LSCF-SDCC Composite Cathode for Low Temperature Solid Oxide Fuel Cell. She graduated from Universiti Tun Hussein Onn Malaysia, Johor, Malaysia in mechanical engineering. She has published 18 papers. Her works have been cited 63 times with an h index of 4 and i10 index of 1 to till date.

Norhayati Ahmad was born in Negeri Sembilan, Malaysia (1970), received the BSc. (Applied Sc. Physics) from Universiti Kebangsaan Malaysia (UKM) in 1993 and MSc. (Engineering Materials) from Universiti Sains Malaysia (USM) in 1998. She obtained her Ph.D degree from Kagoshima University, Japan (sponsored by Hitachi Scholarship Foundation) in 2010. She joined the Department of Materials Engineering, Faculty of Mechanical Engineering, Universiti Teknologi Malaysia (UTM) as a tutor in 1993 and lecturer in 1998. Now she is serving as Associate Professor in Department of Materials, Manufacturing and Industrial Engineering. She is teaching courses of Materials Science, Materials Engineering, Advanced Materials, Materials Technology, Manufacturing Processes, Adv. Tech. for Materials Characterization, Adv. Techniques for Materials Characterization and Materials Selection for more than 23 years at under- and post-graduate levels. She has spent extended periods as a visiting researcher at Kagoshima University, Japan (2011). She has an industrial experience with Kiswire Sdn. Bhd for one year at Kota Kiswire, Johor Bahru (Mac 2013-Mac 2014). Dr Norhayati is a head of Materials Science Laboratory, School of Mechanical Engineering from 2014 until now. The outcomes of her research have been published in international journal (28 high impact factor -Q1/Q2 papers, 13 Scopus journal), 17 non index journal, 37 articles for conference proceedings and 3 articles for Sc. & Tech. Encyclopedia.

Shahid Ali was born in Nowshera, Pakistan, in 1975. Currently, he is working as Associate Professor at National Centre for Physics, Quaid-e-Azam University Campus, Islamabad. He received his M. Sc. and M. Phil. degrees in Physics, in 1998 and 2001 from Gomal University, D. I. Khan and Quaid-e-Azam University, Islamabad, respectively. He served as Research Officer for 4 years at Salam Chair in Physics, GC University Lahore and later earned his Ph. D degree in Theoretical Nonlinear Physics from Umea University, Sweden in 2008. Dr. Ali completed his Post-Doctoral Studies at IST, Lisbon, Portugal during the years 2009-10 and was also appointed as Regular Associate at Abdus Salam International Centre for Theoretical Physics, Italy. His research areas include Nonlinear Plasma Physics, Spacecraft Charging and Space Dynamics, Degenerate Plasmas, Magnetic Fields Generation in Laser-produced Plasmas, etc. He has published more than 120 articles in International Reputed Journals, like Phys. Rev. Lett., Phys. Rev. B, Astrophys. J., Phys. Plasmas, etc., and trained several M. Phil/Ph. D Scholars in the field of Plasma Physics. Remained a regular recipient of the RPA Awards from the Pakistan Council for Science and Technology for many years and received a Gold Medal in Physics from the Pakistan Academy of Sciences in 2016. Dr. Ali also worked as Visiting Associate Professor at University of Science and Technology of China in 2017 and as Visiting Scholar at Khalifa University, UAE, in 2021.

Misbah Khan is a Ph.D. student in the Department of Refrigeration and Cryogenics Engineering, School of Energy and Power Engineering, Xi'an Jiaotong University, Xi'an, China.

Ioannis Kourakis is a Plasma Physicist, currently based at Khalifa University of Science and Technology, Abu Dhabi, UAE, where he is also serving as Acting Local Coordinator for the KU-Mars Program. He has held various academic and research positions in UK, Germany, Belgium and in Brazil. His research focuses on nonlinear dynamics with emphasis on Plasma Physics and materials science. His most recent findings include the formation of nonlinear waves and shocks in Plasma Physics and in Space Science. Earlier work has spanned various areas including the optical instabilities, dissipative solitons, negative-refractive-index materials, Bose-Einstein condensates, hydrodynamics, etc. He has published more than 235 research articles in International refereed journals and conference proceedings.

Santhana Krishnan is a full time post-doctoral researcher in Energy Engineering at the PSU Energy System Research Institute at the Prince of Songkla University, Hatyai, Thailand, since April 2021. He obtained his PhD degree in chemical engineering in 2017 at the University of Malaysia Pahang, Malaysia. In Feb 2018, he went to the University of Technology Malaysia, Malaysia, for the 3 years post doc on Hythane Energy scale up and commercialization. His main areas of research are energy/environment, fermentation, wastewater treatment, bioelectrochemical systems, nanomaterial synthesis for energy and water applications, resource recovery, bioreactors development, and bio-products. He has been involved in national and international research and industrial projects. He has published More than 50 papers in prominent journals in Elsevier's, Springer, Wiley, and has authored 8 book chapters. His works have been cited 700 times with an h index of 14 and an i10 index of 17 to till date. He has been a reviewer for various journals published by Elsevier, Wiley, RSC and Springer. He is an active and lifetime member for Sakura Science Programme, Japan (JST), member of Royal Society of Biology (MRSB), London, and International Water Association (IWA). His ongoing projects involve designing and fabricating pilot scale bioreactors for hydrogen energy production and waste water treatment applications.

Mohd Fadthul Ikmal Misnal was born on 14th November 1991. He graduated from Universiti Teknologi Malaysia (UTM) with a Bachelor of Engineering (Mechanical) in 2014. From 2015 to 2018, he obtained his Master of Philosophy degree in Mechanical Engineering, studying the mathematical model of non-thermal plasma temperature. Currently, he is the Doctor of Philosophy (PhD) candidate in Mechanical Engineering, studying the non-thermal plasma experimental optimization in rice grain decontamination. His research interest are non-thermal plasma modelling, experiment and diagnostic in the field of food treatment and metal surface cleaning.

Norizah Redzuan is Associate Professor in Mechanical Engineering and Program Coordinator of Bachelor Mechanical-Manufacturing Engineering at School of Mechanical Engineering, Universiti Teknologi Malaysia. She also Chairperson for ISO TC 44/ SC 13 Brazing materials and processes. Norizah's focuses research on cold plasma applications in food decontamination and metal cleaning as part of sustainable manufacturing approach. Other interest area are Additive Manufacturing, machining and welding of dissimilar materials. Norizah is the co-author of Book Chapter with the title Additive Manufacturing in Product Design and Development (UTM, 2020) by UTM Press which has been indexed by Web of Science (WOS). She had Norizah received her PhD from the University of Glasgow, UK, where she studied in Cold Plasma Air Decontamination. She had more than 40 publications of journals and articles, also presented numerous international conferences and seminars.

Siti Sarah Safaai is a senior lecturer at University of Teknologi Malaysia's Department of Physics, Faculty of Science (UTM). In 2007, she earned a bachelor's degree in Industrial Physics from Universiti Teknologi Malaysia. She then went on to University Malaya to get a Master of Applied Science degree. She joined UTM in August 2008 while studying for her master's degree. She pursued her studies at the University of Malaya for her Ph. D in a complicated dusty plasma system after obtaining her master's degree in 2010. In 2018, she completed her research. She became interested in plasma technology and fluctuation analysis using a fractal model while studying at the University of Malaya. She utilised the wavelet method to describe the fluctuation time series. Her postgraduate experience, she has always felt, offered one of the finest training grounds for becoming a competent physicist and academician.

Muhammad Asif Shakoori is a PhD student in the School of Energy and Power Engineering, Xi'an Jiaotong University, Shaanxi, China.

Chang Shu Ting is a full-time Master student of Universiti Tunku Abdul Rahman (UTAR) at Kampar, Perak. She did her degree internship in 2018 at Universiti Teknologi Malaysia (UTM), Johor and obtained her bachelor's degree in Microbiology from UTAR. The main area of her current study focuses on biotechnology and molecular biology.

Muhamad Nor Firdaus bin Zainal grew up in Johor, Malaysia and graduated from Universiti Teknologi Malaysia and major in Mechanical Engineering. Currently, he is PhD candidate at Universiti Teknologi Malaysia. His undergoing research related with cold plasma cleaning and his research interest include cold plasma, thermal plasma, design, and manufacturing.

Zarita Zakaria is a full-time lecturer in Department of Biosciences, Faculty of Science, Universiti Technologi Malaysia since 2004. She obtained her PhD degree in Biosciences in 2016 from same university. Her main areas of research are microbiology, environmental science, proteomic, genetic engineering, plant science and biotechnology. She has been involved in national research and industrial projects. She has published papers in Elseviers as well as other publications. She has been a reviewer for various journals published by Elsevier and others. She is an active and lifetime member of Asian Federation of Biotechnology (AFOB). Her ongoing projects involve the microbial protease production and application, microbial biocement optimization, plastic degrading bacteria optimization and genetic engineering.

Ying Zhang is a professor in the School of Energy and Power Engineering, Xi'an Jiaotong University, China.

Index

Q

S

T

W

Premier Reference Source
Emerging GIS Applications for Emergency and Disaster Management

Premier Reference Source
Managerial Strategies and Green Solutions for Project Sustainability

Premier Reference Source
Comparative Approaches to Using R and Python for Statistical Data Analysis

Premier Reference Source
Solutions for High-Touch Communications in a High-Tech World

www.ingramcontent.com/pod-product-compliance
Lightning Source LLC
LaVergne TN
LVHW081519300926
849161LV00006B/1230